Genetics in Rice

Genetics in Rice

Editors

Katsuyuki Ichitani
Ryuji Ishikawa

MDPI • Basel • Beijing • Wuhan • Barcelona • Belgrade • Manchester • Tokyo • Cluj • Tianjin

Editors
Katsuyuki Ichitani
Kagoshima University
Japan

Ryuji Ishikawa
Hirosaki University
Japan

Editorial Office
MDPI
St. Alban-Anlage 66
4052 Basel, Switzerland

This is a reprint of articles from the Special Issue published online in the open access journal *Plants* (ISSN 2223-7747) (available at: https://www.mdpi.com/journal/plants/special_issues/genetics_rice).

For citation purposes, cite each article independently as indicated on the article page online and as indicated below:

LastName, A.A.; LastName, B.B.; LastName, C.C. Article Title. *Journal Name* **Year**, *Article Number*, Page Range.

ISBN 978-3-03936-826-6 (Hbk)
ISBN 978-3-03936-827-3 (PDF)

Cover image courtesy of Katsuyuki Ichitani.

Contents

About the Editors . . . . **vii**

Preface to "Genetics in Rice" . . . . **ix**

Motonori Tomita and Keiichiro Ishimoto
Rice Novel Semidwarfing Gene *d60* Can Be as Effective as Green Revolution Gene *sd1*
Reprinted from: *Plants* **2019**, *8*, 464, doi:10.3390/plants8110464 . . . . **1**

Cuong D. Nguyen, Holden Verdeprado, Demeter Zita, Sachiyo Sanada-Morimura, Masaya Matsumura, Parminder S. Virk, Darshan S. Brar, Finbarr G. Horgan, Hideshi Yasui and Daisuke Fujita
The Development and Characterization of Near-Isogenic and Pyramided Lines Carrying Resistance Genes to Brown Planthopper with the Genetic Background of *japonica* Rice (*Oryza sativa* L.)
Reprinted from: *Plants* **2019**, *8*, 498, doi:10.3390/plants8110498 . . . . **15**

Yu Zhang, Jiawu Zhou, Ying Yang, Walid Hassan Elgamal, Peng Xu, Jing Li, Yasser Z. El-Refaee, Suding Hao and Dayun Tao
Two SNP Mutations Turned off Seed Shattering in Rice
Reprinted from: *Plants* **2019**, *8*, 475, doi:10.3390/plants8110475 . . . . **35**

Hiroki Saito, Yutaka Okumoto, Takuji Tsukiyama, Chong Xu, Masayoshi Teraishi and Takatoshi Tanisaka
Allelic Differentiation at the *E1/Ghd7* Locus Has Allowed Expansion of Rice Cultivation Area
Reprinted from: *Plants* **2019**, *8*, 550, doi:10.3390/plants8120550 . . . . **45**

Buddini Abhayawickrama, Dikkumburage Gimhani, Nisha Kottearachchi, Venura Herath, Dileepa Liyanage and Prasad Senadheera
In Silico Identification of QTL-Based Polymorphic Genes as Salt-Responsive Potential Candidates through Mapping with Two Reference Genomes in Rice
Reprinted from: *Plants* **2020**, *9*, 233, doi:10.3390/plants9020233 . . . . **59**

Takayuki Ogami, Hideshi Yasui, Atsushi Yoshimura and Yoshiyuki Yamagata
Identification of Anther Length QTL and Construction of Chromosome Segment Substitution Lines of *Oryza longistaminata*
Reprinted from: *Plants* **2019**, *8*, 388, doi:10.3390/plants8100388 . . . . **85**

Kumpei Shiragaki, Takahiro Iizuka, Katsuyuki Ichitani, Tsutomu Kuboyama, Toshinobu Morikawa, Masayuki Oda and Takahiro Tezuka
HWA1- and *HWA2*-Mediated Hybrid Weakness in Rice Involves Cell Death, Reactive Oxygen Species Accumulation, and Disease Resistance-Related Gene Upregulation
Reprinted from: *Plants* **2019**, *8*, 450, doi:10.3390/plants8110450 . . . . **99**

Dinh Thi Lam, Katsuyuki Ichitani, Robert J. Henry and Ryuji Ishikawa
Molecular and Morphological Divergence of Australian Wild Rice
Reprinted from: *Plants* **2020**, *9*, 224, doi:10.3390/plants9020224 . . . . **113**

Daiki Toyomoto, Masato Uemura, Satoru Taura, Tadashi Sato, Robert Henry, Ryuji Ishikawa and Katsuyuki Ichitani
Segregation Distortion Observed in the Progeny of Crosses Between *Oryza sativa* and *O. meridionalis* Caused by Abortion During Seed Development
Reprinted from: *Plants* **2019**, *8*, 398, doi:10.3390/plants8100398 . . . . **129**

Hayba Badro, Agnelo Furtado and Robert Henry
Relationships between Iraqi Rice Varieties at the Nuclear and Plastid Genome Levels
Reprinted from: *Plants* **2019**, *8*, 481, doi:10.3390/plants8110481 . . . . **145**

Peterson W. Wambugu, Marie-Noelle Ndjiondjop and Robert Henry
Advances in Molecular Genetics and Genomics of African Rice (*Oryza glaberrima* Steud)
Reprinted from: *Plants* **2019**, *8*, 376, doi:10.3390/plants8100376 . . . . **159**

About the Editors

Katsuyuki Ichitani, Associate Professor in Kagoshima University.
Born in 1970
Graduated from Kyoto University, Japan, in 1993.
Lecturer of Kagoshima University, Japan, from 1998.
The degree of doctor was awarded in 1998.
Associate Professor of Faculty of Agriculture, Kagoshima University, Japan, from 2004.
Major academic field is the genetics of agronomic traits and reproductive barrier in rice

Ryuji Ishikawa, Professor in Hirosaki University
Born in 1962
Graduated from Hokkaido University, Japan in 1985.
1988 Assistant Professor in Faculty of Agriculture, Hirosaki University.
The degree of doctor was awarded in 1993.
1993 Associate Professor in Faculty of Agriculture, Hirosaki University.
1997 Associate Professor in Faculty of Agriculture and Life Science, Hirosaki University.
2008 Professor in Faculty of Agriculture and Life Science, Hirosaki University.
Major academic field is rice genetics, evolutionary genetics.

Preface to "Genetics in Rice"

Rice feeds more than half of the world population. Its small genome size and ease in transformation have made rice the model crop in plant physiology and genetics. Molecular as well as Mendelian, forward as well as reverse genetics, collaborate with each other to expand rice genetics. The syntety of rice with other grasses, such as wheat, barley and maize, has helped accelerate their genomic studies.

The wild relatives of rice belonging to the genus *Oryza* are distributed in Asia, Africa, Latin America and Oceania. Phenotypic and genetic diversity among them contributes to their adaptation to a wide range of environments. They are good sources for the study of domestication and adaptation.

Rice was the first crop to have its entire genome sequenced. With the help of the reference genome of Nipponbare and the advent of the next generation sequencer, the study of the rice genome has been accelerated. Now, 3000 (3K) cultivar genome information, the pangenome information comprising the whole genes among rice as a species, and the genomes of wild relatives of rice are available.

The mining of DNA polymorphism has permitted map-based cloning, QTL (quantitative trait loci) analysis, GWAS (genome-wide association study), and the production of many kinds of experimental lines, such as recombinant inbred lines, backcross inbred lines, and chromosomal segment substitution lines. The genetics of agronomic traits and pest resistance has led to the breeding of elite rice cultivars.

Inter- and intraspecific hybridization among *Oryza* species has opened the door to various levels of reproductive barriers ranging from prezygotic—e.g., hybrid sterility, male sterility—to postzygotic—e.g., hybrid weakness, hybrid breakdown.

This Special Issue of Plants, Genetics in Rice https://www.mdpi.com/journal/plants/special_issues/genetics_rice, contains eleven papers on genetic studies of rice and its relatives utilizing the rich genetic resources and/or rich genome information described above.

Katsuyuki Ichitani, Ryuji Ishikawa
Editors

plants

MDPI

Article

Rice Novel Semidwarfing Gene *d60* Can Be as Effective as Green Revolution Gene *sd1*

Motonori Tomita [1,*] and Keiichiro Ishimoto [2]

1 Research Institute of Green Science and Technology, Shizuoka University, 836 Ohya, Suruga-ku, Shizuoka City, Shizuoka 422-8529, Japan
2 Faculty of Agriculture, Tottori University, 4-101 Koyama Minami, Tottori 680-8550, Japan; kei-ishimoto@pref.nagasaki.lg.jp
* Correspondence: tomita.motonori@shizuoka.ac.jp

Received: 3 September 2019; Accepted: 7 October 2019; Published: 30 October 2019

Abstract: Gene effects on the yield performance were compared among promising semidwarf genes, namely, novel gene *d60*, representative gene *sd1* with different two source IR8 and Jukkoku, and double dwarf combinations of *d60* with each *sd1* allele, in a Koshihikari background. Compared with the culm length of variety Koshihikari (mean, 88.8 cm), that of the semidwarf or double dwarf lines carrying Jukkoku_*sd1*, IR8_*sd1*, *d60*, Jukkoku_*sd1* plus *d60*, or IR8_*sd1* plus *d60* was shortened to 71.8 cm, 68.5 cm, 65.7 cm, 48.6 cm, and 50.3 cm, respectively. Compared with the yield of Koshihikari (mean, 665.3 g/m^2), that of the line carrying Jukkoku_*sd1* allele showed the highest value (772.6 g/m^2, 16.1% higher than Koshihikari), while that of IR8_*sd1*, *d60* and IR8_*sd1* plus *d60*, was slightly decreased by 7.1%, 5.5%, and 9.7% respectively. The line carrying Jukkoku_*sd1* also showed the highest value in number of panicles and florets/panicle, 16.2% and 11.1% higher than in Koshihikari, respectively, and these effects were responsible for the increases in yield. The 1000-grain weight was equivalent among all genetic lines. Except for the semidwarf line carrying Jukkoku_*sd1*, semidwarf line carrying *d60* was equivalent to line carrying IR8_*sd1*in the yield of unpolished rice, and yield components such as panicle length, panicle number, floret number /panicle. Therefore, the semidwarfing gene *d60* is one of the best possible choices in practical breeding.

Keywords: rice; semidawarf gene; *d60*; *sd1*; yield component; phenotyping; growth

1. Introduction

Semidwarfing prevents plants from lodging at their full-ripe stage, making them lodging-resistant to wind and rain, enhances their adaptability for heavy manuring and markedly improved the global productivity of rice and wheat between 1960–1990 (up to double yields of rice and quadruple yields of wheat) [1,2]. Semidwarf rice contributes stable production in the monsoonal regions of Asia, where typhoons frequently occur during the yielding season and also brings benefits such as erect leaf angle, reduced photoinhibition, and possibility to plant at higher densities to japonica varieties grown in California and also in South America [3]. However, gene source of semidwarfness is limited. The International Rice Research Institute (IRRI) developed a semidwarf rice variety IR8 in 1966 by using Taiwanese native semidwarf variety Dee-geo-woo-gen (DGWG). IR8 called as Miracle Rice, has been improved with lodging resistance and light-reception attitude, and it brought the Green Revolution in tropical Asia [2]. In Japan, semidwarf cultivars in the Kyushu region were developed in the 1960s using the native semidwarf variety Jukkoku [4]. In the Tohoku region, semidwarf cultivars were developed in the 1970s using the semidwarf mutant Reimei induced by Fujiminori-gamma-ray irradiated [5]. In the United States, Calrose 76 was developed in 1976 by Calrose-gamma-ray irradiated [6,7].

Genetic study has also been devoted on the genes responsible for semidwarfism in rice. First, a recessive semidwarf gene *d47* was identified in DGWG, the parental line of IR8 [8,9]. Next, the semidwarf gene *sd1* in Calrose 76 was shown to be allelic to *d47* [10,11]. Finally, semidwarf genes in Taichung Native 1 descend from DGWG, Shiranui from Jukkoku, and *d49* in the mutant cultivar Reimei were attributed to the same allele by allelism examination [12–14]. Therefore, only a single semidwarf gene, *sd1*, has been commonly used across the world. A little genetic source of current semidwarf rice cultivars have a risk for environmental change. Thus, it is an emerging subject to find a novel semidawrf gene to replace *sd1* and to utilize it to diversify genetic variations of semidwarf rice worldwide.

A novel semidwarf gene, *d60*, which was found in the mutant Hokuriku 100 induced by irradiation of 20 kR of gamma-ray to Koshihikari, is thus of particular importance [15]. While *sd1* is on rice chromosome 1 [16,17], *d60* is located on chromosome 2 (Tomita et al., submitted to Genes). *sd1* is a defective allele encoding GA20-oxidase gene in a late step in the GA biosynthesis pathway [18–20]. Moreover, unlike *sd1*, *d60* complements the gamete lethal gene, *gal*. Therefore, in the cross between Hokuriku 100 (*d60d60GalGal*) and Koshihikari (*D60D60galgal*), male and female gametes, in which *gal* and *d60* coexistent, become lethal and the pollen and seed fertility in the F_1 (genotype *D60d60Galgal*) breakdown to 75%. As a result, the F_2 progeny exhibits a unique genotype ratio of 6 fertile long-culm (4*D60D60*:2*D60d60GalGal*: 2 partially fertile long-culm (*D60d60Galgal* = F_1 type):1 dwarf(*d60d60GalGal*) [15].

Although there are multiple alleles in *sd1* locus of DGWG, Jukkoku, Reimei, and Calrose 76, the differences in their influences on the yield performance have not been reported. Therefore, investigating the differences in phenotypic traits among the different *sd1* allele-carrying plants, *d60*-carrying plant and their double dwarf plants, will be beneficial for practical selection of *d60* and *sd1*alleles. In this study, semidwarf or double dwarf lines, which were integrated with *sd1* of Jukkoku, *sd1* of IR8, *d60*, or both gene combinations in the genetic background of Koshihikari, were used for investigating the influence of these semidwarf genes on phenotypic traits, especially related to yield performance.

2. Results

2.1. Effects of Semidwarf and Double Dwarf Genes on Growth

The trends in full-length growth, depicted by growth curves, were comparable among all lines. (Figure 1). The full length in lines carrying one or two semidwarf genes was already shorter than that of Koshirikari lines at the time of transplanting (June 7, 28 days after sowing), and the differences became prominent around 64–70 days after sowing (July 13 and 19) (Figure 1, Table 1). The full length of *d60*-carrying line was longer than that of *sd1*-carrying lines at the time of transplanting. However, the full length of line carrying Jukkoku_*sd1* and that of line carrying IR8_*sd1* exceeded that of line carrying *d60* on June 23 (43 days after sowing) and on July 13 (64 days after sowing), respectively: full length in lines carrying either Jukkoku_*sd1* or IR8_*sd1* was longer than that in line carrying *d60* at the time of final measurement (August 23, 103 days after sowing). Days to heading ranged from 86.5 days of line carrying IR8_*sd1* to 90.5 days of those carrying *d60*. Such a four-day difference was thought to be little. Therefore, the differences appeared in morphological traits, such as culm length and panicle length, were attributed to genetic reason.

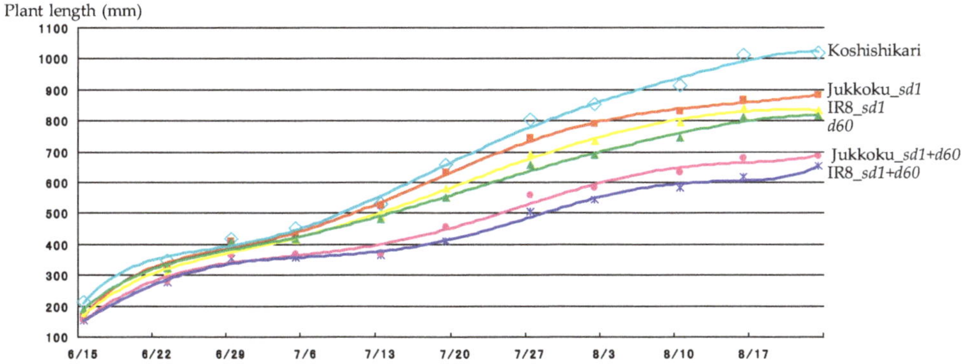

Figure 1. Effect of growth of semidwarf and double dwarf gene lines. Ten plants were randomly selected, and the distance between the ground and the highest standing point (i.e., the full length) was measured every week for approximately three months until the panicle emerged. The full length of *d60*-carrying line was longer than that of *sd1*-carrying lines at the time of transplanting. However, the full length of line carrying Jukkoku_*sd1* and that of line carrying IR8_*sd1* exceeded that of line carrying *d60* on June 23 (43 days after sowing) and on July 13 (64 days after sowing), respectively: full length in lines carrying either Jukkoku_*sd1* or IR8_*sd1* was longer than that in line carrying *d60* at the time of final measurement (August 23, 103 days after sowing).

Table 1. Plant length of semidwarf and double dwarf gene lines.

Days after Sowing	35	43	49	55	64	70	78	82	90	96	103
Koshihikari	212.6	345.0	415.4	452.4	529.5	657.0	805.6	852.6	913.1	1011.6	1020.3
Jukkoku_*sd1*	166.8 *	330.2	408.9	430.7	523.8	631.0 *	743.0 *	791.7 *	828.7 *	865.9 *	880.8 *
IR8_*sd1*	170.3 *	308.9 *	393.2	414.9 *	488.8 *	578.2 *	693.2 *	734.4 *	794.6 *	837.9 *	831.0 *
d60	189.1 *	321.0 *	406.5	415.9 *	480.5 *	550.4 *	656.8 *	689.3 *	744.1 *	810.6 *	814.3 *
Jukkoku_*sd1+d60*	153.1 *	275.4 *	352.8 *	357.8 *	364.1 *	409.2 *	506.4 *	543.4 *	581.9 *	616.6 *	651.7 *
IR8_*sd1+d60*	153.7 *	279.2 *	364.8 *	368.5 *	369.1 *	454.5 *	558.4 *	580.9 *	630.9 *	679.2 *	684.2 *

The full length in lines carrying one or two semidwarf genes was already shorter than that of Koshirikari lines at the time of transplanting 28 days after sowing, and the differences became prominent around 64–70 days after sowing. Finally, the full length of semidwarf and double dwarf lines were significantly shorter than Koshihaikri. Color in the boxes of genetic lines coincide the color of growth curve in Figure 1. *: statistically significant at the 5% level.

Integration of a semidwarf gene (or genes) resulted in a reduction in culm length: the mean culm length of Koshihikari was 88.8 cm, while that of lines carrying Jukkoku_*sd1*, IR8_sd1, *d60*, Jukkoku_*sd1* plus *d60*, or IR8_*sd1* plus *d60* was 71.8 cm, 68.5 cm, 65.7 cm, 48.6 cm, or 50.2 cm, respectively. Leaf length was shorter in line carrying Kinuhikari_*sd1* (9–16% reduction compared with Koshihikari) or *d60* (9–18% reduction compared with Koshihikari) than in those carrying Jukkoku_*sd1* (1–9% reduction compared with Koshihikari (Figure 2). Furthermore, leaves of the semidwarf and double dwarf lines were slightly shorter and straighter (pointing upwards) than in Koshihikari (Figure 3), indicating improved light-reception attitude by the integration of semidwarf gene (or genes). Panicle length was slightly longer (by 2.5%) in line carrying Jukkoku_*sd1* and slightly shorter in lines carrying Kinuhikari_*sd1* (by 2.4%) or *d60* (by 3.0%), compared with Koshihikari (Table 2). However, the reduction in panicle length was quite less than that in culm length (22.8% decrease in lines carrying Kinuhikari_*sd1* vs a 26.1% decrease in lines carrying *d60*). Therefore, the negative effects of semidwarf genes *sd1* and *d60* on panicle length were negligible.

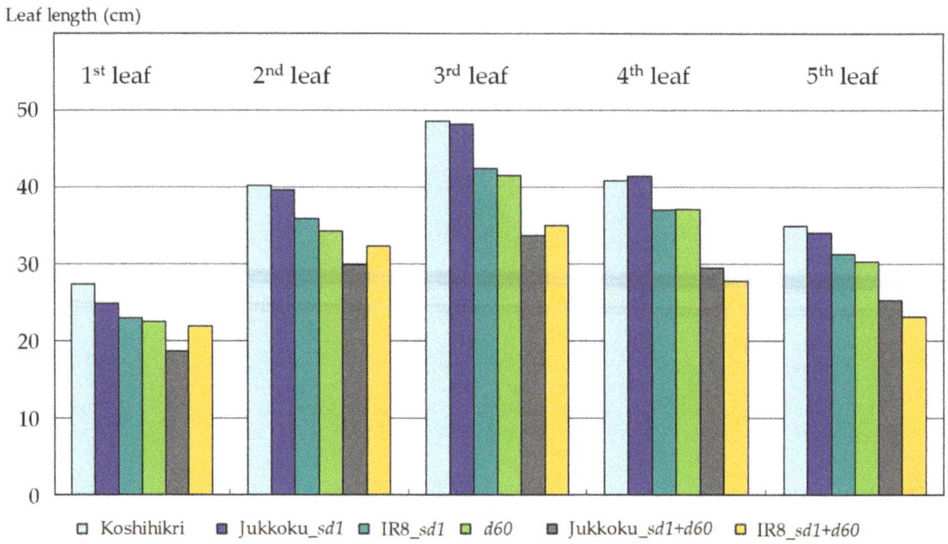

Figure 2. Effect of semidwarf and double dwarf genes to leaf length. Upper five leaves, arising from the main culm, were measured. Except for Jukkoku_*sd1* line. leaves of the semidwarf and double dwarf lines were slightly shorter than that of Koshihikari.

Figure 3. Plant phenotype of semidwarf and double dwarf gene lines. Leaves of the semidwarf and double dwarf lines were straighter (pointing upwards) than in Koshihikari, indicating improved light-reception attitude by the integration of semidwarf gene (or genes).

2.2. Effects of Semidwarf and Double Dwarf Genes on Yield

The yield components of each genotype are summarized in Table 2. The weight of unpolished rice/1000 grains and the proportions of fertile florets differed only slightly between lines. The effect of these genes on the proportion of fertile florets and the weight of unpolished rice/1000 grains were negligible. The number of panicles/plants was 17.9 in Koshihikari: 20.8 in line carrying Jukkoku_*sd1* (+16.2% vs Koshihikari), and 15.4 in Jukkoku DW line (−14.0% vs Koshihikari) (Figure 4, Table 3). In addition, the floret number/panicle was 87.3 in line carrying Jukkoku_*sd1* (+11.1% vs Koshihikari) and 72.1 in Jukkoku DW line (−8.3% vs Koshihikari) (Figure 5, Tables 2 and 4). The number of panicles was larger in line carrying Jukkoku_*sd1*, while floret density was larger in all semidwarf varieties than in Koshihikari (Figure 6, Table 5). Thus, an increase in both the number of panicles/plant and the floret number/panicle resulted in an increase in the number of panicles/m^2 and a consequent increase in yield (Figure 7, Table 6).

Table 2. Effect of semidwarf and double dwarf genes to yield components.

	Koshihikari	**Jukkoku_*sd1***	**IR8_*sd1***	***d60***	**Jukkoku _*sd1*+*d60***	**IR8_*sd1*+*d60***
Weight of unpolished rice/1000 grains (g)	22.1	20.2	20.4	20.9	21.3	20.5
Panicles /m^2	397.4	461.8 *	391.8	397.4	341.9 *	399.6
Floret number/panicle	78.6	87.3 *	82.8	79.3	72.1 *	76.2
Seed fertility (%)	96.6	94.9	93.4	95.4	95.6	96.2
Yield of unpolished rice (g/m^2=kg/a)	665.3	772.6 *	617.9 *	628.5	502.1 *	600.5 *

Compared with the yield of Koshihikari (mean, 665.3 g/m^2), that of the line carrying Jukkoku_*sd1* was highest value 772.6 g/m^2 increased by 16.1%, while that of IR8_*sd1*, *d60* and IR8_*sd1* plus *d60*, was slightly decreased by 7.1%, 5.5%, and 9.7%, respectively. The line carrying Jukkoku_*sd1* also showed highest value in number of panicles and florets/panicle, each 16.2% and 11.1% higher than in Koshihikari, which were responsible for the increases in yield. The weight of rice/1000 grains was equivalent among all genetic lines. Except for the semidwarf line carrying Jukkoku_*sd1*, semidwarf line carrying *d60* was equivalent to line carrying IR8_*sd1*in the yield of unpolished rice, and yield components such as panicles/m^2, floret number /panicle. *: statistically significant at the 5% level.

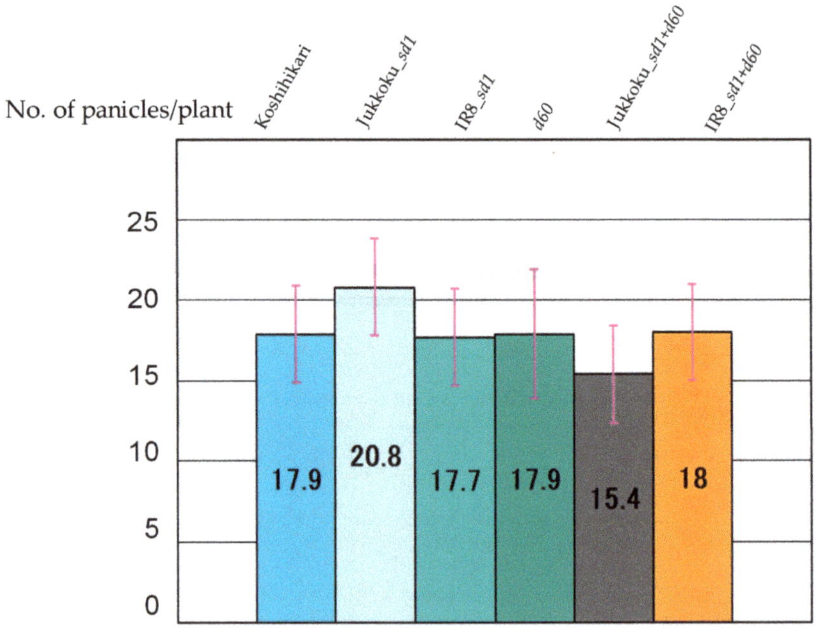

Figure 4. No. of panicles/plant in semidwarf and double dwarf gene lines. The number of panicles/plants was highest at 20.8 in line carrying Jukkoku_*sd1* (+10.2% vs Koshihikari).

Table 3. Effect of semidwarf and double dwarf genes to No. of panicles/plant.

	Koshihikari	Jukkoku_*sd1*	IR8_*sd1*	*d60*	Jukkoku _*sd1+d60*	IR8_*sd1+d60*
No. of panicles/plant	17.9	20.8 *	17.7	17.9	15.4 *	18.0
Percent change (%)	-	+16.2	−1.4	±0	−14.0	+0.6

The number of panicles/plants in line carrying *d60* (17.9) was comparable to that in line carrying IR8_*sd1* (17.7). *: statistically significant at the 5% level.

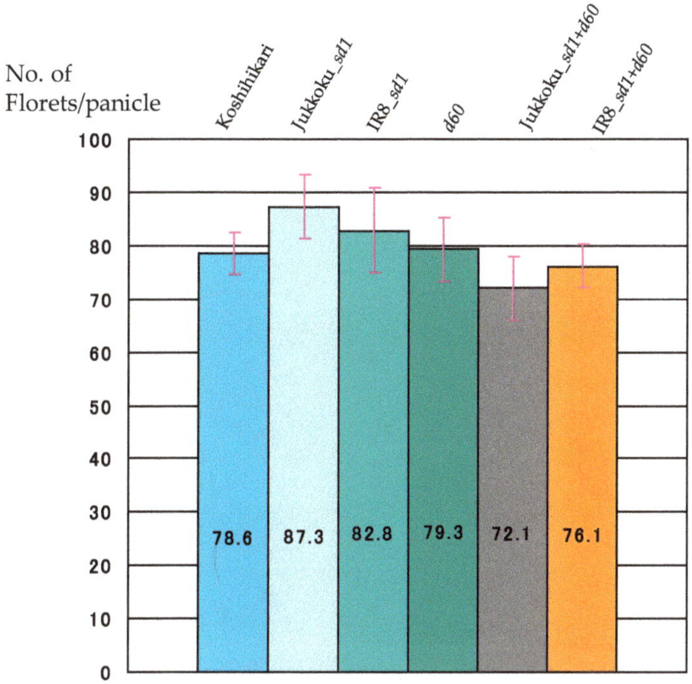

Figure 5. Floret number/panicle of semidwarf and double dwarf gene lines. The number of panicles/plants in line carrying *d60* was comparable to that in line carrying IR8_*sd1*.

Table 4. Effect of semidwarf and double dwarf genes to floret number/panicle.

	Koshihikari	Jukkoku_*sd1*	IR8_*sd1*	*d60*	Jukkoku _*sd1*+*d60*	IR8_*sd1*+*d60*
No. of Florets/panicle	78.6	87.3 *	82.8	79.3	72.1 *	76.1
Percent change (%)	-	+11.1	+5.3	+0.9	−8.3	−3.1

The floret number/panicle was highest at 87.3 in line carrying Jukkoku_*sd1* (+11.1% vs Koshihikari). *: statistically significant at the 5% level.

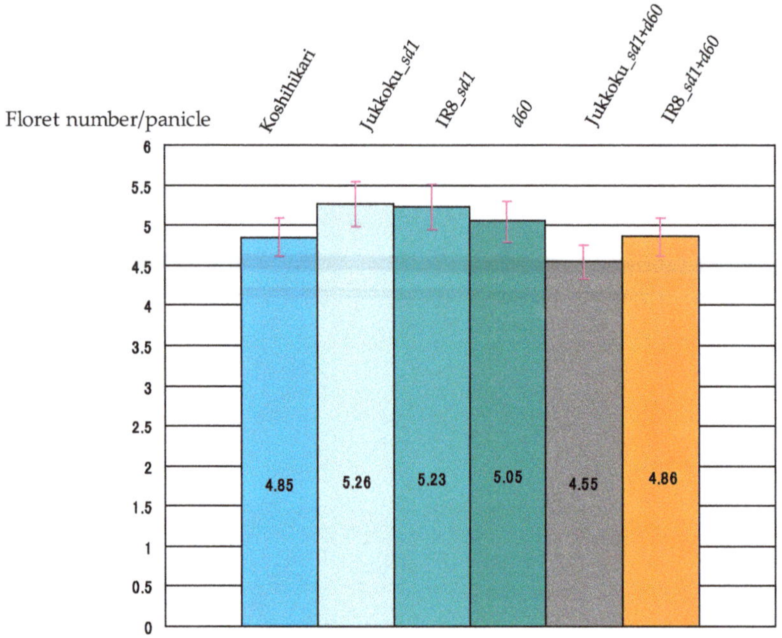

Figure 6. Effect of semidwarf and double dwarf genes to floret density. The floret density was larger in all lines carrying one semidwarf gene than that of Koshihikari.

Table 5. Effect of semidwarf and double dwarf genes to panicle.

	Koshihikari	**Jukkoku_*sd1***	**IR8_*sd1***	***d60***	**Jukkoku _*sd1*+*d60***	**IR8_*sd1*+*d60***
Floret number/panicle	78.6	87.3 *	82.8	79.3	72.1 *	76.1
Panicle length (cm)	16.2	16.6	15.8	15.7	15.8	15.7
Floret density (/cm)	4.85	5.26 *	5.23 *	5.05	4.55 *	4.86
Percent change of floret density (%)	-	+8.53	+7.95	+4.11	−6.15	+0.32

Panicle length was slightly longer (by 2.5%) in line carrying Jukkoku_*sd1* and slightly shorter in lines carrying IR8_*sd1* (by 2.4%) or *d60* (by 3.0%), compared with Koshihikari. The reduction in panicle length was quite less than that in culm length (22.8% decrease in line carrying IR8_*sd1* vs a 26.1% decrease in line carrying *d60*).

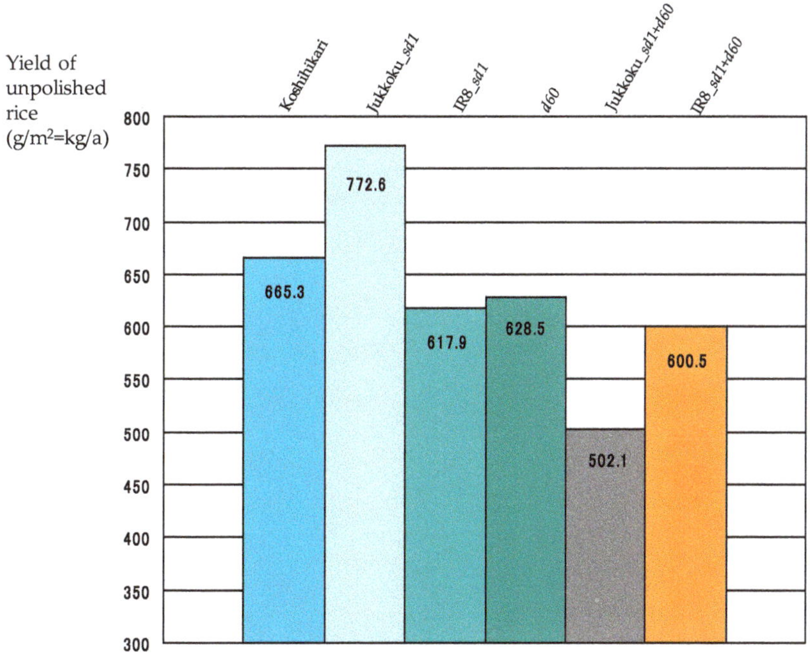

Figure 7. Yield of semidwarf and double dwarf gene lines. The yield of unpolished rice was 665.3 g/m^2 in Koshihikari, 772.6 g/m^2 in line carrying Jukkoku_*sd1* (+15.9% vs Koshihikari), 617.9 g/m^2 in line carrying Kinuhikari_*sd1* (−7.1% vs Koshihikari), and 628.5 g/m^2 in line carrying *d60* (−5.5% vs Koshihikari).

Table 6. Effect of semidwarf and double dwarf genes to yield.

	Koshihikari	Jukkoku_*sd1*	IR8_*sd1*	*d60*	Jukkoku _*sd1+d60*	IR8_*sd1+d60*
Yield of unpolished rice (g/m^2=kg/a)	665.3	772.6 *	617.9 *	628.5	502.1 *	600.5 *
Percent change (%)	-	+16.1	−7.1	-5.5	−24.5	−9.7

The yield of line carrying *d60* was comparable to that in line carrying IR8_*sd1*. *: statistically significant at the 5% level.

The yield of unpolished rice was 665.3 g/m^2 in Koshihikari, 772.6 g/m^2 in line carrying Jukkoku_*sd1* (+15.9% vs Koshihikari), 617.9 g/m^2 in line carrying Kinuhikari_*sd1* (−7.1% vs Koshihikari), and 628.5 g/m^2 in line carrying *d60* (−5.5% vs Koshihikari) (Figure 7, Table 6). The introduction of Kinuhikari_*sd1* or *d60* into Koshihikari appears to cause a slight reduction in yield. On the other hand, the yield of DW lines was markedly lower than that of Koshihikari: for Jukkoku DW line (−24.5% vs Koshihikari) and 600.5 g/m^2 for IR8 DW line (−9.7% vs Koshihikari) (Figure 7, Table 6). When using the alternative equation, the yield index was higher in all semidwarf-gene-carrying lines than in Koshihikari (Figure 8). The high yield index and lodging resistance of semidwarf varieties suggest that introduction of *sd1* and *d60* into non-dwarf genomes will be beneficial for increasing crop yield. Moreover, only minor differences in the grain appearance were observed among lines including Koshihikari, indicating that the grain quality of semidwarf lines is equivalent to that of Koshihikari. Taken together, semidwarf genes *sd1* and *d60* are useful in the agricultural industry.

The yield index was higher in line carrying IR8_*sd1* than in those carrying Jukkoku_*sd1* (Figure 8, Table 7). Although line carrying Jukkoku_*sd1* gave a higher yield than those carrying IR8_*sd1*, the higher yield index associated with IR8_*sd1* than Jukkoku_*sd1* suggests that the efficiency of the distribution to sink organs (e.g., seeds) is higher. Thus, the yield favors a gain in dry matter, which may be also higher in plants carrying IR8_*sd1* than in those carrying Jukkoku_*sd1*. Furthermore, the yield index for Jukkoku DW and IR8 DW are high—-45.6% and 46.5% higher, respectively than that of Koshihikari (Figure 8, Table 7). In order to increase a markedly low yield in DW lines, the use of conditions that favor a gain in dry matter, such as intensive cultivation with heavy fertilization to increase the number of tillers, may be effective.

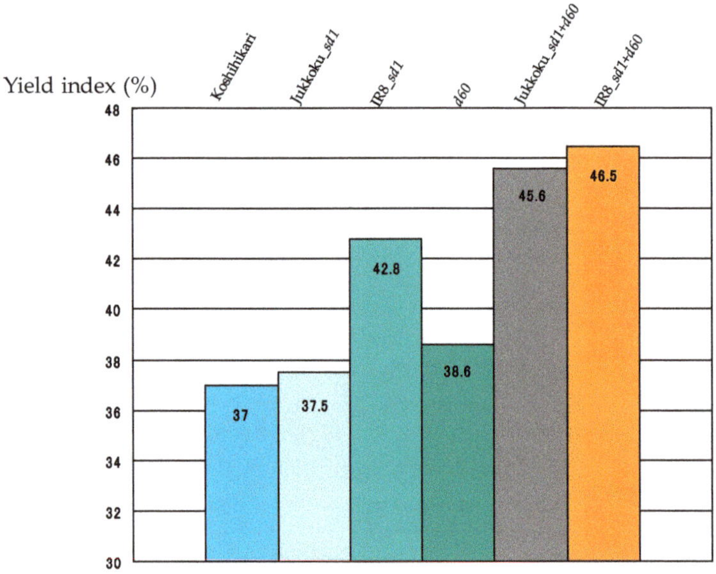

Figure 8. Yield index of semidwarf and double dwarf gene lines. The yield index was higher in line carrying IR8_*sd1* than in those carrying Jukkoku_*sd1*. The yield index for Jukkoku DW and IR8 DW are 45.6% and 46.5%, respectively higher than that of Koshihikari.

Table 7. Effect of semidwarf and double dwarf genes to yield index.

	Koshihikari	Jukkoku_*sd1*	IR8_*sd1*	*d60*	Jukkoku _*sd1+d60*	IR8_*sd1+d60*
Weight of winnowed paddy (g)	26.9	27.3	26.4	22.7 *	20.3 *	23.9 *
Weight of the plant part above the ground (g)	72.7	72.8	61.5 *	58.9 *	44.6 *	51.4 *
Yield index (%)	37.0	37.5	42.8 *	38.6	45.6 *	46.5 *

The high yield index and lodging resistance of semidwarf varieties suggest that introduction of *sd1* and *d60* into non-dwarf genomes will be beneficial for increasing crop yield. *: statistically significant at the 5% level.

3. Discussion

As exemplified by IR8, which was the variety behind the Green Revolution, many of the rice varieties cultivated worldwide commonly carry the semidwarf gene *sd1*. Another semidwarf gene *d60* is non-allelic to *sd1* and is of particular interest as a different source of semidwarfism to give genetic diversity among the semidwarf varieties. In this study, semidwarf lines, namely Jukkoku (Jukkoku _sd1), sd1 of Kinuhikari (IR8_sd1: Kinuhikari maintains sd1 of IR8 origin), d60 or sd1 plus d60 into the Koshihikari background, were used to investigate influence of these semidwarf genes on phenotypic traits, in relation to yield.

This study showed that all tested semidwarf lines had shorter culm lengths than Koshihikari, indicating improved lodging resistance. The effect on culm length carrying *d60*(65.7 cm) is slight shorter than in those carrying *sd1* (Jukkoku_*sd1*, 71.8 cm; IR8_*sd1*, 68.5 cm), Among the genetic lines, line carrying Jukkoku_*sd1*showed the highest yield of unpolished rice 772.6 g/m^2, which is 16.1% higher than in Koshikikari. The Jukkoku_*sd1* line also showed highest value in the number of panicles, the number of florets per panicle than in Koshihikari. Therefore, it was highly possible that the increasing yield of Jukkoku_*sd1* line was ascribed to the increasing numbers of panicles and florets. Although the yield of unpolished rice of *d60* line, 628.5 g/m^2 was 5.5% lower than that of Koshihikari (665.3 g/m^2), but this is almost equivalent yield performance of IR8_*sd1* (617.9 g/m^2). Ogi et al. (1993) [21] and Murai et al. (2004) [22] reported characteristics of isogenic line carrying *sd1* derived from DGWG, the source of IR8 *sd1*. These isogenic lines showed almost same number of panicles as that of original varieties, 'Norin 29' and 'Shiokari'. Therefore, it was concluded that Jukkoku_*sd1* especially has potential increasing panicle numbers compared to IR8_*sd1*. Hence, Jukkoku_*sd1* appears to confer a pleiotropic effect of increasing panicle number very well in the Koshihikari genetic background. The difference of such as effect between *sd1* alleles may be ascribed to that IR8_*sd1* suffered 383 bp deficit in the region exon 1-2 of *GA20-ox* [18], whereas, Jukkoku_*sd1* has only a SNP against the wild type *GA20-ox* [18] and the transcripts existed [23].

This study demonstrated that *d60* confers slightly shorter culm length than IR8_*sd1*, but almost equivalent yield performance with IR8_*sd1* together with effects on yield-related phenotypic traits comparable to IR8_*sd1*, which actually contributed to green revolution [2]. Although many dwarf genes are accompanied with a reduction in panicle length, yield of unpolished rice, and grain thresh ability (which is likely attributed to excessive dwarfing effects), *d60* does not exert such negative effects on yield-related phenotypic traits of rice plant. In conclusion, *d60* is applicable to practical breeding and one of choice for expanding genetic diversity of rice varieties.

4. Materials and Methods

4.1. Genetic Lines

The following rice semidwarf or double dwarf lines, Koshihikari, Koshihikari carrying Jukkoku_*sd1* [Koshihikari*6//(Kanto 79/Jukkoku F_4) B_6F_4], Koshihikari carrying IR8_*sd1* [Koshihikari/Kinuhikari F_5)], Koshihikari carrying *d60* [Koshihikari Koshihikari*7//(Koshihikari/Hokuriku 100) B_7F_3], Koshihikari carrying *d60* and Jukkoku_*sd1* [Jukkoku_DW, Koshihikari carrying Jukkoku_*sd1*/Koshihikari carrying_*d60* F_7), and Koshihikari carrying *d60* and IR8_*sd1* (IR8 DW: Koshihikari carrying IR8_*sd1*/ Koshihikari carrying_*d60* F_7) were used in this study. Koshihikari carrying Jukkoku_*sd1* was developed by six times of backcrosses with Koshihiakri as a recurrent parent using the short stemmed *sd1* homozygous fixed F_4 strain in Kanto No. 79 × Jukkoku (*sd1*) as a non-recurrent parent [24]. Koshihikari carrying *d60* was developed by seven times of backcrosses with Koshihikari as the recurrent parent using the short-stemmed F_2 plant as a non-recurrent parent [15]. Koshihikari carrying IR8__*sd1* was the short stemmed *sd1* homozygous fixed F_7 strain derived from Koshihikari × Kinuhikari. Koshihikari carrying *d60* and Jukkoku_*sd1*(Jukkoku DW) was double dwarf *d60sd1* homozygous fixed F_7 strain derived from Koshihikari carrying Jukkoku_*sd1* × Koshihikari carrying_*d60* [15]. Koshihikari carrying *d60* and IR8_*sd1*(IR8 DW) was double dwarf *d60sd1* homozygous fixed F_7 strain derived from Koshihikari carrying IR8_*sd1* × Koshihikari carrying_*d60* [15]. Kinuhikari has *sd1* derived from IR8 [25], which suffer 383 bp deficit in the region exon 1-2 from wild type *GA20-ox* [18]. Koshihikari carrying Jukkoku_*sd1*was plant-variety registered via further 7–8th backcrosses and it was designated as Hikarishinseiki [24], whose *sd1* has only a SNP against wild type *GA20-ox* [18,26]. Genomic *sd1*allele and the RNA transcript in Hikarishinseiki are detectable by diagnosis targeting the SNP [23,27].

4.2. Cultivation

Rice seeds were taken from stocks kept in a refrigerator. Seeds of each line were immersed in enough water just to cover the seeds. Water was exchanged every day for seven days (May 2 to May 8) during seed soaking and stimulation of germination. Seedlings were grown in nursery boxes (30 × 15 × 3 cm) for approximately 20 days. Seedlings were then individually transplanted into a paddy field (120 m^2: 6.0 × 20.0 m) of the University Farm on June 8. Two 4-m^2 plots (2 × 2 m) with transplanting densities 22.2 seedlings/m^2 (one seedling per 30 × 15 cm, 78 seedlings per field) were prepared for each genetic line (two instances). The paddy field was fertilized by 4.0 kg of basal fertilizer containing nitrogen, phosphorus, and potassium (weight ratio, nitrogen:phosphorus:potassium = 2.6:3.2:2.6) at the rate with 4.3 g/m^2 nitrogen, 5.3 g/m^2 phosphorus, and 4.3 g/m^2 potassium evenly across the field. A herbicide (Joystar L Floable, Kumiai Chemical Industry, Tokyo, Japan) was applied on June 20 to kill weeds grown uncontrollably, and the water level was then kept at a high enough level to cover the weeds for a week.

4.3. Growth Analysis

Ten seedlings were randomly selected for each line at the time of transplantation, and the full length was measured individually. After transplantation, ten plants were randomly selected, and the distance between the ground and the highest standing point (i.e., the full length) was measured every week for approximately three months until the panicle emerged. The time when the tip of the panicle first emerged from the flag leaf sheath was recorded as the heading time for all plants.

4.4. Plant Phenotyping

After ripening, ten plants typical of each genotype were sampled twice. Sampled plants were air-dried, and were assessed or measured the following traits. Culm length: the length between the ground surface and the panicle base of the main culm was measured at the time of sampling. Leaf length: the lengths of the upper five leaves, arising from the main culm, were individually measured. Length and weight of panicle: the length between the panicle base and the tip of the panicle, and the weight of the panicle, were measured. Total panicle number: the number of panicles were counted by sampled individuals and panicle numbers per 1 m^2 area (panicles/m^2) were counted twice in each plot of the paddy field. Total floret number: florets were counted to obtain total floret number. Floret number/panicle: the total floret number (including both sterile and fertile florets) was divided by the total panicle number. Proportion of fertile florets: each floret was assessed to determine its fertility. Floret density (floret number/cm): the number of florets per panicle was divided by the length of the panicle. Presence of awns: florets with an awn were counted when counting the florets. Grain threshability: was manually tested during examination of phenotypic traits. Appearance of grains: the size, color, and presence of an awn were observed for assessment of grain quality. Total weight of winnowed paddy: the total weight of winnowed paddy was weighed after grain selection using the salt solution (salt content of 1.06 g/m^3). Weight of sieved unpolished rice/1000 grains: obtained by multiplying the total weight of winnowed paddy by 0.84. Weight of plant parts above the ground: the weight of the plant parts above the ground was measured. Yield index: the winnowed paddy weight was divided by the weight of the plant part above the ground to obtain the yield index. The means of traits were statistically compared using the *t*-test.

4.5. Yield

Yield of unpolished rice was calculated using the following equation.

- Yield of unpolished rice (g/m^2) = (number of panicles/m^2) × (floret number/panicle) × (proportion of fertile florets) × (weight of unpolished rice/grain)

The following alternative equation (see below) was also used to calculate the comparison of yields:

- Yield = (yield index) × (weight of the plant parts above the ground)

Author Contributions: Conceptualization—M.T.; methodology—M.T.; investigation—K.I., M.T.; resources—M.T.; writing—original draft preparation—M.T.; writing—review and editing—M.T.; project administration, M.T.; funding acquisition—M.T.

Funding: This work is founded by Adaptable and Seamless Technology Transfer Program (A-STEP) through Target-driven R&D (high-risk challenge type) by Japan Science and Technology Agency (JST) to Motonori Tomita, whose project ID14529973 was entitled "NGS genome-wide analysis-based development of rice cultivars with super high-yield, large-grains, and early/late flowering suitable for the globalized world and global warming", since 2014 to 2018.

Conflicts of Interest: The authors declare no conflict of interest.

References

1. Khush, G.S. Green revolution: Preparing for the 21st century. *Genome* **1999**, *42*, 646–655. [CrossRef] [PubMed]
2. Hergrove, T.; Coffman, W.R. Breeding history. In *Rice that Changed the World: Cerebrating 50 Years of IR8*; Special supplement focusing on IR8; International Rice Research Institute: Los Baños, Philippines, 2016; pp. 6–10. Available online: http://books.irri.org/RT_Supplement-IR8.pdf (accessed on 25 September 2019).
3. Jennings, P. Rice revolutions in Latin America. In *Rice that Changed the World: Cerebrating 50 Years of IR8*; Special supplement focusing on IR8; International Rice Research Institute: Los Baños, Philippines, 2016; p. 19. Available online: http://books.irri.org/RT_Supplement-IR8.pdf (accessed on 25 September 2019).
4. Okada, M.; Yamakawa, Y.; Fujii, K.; Nishiyama, H.; Motomura, H.; Kai, S.; Imai, T. On the new varieties of paddy rice, 'Hoyoku', 'Kokumasari' and 'Shiranui'. *J. Agric. Exp. Stn. Kyushu* **1967**, *2*, 187–224.
5. Futuhara, Y. Breeding of new rice variety Reimei by gamma ray irradiation. *Gamma. Field Symp.* **1968**, *7*, 87–109.
6. Rutger, J.N.; Peterson, M.L.; Hu, C.H.; Lehman, W.F. Induction of usefil short stature and early maturing mutants in two japonica rice cultivars. *Crop Sci.* **1976**, *16*, 631–635. [CrossRef]
7. Rutger, J.N.; Peterson, M.L.; Hu, C.H. Registration of Calrose76 rice. *Crop Sci.* **1977**, *17*, 978. [CrossRef]
8. Aquino, R.C.; Jennings, P.R. Inheritance and significance of dwarfism in an indica rice variety. *Crop Sci.* **1966**, *6*, 551–554. [CrossRef]
9. Heu, M.; Chang, T.T.; Cabanilla, V.L. The inheritance of culm length, panicle length, duration to heading and bacterial leaf blight reaction in rice cross: Sigadis×Taichung (Native)1. *Jpn. J. Breed.* **1968**, *18*, 7–11.
10. Foster, K.W.; Rutger, J.N. Inheritance of semidwarfism in rice, *Oryza sativa* L. *Genetics* **1978**, *88*, 559–574.
11. Mackill, D.J.; Rutger, J.N. The inheritance of induced-mutant semidwarfing genes bin rice. *J. Hered.* **1979**, *70*, 335–341. [CrossRef]
12. Ikehashi, H.; Kikuchi, F. Genetic analysis of semidwarfness and their significance for breeding of high-yielding varieties in rice. *JARQ* **1982**, *15*, 231–235.
13. Kikuchi, F.; Itakura, N.; Ikehashi, H.; Yokoo, M.; Nakane, A.; Maruyama, K. Genetic analysis of semidwarfism in high-yielding rice varieties in Japan. *Bull. Natl. Inst. Agr. Sci. Ser. D* **1985**, *36*, 125–145.
14. Kikuchi, F.; Futsuhara, Y. Inheritance of morphological characters. 2. Inheritance of semidwarf. In *Science of the Rice Plant*, 3rd ed.; Matsuo, T., Shimizu, S., Tsunoda, S., Murata, Y., Kumazawa, K., Futsuhara, Y., Hoshikawa, K., Yamaguchi, H., Kikuchi, F., Eds.; Tokyo Food and Agricultural Policy Research Center: Tokyo, Japan, 1997; Volume 3, pp. 309–317.
15. Tomita, M. Combining two semidwarfing genes *d60* and *sd1* for reduced height in 'Minihikari', a new rice germplasm in the 'Koshihikari' genetic background. *Genet. Res. Camb.* **2012**, *94*, 235–244. [CrossRef] [PubMed]
16. Suh, H.S.; Heu, M.H. The segregation mode of plant height in the cross of rice varieties. VI. Linkage analysis of the semi-dwarfness of the rice variety Tongil. *Korean J. Breed.* **1978**, *10*, 1–6.
17. Maeda, H.; Ishii, T.; Mori, H.; Kuroda, J.; Horimoto, M.; Takamure, I.; Kinoshita, T.; Kamijima, O. High density molecular map of semidwarfing gene, *sd-1*, in rice (*Oryza sativa* L.). *Breed. Sci.* **1997**, *47*, 317–320. [CrossRef]
18. Sasaki, A.; Ashikari, M.; Ueguchi-Tanaka, M.; Itoh, H.; Nishimura, A.; Swapan, D.; Ishiyama, K.; Saito, T.; Kobayashi, M.; Khush, G.S.; et al. Green revolution: A mutant gibberellin-synthesis gene in rice. *Nature* **2002**, *416*, 701–702. [CrossRef] [PubMed]

19. Monna, L.; Kitazawa, N.; Yoshino, R.; Suzuki, J.; Masuda, H.; Maehara, Y.; Tanji, M.; Sato, M.; Nasu, S.; Minobe, Y. Positional cloning of rice semidwarfing gene, *sd-1*: Rice "Green revolution gene" encodes a mutant enzyme involved in gibberellin synthesis. *DNA Res.* **2002**, *9*, 11–17. [CrossRef]
20. Spielmeyer, W.; Ellis, M.H.; Chandler, P.M. Semidwarf (*sd-1*), "green revolution" rice, contains a defective gibberellin 20-oxidase gene. *Proc. Natl. Acad. Sci. USA* **2002**, *99*, 9043–9048. [CrossRef]
21. Ogi, Y.; Kato, H.; Maruyama, K.; Kikuchi, F. The effects of the culm length andother agronomic characters cause by semidwarfing genes at the sd-1 locus inrice. *Jpn. J. Breed.* **1993**, *43*, 267–275. [CrossRef]
22. Murai, M.; Komazaki, T.; Sato, S. Effects of *sd1* and *Ur1* (Undulate rachis-1) on lodging resistance and related traits in rice. *Breed. Sci.* **2004**, *54*, 333–340. [CrossRef]
23. Tomita, M.; Matsumoto, S. Transcription of rice Green Revolution *sd1* gene is clarified by comparative RNA diagnosis using the isogenic background. *Genom. Appl. Biol.* **2011**, *2*, 29–35.
24. Tomita, M. Introgression of Green Revolution *sd1* gene into isogenic genome of rice super cultivar Koshihikari to create novel semidwarf cultivar 'Hikarishinseiki' (Koshihikari-sd1). *Field Crops Res.* **2009**, *114*, 173–181. [CrossRef]
25. Tabuchi, H.; Hashimoto, N.; Takeuchi, A.; Terao, T.; Fukuta, Y. Genetic analysis of semidwarfism of the japonica rice cultivar Kinuhikari. *Breed. Sci.* **2000**, *50*, 1–7. [CrossRef]
26. Tomita, M.; Ishii, K. Genetic performance of the semidwarfing allele *sd1* derived from a Japonica rice cultivar and minimum requirements to detect its single-nucleotide polymorphism by MiSeq whole-genome sequencing. *BioMed Res. Int.* **2018**, *2018*, 4241725. [CrossRef] [PubMed]
27. Naito, Y.; Tomita, M. Identification of an isogenic semidwarf rice cultivar carrying the Green Revolution *sd1* gene using multiplex codominant ASP-PCR and SSR markers. *Biochem. Genet.* **2013**, *51*, 530–542. [CrossRef] [PubMed]

plants

MDPI

Article

The Development and Characterization of Near-Isogenic and Pyramided Lines Carrying Resistance Genes to Brown Planthopper with the Genetic Background of *japonica* Rice (*Oryza sativa* L.)

Cuong D. Nguyen [1,2], Holden Verdeprado [3], Demeter Zita [4], Sachiyo Sanada-Morimura [5], Masaya Matsumura [5], Parminder S. Virk [3,6], Darshan S. Brar [3,7], Finbarr G. Horgan [8], Hideshi Yasui [9] and Daisuke Fujita [1,3,4,9,*]

1 United Graduate School of Agricultural Sciences, Kagoshima University, Kagoshima 890-0065, Japan; cuongdhqt2000@gmail.com
2 College of Food Industry, 101B Le Huu Trac Street, Son Tra District, Da Nang City 550000, Vietnam
3 International Rice Research Institute, DAPO Box 7777, Metro Manila 1301, Philippines; h.verdeprado@irri.org (H.V.); p.virk@cgiar.org (P.S.V.); darshanbrar@pau.edu (D.S.B.)
4 Faculty of Agriculture, Saga University, Saga 840-8502, Japan; demeterzita2012@gmail.com
5 NARO Kyushu Okinawa Agricultural Research Center, 2421 Suya, Koshi, Kumamoto 861–1192, Japan; sanadas@affrc.go.jp (S.S.-M.); mmasa@affrc.go.jp (M.M.)
6 International Center for Tropical Agriculture, A.A, 6713 Cali, Colombia
7 School of Agricultural Biotechnology, Punjab Agricultural University, Ludhiana 141027, India
8 EcoLaVerna Integral Restoration Ecology, Bridestown, Kildinan, Co. Cork, T56 CD39, Ireland; f.g.horgan@gmail.com
9 Plant Breeding Laboratory, Graduate School, Kyushu University, Fukuoka 812-8581, Japan; hyasui@agr.kyushu-u.ac.jp
* Correspondence: dfujita@cc.saga-u.ac.jp; Tel.: +81-952-28-8724

Received: 18 September 2019; Accepted: 7 November 2019; Published: 12 November 2019

Abstract: The brown planthopper (BPH: *Nilaparvata lugens* Stål.) is a major pest of rice, *Oryza sativa*, in Asia. Host plant resistance has tremendous potential to reduce the damage caused to rice by the planthopper. However, the effectiveness of resistance genes varies spatially and temporally according to BPH virulence. Understanding patterns in BPH virulence against resistance genes is necessary to efficiently and sustainably deploy resistant rice varieties. To survey BPH virulence patterns, seven near-isogenic lines (NILs), each with a single BPH resistance gene (*BPH2*-NIL, *BPH3*-NIL, *BPH17*-NIL, *BPH20*-NIL, *BPH21*-NIL, *BPH32*-NIL and *BPH17-ptb*-NIL) and fifteen pyramided lines (PYLs) carrying multiple resistance genes were developed with the genetic background of the *japonica* rice variety, Taichung 65 (T65), and assessed for resistance levels against two BPH populations (Hadano-66 and Koshi-2013 collected in Japan in 1966 and 2013, respectively). Many of the NILs and PYLs were resistant against the Hadano-66 population but were less effective against the Koshi-2013 population. Among PYLs, *BPH20+BPH32*-PYL and *BPH2+BPH3+BPH17*-PYL granted relatively high BPH resistance against Koshi-2013. The NILs and PYLs developed in this research will be useful to monitor BPH virulence prior to deploying resistant rice varieties and improve rice's resistance to BPH in the context of regionally increasing levels of virulence.

Keywords: rice (*Oryza sativa* L.); brown planthopper; near-isogenic lines; pyramided lines; resistance; virulence

1. Introduction

The brown planthopper (BPH: *Nilaparvata lugens* Stål.) is a major pest of rice (*Oryza sativa* L.) in tropical and subtropical Asia [1]. BPH damages rice by sucking phloem from the plants (mechanical damage) or by transmitting viruses such as rice grassy stunt virus (RGSV), rice ragged stunt phytoreovirus (RRSV) and rice wilted stunt virus (RWSV) [2,3]. At high planthopper densities, rice crops display patches of desiccated rice known as 'hopperburn.' Insecticides have been widely used to reduce BPH populations [4]. However, insecticides are damaging to human health and the environment, and are increasingly recognized as contributing to BPH outbreaks through physiological and ecological pest resurgence mechanisms [1]. Host plant resistance is considered a potentially effective alternative to harmful insecticides, that reduces BPH damage without detrimental effects on the natural enemies of BPH [5].

To date, more than 34 BPH resistance genes have been identified from rice cultivars and wild rice species. Seven genes: *BPH9, BPH14, BPH17, BPH18, BPH26, BPH29* and *BPH32* have been cloned and characterized for different BPH resistance levels [6–12]. Four gene clusters (chromosomal regions) strongly associated with BPH resistance have been identified. These occur on chromosomes 4S (short arm), 4L (long arm), 6S and 12 [3,13]. Four genes (*BPH12* from B14, *BPH15, BPH17* and *BPH20*) and six quantitative trait loci (QTLs) (*QBph4, QBPH4.1, QBPH4.2, QBph4.2 qBph4.3* and *qBph4.4*) for BPH resistance have been identified on chromosome 4S [13–20]. Among those QTLs, *QBph4* (6.7–6.9 Mbp) from IR02W101 (*Oryza officinalis*) and *QBph4.2* (6.6–6.9 Mbp) from IR65482-17 (*Oryza australiensis*) were identified at similar locations based on physical distance; *QBPH4.1* (5.8–7.8 Mbp) and *QBPH4.2* (15.2–17.2 Mbp) were identified from Rathu Heenati; *qBph4.3* (0.2–0.7 Mbp) and *qBph4.4* (0.7–13.1 Mbp) were detected in Salkathi. On chromosome 4L, five genes for BPH resistance—*BPH6, BPH12(t)* from GSK185-2, *BPH18(t)* from BPH2183, *BPH27* from GX2183 and *BPH27(t)* from Balamawee have been identified [21–25]. Six genes/QTLs have been detected on the short arm of chromosome 6: *BPH3, BPH4, BPH25, BPH32, Qbph6* and *qBPH6(t)* [12,26–30]. Eight genes have been identified on the long arm of chromosome 12—*BPH1, BPH2, BPH7, BPH9, BPH10* and *BPH18* from IR65482-7-216-1-2, and *BPH21* and *BPH26* [17,29,31–38]. The BPH resistance genes on chromosome 12 were classified into four allelic types based on their amino acid sequences and different resistance levels: type 1—*BPH1, BPH10, BPH18* and *BPH21*; type 2—*BPH2* and *BPH26*; type 3—*BPH7*; and type 4—*BPH9* [6]. A number of highly resistant rice breeding lines and varieties contain several genes with resistance to BPH and other phloem feeding Hemiptera. These include IR71033-121-15 introgressed from *Oryza minuta* carrying *BPH20, BPH21* and *qBPH6(t)* [17,39]; Rathu Heenati carrying *BPH3, BPH17, QBPH4.1* and *QBPH4.2* [8,13,27]; and PTB33 carrying *BPH2* and *BPH32* [12,40,41].

Monogenic resistance is vulnerable to rapid adaptation by BPH populations. Research indicates that BPH populations have sufficient genetic variability to enable them to overcome specific resistance genes when selected on a resistant host over multiple generations [4,42,43]. In the late 1970s, BPH populations adapted to varieties carrying the *BPH1* and/or *BPH2* genes after these were widely deployed in rice varieties across Asia [4,42,44]. A recent multi-national study has indicated that BPH populations across much of Asia have adapted to feed on rice carrying the *BPH1, BPH2, BPH5, BPH7, BPH8, BPH9, BPH10* and *BPH18* genes [45]. Under laboratory conditions, BPH populations continually reared for between seven to 15 generations on resistant rice varieties were capable of adapting to resistance from a range of genes, including *BPH1, BPH2, BPH3, BPH8, BPH9, BPH10* and *BPH32* [43,46–49]. Through adaptation to resistance genes, BPH acquires stronger virulence against resistance genes and BPH virulence remains stable for several decades [4,50]. Therefore, it is important to preserve the effects of resistance genes by preventing BPH adaptation.

To prevent further adaptation by BPH populations to available resistance genes, a strategy for deploying resistance based on insect virulence is necessary [4]. However, BPH virulence varies under different environments depending on the predominant rice cultivars, BPH migration routes, and the length of population exposure to different resistance genes [4]. Without exposing resistance genes to BPH populations under controlled conditions prior to deployment, the potential effectiveness of the resistance genes for target regions is difficult to predict. In previous studies, the virulence of BPH has been characterized using resistant varieties [45,48,51–53]. However, many BPH-resistant varieties have multiple resistance genes, such that the effects of any single resistance gene cannot be assessed using these varieties. In contrast, the effects of any single resistance gene may be revealed in detail by using near-isogenic lines (NILs) that carry the gene on the genetic background of a susceptible variety. Recently, more than 16 NILs with BPH resistance genes (*BPH3, BPH4, BPH6, BPH9, BPH10, BPH12, BPH14, BPH15, BPH17, BPH18, BPH20, BPH21, BPH25, BPH26, BPH30* and *BPH32*) have been developed on the genetic backgrounds of several *indica* and *japonica*-susceptible varieties. These NILs have been evaluated against different BPH populations from China, the Philippines and Japan [14,54–58].

Because the resistance of rice varieties carrying single genes is weaker and less durable (i.e., allowing rapid BPH adaptation) to BPH than varieties with multiple resistance genes, several researchers have proposed the pyramiding of two or more genes to enhance resistance levels and thereby avoid pest adaptation [59]. Combinations of multiple BPH resistance genes have been reported to increase levels of plant resistance to BPH. For example, a pyramided line (PYL) with *BPH14* and *BPH15* enhanced resistance against BPH from China compared to monogenic NILs with either *BPH14* or *BPH15* alone [60]. Similarly, the pyramided lines *BPH6* + *BPH12* PYL and *BPH3* + *BPH27* PYL exhibited greater resistance levels in bulk seedling tests than monogenic lines with each of the genes present alone [14,55], and *BPH17* + *BPH21* PYL had greater resistance against BPH in the Philippines than lines with either gene alone [57]. Pyramiding the *BPH25* and *BPH26* genes into a single rice line was reported to have positive epistatic effects against BPH populations collected in Vietnam, the Philippines and Japan [51,61]. Therefore, the development of rice varieties carrying multiple BPH resistance genes might be an effective way to enhance BPH resistance.

In this study, seven NILs with BPH resistance genes (*BPH2, BPH3, BPH17, BPH20, BPH21, BPH32* and *BPH17-ptb*) and a *japonica* rice genetic background were developed to evaluate the effects of different resistance genes on BPH populations. Based on the NILs developed, 15 pyramided lines (PYLs) carrying two or three resistance genes were developed to enhance levels of resistance against BPH. Additionally, using the NILs and PYLs we developed, the study compared resistance against two BPH populations collected in Japan: the first was collected in 1966 (before resistant varieties were widely released) and the second was collected in 2013 (recently migrated from China to Japan). Comparisons of the reactions by BPH from each population to the NILs and PYLs indicates the utility of resistance genes and their different combinations (some with epistatic effects) against modern BPH populations.

2. Results

2.1. Development of Seven NILs for BPH Resistance

Seven NILs with BPH resistance genes from three donor parents on the genetic background of Taichung 65 (T65) were developed through marker-assisted selection (MAS) and backcrossing (Tables 1 and 2). For three donor parents, IR71033-121-15 has *BPH20* and *BPH21*; Rathu Heenati contains *BPH3* and *BPH17*; and PTB33 carries *BPH2, BPH17-ptb* and *BPH32* based on previous studies. For PTB33, there has been no previous report of a BPH resistance gene on chromosome 4S. However, amino acid sequences for the *BPH17* locus in PTB33 were identical to those of Rathu Heenati [8]. Thus, we assume that PTB33 contains a gene for BPH resistance on chromosome 4S and tentatively named this as *BPH17-ptb*. The substituted chromosomal segments of the NILs were detected by polymorphic simple sequence repeat (SSR) markers that were equally distributed across the whole genome (Table 3; Figure 1). The genetic background of *BPH2*-NIL was analyzed using 203 polymorphic SSR markers. The ratio of substituted segments from PTB33 on *BPH2*-NIL was 9.1–14.8% (total 33.9–55.0 Mbp). One substituted segment with a size of 21.3–25.4 Mbp encompassing *BPH2* was detected between RM247 and RM5479 on chromosome 12. The other three segments were detected between RM5426 and

RM248 on chromosome 7 with a size of 3.4–4.2 Mbp, between RM5688 and RM444 on chromosome 9 with a size of 4.2–9.2 Mbp and between RM7492 and RM216 on chromosome 10 with a size of 5.0–16.2 Mbp.

The genetic background of *BPH3*-NIL was confirmed using 195 polymorphic SSR markers, and the ratio of substituted segment from Rathu Heenati was 1.0–3.0% (total 3.8–11.3 Mbp). One segment with a size of 1.6–1.8 Mbp including *BPH3* was detected between MSSR1 and RM1369 on the short arm of chromosome 6. The other substituted segments were detected between RM1359 and RM1155 on chromosome 4 with a size of 0.5–4.2 Mbp and between RM1345 and RM3155 on chromosome 8 with a size of 1.8–4.9 Mbp.

The genetic background of *BPH17*-NIL was surveyed using 173 polymorphic SSR markers. The ratio of substituted segments was 1.0–4.8% (total 3.8–17.6 Mbp) containing one segment located between RM8213 and B40 on chromosome 4, including the *BPH17* region.

The genetic background of *BPH17-ptb*-NIL was analyzed using 229 polymorphic SSR markers, and the ratio of substituted segments from PTB33 on *BPH17-ptb*-NIL was 2.8–7.6% (total 10.5–28.1 Mbp). One substituted segment with 5.8–13.1 Mbp encompassing *BPH17-ptb* was detected between C61009 and B40 on chromosome 4. Two other substituted segments were detected at RM3126 on chromosome 3 and between RM7048 and RM6971 on chromosome 9 (4.7–11.5 Mbp).

The genetic background of *BPH20*-NIL was confirmed using 237 polymorphic SSR markers and the ratio of substituted segments of IR71033-121-15 was 5.6–9.6% (total 20.6–35.5 Mbp). One segment with a size of 13.8–19.9 Mb containing *BPH20* was detected between RM335 and RM5900 on chromosome 4. Two other substituted segments were detected between RM224 and RM5926 on chromosome 11 (1.5–7.7 Mbp) and between RM7315 and RM3103 on chromosome 12 (5.3–7.9 Mbp).

The genetic background of *BPH21*-NIL was surveyed using 229 polymorphic SSR markers, and the ratio of substituted segments from IR71033-121-15 was 7.1–11.6% (total 26.4–43.1 Mbp). One segment with a size of 22.6–23.7 Mbp, including *BPH21*, was detected between RM1880 and RM28493 on chromosome 12. Three other segments were detected between RM6841 and RM3348 on chromosome 5 (2.3–7.2 Mbp), around RM1328 on chromosome 9 and between RM224 and RM5926 on chromosome 11 (1.5–7.7 Mbp).

The genetic background of *BPH32*-NIL was confirmed using 233 polymorphic SSR markers. The ratio of substituted segments of PTB33 on *BPH32*-NIL was 1.9–4.1% (total 7.1–15.1 Mbp). One segment with a size of 1.6–3.2 Mbp containing *BPH32* was detected between RM6775 and RM190 on chromosome 6. Three other segments from the donor parent were detected between RM5755 and RM3280 on chromosome 3 with a size of 4.9–8.1 Mbp, between RM1306 and RM248 on chromosome 7 with a size of 0.4–3.2 Mbp and between RM5349 and RM5961 on chromosome 11 with a size of 0.2–0.6 Mbp.

Table 1. The SSR markers used for maker-assisted selection of nine genes for resistance to the brown planthopper.

Marker	Resistance Gene Tagged	Chromosome	Forward Primer Sequence (5′→3′)	Reverse Primer Sequence (5′→3′)	Physical Position (Mbp) *	Predicted Size (bp) **
RM1246 [a]	*BPH2, BPH21*	12	GGCTCACCTCGTTCTCGATCC	CATAAATAAATAGGGCGCCACACC	19.16	195
RM28493 [b]	*BPH2, BPH21*	12	ACCGTTAGATGACACAAGCAACG	GGTTAGCAAGACTGGAGGAGACG	23.28	259
RM508 [c]	*BPH3, BPH32*	6	AGAAGCCGGTTCATAGTTCATGC	ACCCGTGAACCACAAAGAACG	0.44	158
RM588 [c]	*BPH3, BPH32*	6	TCTTGCTGTGCTGTTAGTGTACG	GCAGGACATAAATACTAGGCATGG	1.61	97
RM8213 [a]	*BPH17, BPH17-ptb*	4	TGTTGGGTGGGTAAAGTAGATGC	CCCAGTGATACAAAGATGAGTTGG	4.42	178
MS10 [d]	*BPH17, BPH17-ptb, BPH20*	4	CAATACGAGAAGCCCCTCAC	CTGAAGGAACACGCGGTAGT	8.08	167
RM5900 [a]	*BPH20*	4	TTCTACGTTTGACCGTCA	TCTAGGAGCGTTTGTAGGAG	13.77	248
S00310 [e]	*BPH25*	6	CAACAAGATGGACGGCAAGG	TTGGAAGAAAAGGCAGGCAC	0.21	215
MSSR1 [e]	*BPH25*	6	CTAGCTGCTCTGCTCTGCTG	CGGCAATCTCTCCGAATC	0.22	114
RM309 [f]	*BPH26*	12	CACGCACCTTTCTGGCTTTCAGC	AGCAACCTCCGACGGGAGAAGG	21.52	177
RM28438 [b]	*BPH26*	12	GTTCGTGAGCCACAACAAATCC	GTTAAATGCTCCACCAAACACACC	22.59	216
InD14 [g]	*BPH26*	12	GGCCGAGTAGGATACTCTAGAAA	CTGCGAGAAAGGAGAGGTGG	22.87	387
RM28466 [b]	*BPH26*	12	CCGACGAAGAAGACGAGGAGTAGCC	AGGCCGGAGAGCAATCATGTCG	22.98	93
RM28481 [b]	*BPH26*	12	GTCAATTAACCATTGCCCATGC	TTCACGTGGGAACTACTCATGC	23.14	242
MSSR2 [e]	*BPH26*	12	CATGTCGAAGAGGTTGCAGA	GGTTTCATCCAAGTCCACGA	25.03	265

Primer sequence information was obtained from: [a]: McCouch et al. (2002) [62], [b]: International Rice Genome Sequencing Project (IRGSP) (2005) [63], [c]: Temnykh et al. (2001) [64], [d]: Rahman et al. (2009) [17], [e]: Myint et al. (2012) [29], [f]: Temnykh et al. (2000) [65] and [g]: Zhao et al. (2016) [6]. * The physical position of marker was the physical location of forward primer for each marker obtained from The Rice Annotation Project Database [66]. ** Physical distance and predicted size were estimated on the basis of the Nipponbare genome sequence.

Table 2. Details of the seven near-isogenic lines and 15 pyramided lines carrying brown planthopper resistance genes.

Entry	Gene (Donor Parent)						Generation
BPH2-NIL	***BPH2***	**(PTB33)**					BC_4F_3
BPH3-NIL	*BPH3*	(Rathu Heenati)					BC_4F_4
BPH17-NIL	*BPH17*	(Rathu Heenati)					BC_4F_4
BPH20-NIL	*BPH20*	(IR71033-121-15)					BC_4F_5
BPH21-NIL	*BPH21*	(IR71033-121-15)					BC_4F_5
BPH32-NIL	*BPH32*	(PTB33)					BC_4F_4
BPH17-ptb-NIL	*BPH17-ptb*	(PTB33)					BC_4F_3
BPH2+BPH17-PYL	*BPH2*	(PTB33)	*BPH17*	(Rathu Heenati)			BC_4F_3 equivalent
BPH2+BPH25-PYL	*BPH2*	(PTB33)	*BPH25*	(ADR52)			BC_4F_3 equivalent
BPH2+BPH32-PYL	*BPH2*	(PTB33)	*BPH32*	(PTB33)			BC_4F_3 equivalent
BPH2+BPH17-ptb-PYL	*BPH2*	(PTB33)	*BPH17-ptb*	(PTB33)			BC_4F_3 equivalent
BPH3+BPH17-PYL	*BPH3*	(Rathu Heenati)	*BPH17*	(Rathu Heenati)			BC_4F_4 equivalent
BPH17+BPH21-PYL	*BPH17*	(Rathu Heenati)	*BPH21*	(IR71033-121-15)			BC_4F_3 equivalent
BPH20+BPH21-PYL	*BPH20*	(IR71033-121-15)	*BPH21*	(IR71033-121-15)			BC_3F_8 equivalent
BPH20+BPH32-PYL	*BPH20*	(IR71033-121-15)	*BPH32*	(PTB33)			BC_4F_3 equivalent
BPH21+BPH25-PYL	*BPH21*	(IR71033-121-15)	*BPH25*	(ADR52)			BC_4F_3 equivalent
BPH21+BPH17-ptb-PYL	*BPH21*	(IR71033-121-15)	*BPH17-ptb*	(PTB33)			BC_4F_3 equivalent
BPH25+BPH17-ptb-PYL	*BPH25*	(ADR52)	*BPH17-ptb*	(PTB33)			BC_4F_3 equivalent
BPH32+BPH17-ptb-PYL	*BPH32*	(PTB33)	*BPH17-ptb*	(PTB33)			BC_3F_8 equivalent
BPH2+BPH3+BPH17-PYL	*BPH2*	(PTB33)	*BPH3*	(Rathu Heenati)	*BPH17*	(Rathu Heenati)	BC_4F_3 equivalent
BPH2+BPH32+BPH17-ptb-PYL	*BPH2*	(PTB33)	*BPH32*	(PTB33)	*BPH17-ptb*	(PTB33)	BC_4F_3 equivalent
BPH20+BPH21+BPH32-PYL	*BPH20*	(IR71033-121-15)	*BPH21*	(IR71033-121-15)	*BPH32*	(PTB33)	BC_4F_3 equivalent

Table 3. Background survey analysis of seven near-isogenic lines using SSR polymorphic markers.

NIL	Donor	No. of SSR Markers			Genome Ratio (%)		Total Physical Distance of Donor Segment (Mbp) *	
		T65	Donor	Total	T65	Donor		
bph2-NIL	PTB33	183	20	203	85.2–90.9	9.1–14.8	33.9	55.0
Bph3-NIL	Rathu Heenati	181	14	195	97.0–99.0	1.0–3.0	3.8	11.3
Bph17-NIL	Rathu Heenati	170	3	173	95.2–99.0	1.0–4.8	3.8	17.6
BPH17-ptb-NIL	PTB33	219	10	229	92.4–97.2	2.8–7.6	10.5	28.1
Bph20-NIL	IR71033-121-15	224	13	237	90.4–94.4	5.6–9.6	20.6	35.5
Bph21-NIL	IR71033-121-15	210	19	229	88.4–92.9	7.1–11.6	26.4	43.1
BPH32-NIL	PTB33	220	13	233	95.9–98.1	1.9–4.1	7.1	15.1

* The minimum physical distance of donor segment was calculated by the distance between two markers delimiting the substituted segment and the maximum amount was calculated by two flanking markers of substituted segments.

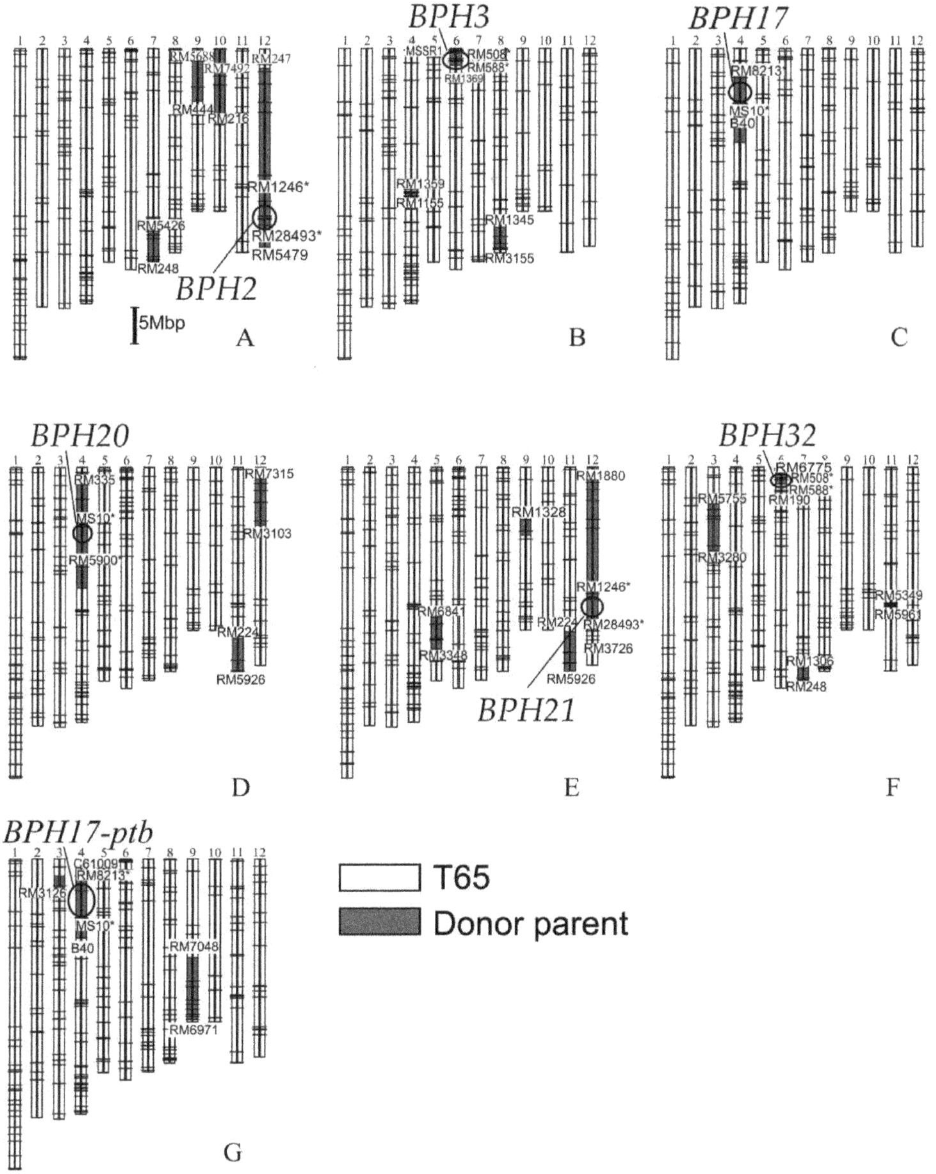

Figure 1. Graphical genotypes of *BPH2*-NIL (**A**), *BPH3*-NIL (**B**), *BPH17*-NIL (**C**), *BPH20*-NIL (**D**), *BPH21*-NIL (**E**), *BPH32*-NIL (**F**) and *BPH17-ptb*-NIL (**G**). The 12 bars indicate 12 chromosomes of rice. Horizontal lines across the chromosomes show the positions of polymorphic SSR markers. Circles indicate the approximate positions of brown planthopper resistant genes. The asterisks (*) indicate SSR markers that were used for marker-assisted selection.

2.2. Development of 15 PYLs Carrying Two or Three BPH Resistance Genes

Twelve PYLs carrying two BPH resistance genes (BPH2 + BPH17-PYL, BPH2 + BPH25-PYL, BPH2 + BPH32-PYL, BPH2 + BPH17-ptb-PYL, BPH3 + BPH17-PYL, BPH17 + BPH21-PYL, BPH20 + BPH21-PYL, BPH20 + BPH32-PYL, BPH21 + BPH25-PYL, BPH21 + BPH17-ptb-PYL, BPH25 + BPH17-ptb-PYL and BPH32 + BPH17-ptb-PYL) and three PYLs containing three BPH resistance genes (BPH2 + BPH3 + BPH17-PYL, BPH2 + BPH32 + BPH17-ptb-PYL and BPH20 + BPH21 + BPH32-PYL) were developed using NILs and PYLs with BPH resistance gene(s) (Table 2). The PYLs were confirmed for resistance genes through foreground selection using flanking SSR markers tightly linked to each resistance gene. Most of PYLs were selected from the BC_4F_3 equivalent generation, except BPH3 + BPH17-PYL from the BC_4F_4 equivalent generation, BPH20+BPH21-PYL from the BC_3F_8 generation and BPH32 + BPH17-ptb-PYL from the BC_3F_8 generation.

2.3. Comparison of Resistance Levels against Hadano-66 by Modified Seedbox Screening Test (MSST)

T65 was highly damaged (damage score (DS) = 8.2) by the Hadano-66 population (Figure 2A). The DSs of the donor parents were significantly lower (0.7 for IR71033-121-15, 0.7 for PTB33 and 0.2 for Rathu Heenati) than that of T65. The donor parents also had higher levels of resistance compared with their respective NILs and PYLs. Among the NILs, *BPH2*-NIL (DS: 3.0) and *BPH17*-NIL (3.2) showed the highest resistance levels. The other NILs *BPH3*-NIL (6.0), *BPH20*-NIL (6.0), *BPH21*-NIL (6.5), *BPH25*-NIL (6.7), *BPH26*-NIL (4.8), *BPH32*-NIL (6.7) and *BPH17-ptb*-NIL (5.7), had lower DSs than T65's but were not significantly different from the T65. Damage scores across the 15 PYLs ranged from 2.3 to 6.0. Among PYLs, the DSs of 10 PYLs—*BPH2* + *BPH17*-PYL (2.7), *BPH2* + *BPH25*-PYL (2.5), *BPH2* + *BPH32*-PYL (3.0), *BPH2* + *BPH17-ptb*-PYL (3.0), *BPH17* + *BPH21*-PYL (2.3), *BPH20* + *BPH21*-PYL (2.3), *BPH21* + *BPH25*-PYL (3.3), *BPH21* + *BPH17-ptb*-PYL (2.7), *BPH2* + *BPH3* + *BPH17*-PYL (3.0) and *BPH20* + *BPH21* + *BPH32*-PYL (2.3), were equal to or less than 3.3, while the DSs of five PYLs—*BPH3* + *BPH17*-PYL (5.0), *BPH20* + *BPH32*-PYL (5.3), *BPH25* + *BPH17-ptb*-PYL (6.0), *BPH32* + *BPH17-ptb* (5.0) and *BPH2* + *BPH32* + *BPH17-ptb*-PYL (4.3), were more than 4.3. Although the DSs between NILs and PYLs were not significantly different, the resistance levels of the PYLs tended to be higher than those of the NILs.

Additionally, fresh biomass reduction rates (FBRRs) of the NILs and PYLs were calculated as an indicator of resistance (Figure 2B). T65 had the highest FBRR (89.0%) and was significantly different from the donor parents: IR71033-121-15 (35.7%), PTB33 (39.2%) and Rathu Heenati (20.4%). Among the NILs, *BPH17*-NIL (58.7%) had the lowest FBRR and was significantly different from T65. The other NILs, *BPH2*-NIL (68.6%), *BPH3*-NIL (82.4%), *BPH20*-NIL (77.3%), *BPH21*-NIL (84.3%), *BPH25*-NIL (85.3%), *BPH26*-NIL (73.8%), *BPH32*-NIL (86.7%) and *BPH17-ptb*-NIL (77.6%), had lower FBRRs than T65; however, the differences were not significant. The FBRRs of four PYLs—*BPH2* + *BPH32*-PYL (59.1%), *BPH2* + *BPH17-ptb*-PYL (56.7%), *BPH21* + *BPH17-ptb*-PYL (50.1%) and *BPH2* + *BPH3* + *BPH17*-PYL (57.6%), were less than 60%. The FBRRs of five PYLs—*BPH2* + *BPH17*-PYL (64.5%), *BPH2* + *BPH25*-PYL (68.4%), *BPH20* + *BPH21*-PYL (62.3%), *BPH2* + *BPH32* + *BPH17-ptb*-PYL (64.0%) and *BPH20* + *BPH21* + *BPH32*-PYL (63.2%), ranged from 60% to 70%; and the FBRRs of six PYLs—*BPH3* + *BPH17*-PYL (79.3%), *BPH17* + *BPH21*-PYL (70.9%), *BPH20* + *BPH32*-PYL (71.6%), *BPH21* + *BPH25*-PYL (70.3%), *BPH25* + *BPH17-ptb*-PYL (78.9%) and *BPH32* + *BPH17-ptb*-PYL (74.4%), ranged from 70% to 80%. Additionally, DSs and FBRRs were positively correlated (Pearson's C = 0.89; $p < 0.001$).

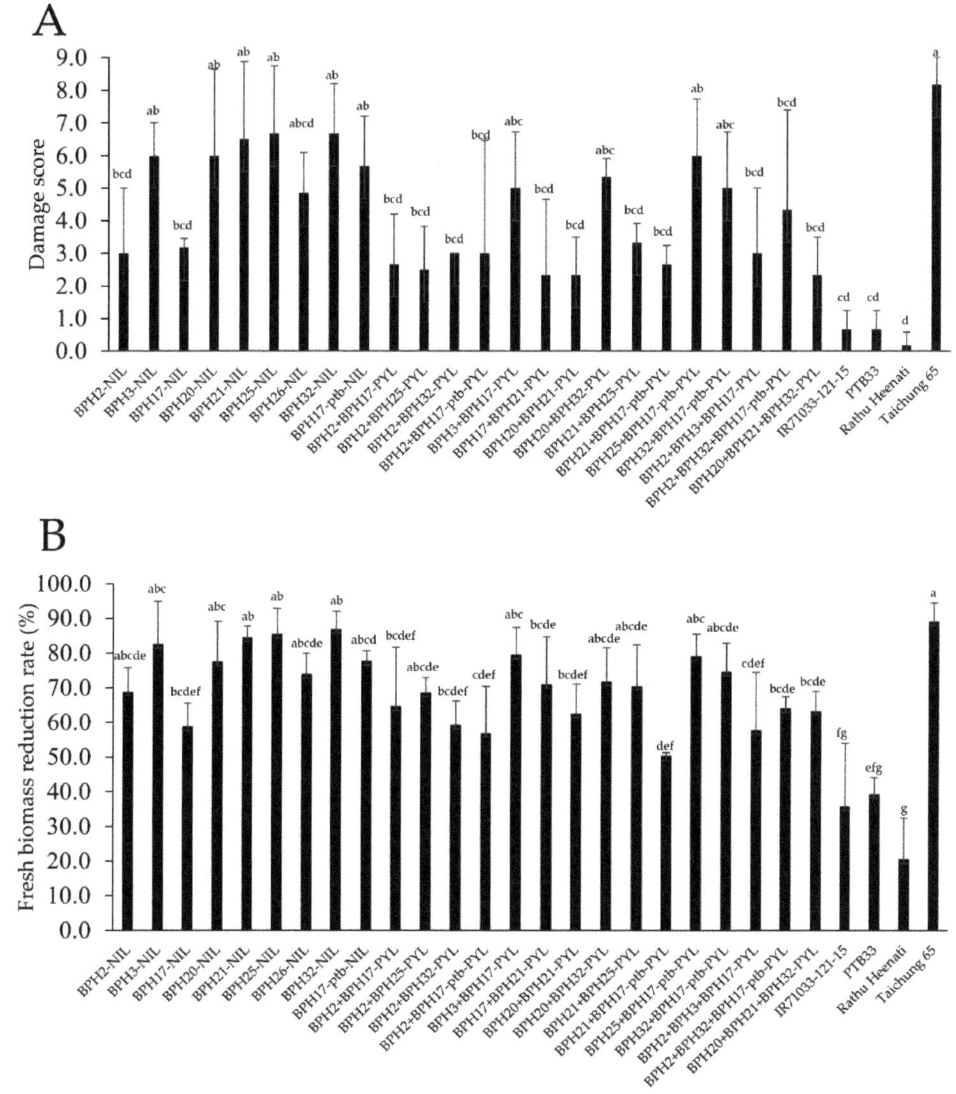

Figure 2. Damage scores (**A**) and fresh biomass reduction rates (**B**) of near-isogenic lines and pyramided lines infested with the Hadano-1966 *Nilaparvata lugens* population using the modified seedbox screening test at the seedling stage. The lower damage scores and fresh biomass reduction rates indicate higher resistance levels.

2.4. Comparison of Adult Mortality (ADM) of the Hadano-66 Population to the NILs and PYLs

Levels of adult mortality (ADM) of the donor parents IR71033-121-15, PTB33 and Rathu Heenati (100%) were significantly higher than that of T65 (17.6%) (Table 4). Among the NILs, *BPH2*-NIL and *BPH17*-NIL had the highest ADM rates, 68.9% and 59.0%, respectively. The ADM rates of other NILs were not significantly different from that of T65. The PYLs carrying the *BPH2*—*BPH2* + *BPH17*-PYL (75.0%), *BPH2* + *BPH25*-PYL (87.5%), *BPH2* + *BPH32*-PYL (84.0%) and *BPH2* + *BPH17-ptb*-PYL (84.0%), showed the highest ADM rates among PYLs with two genes and were higher than any of the corresponding NILs. The ADM rates of seven PYLs—*BPH3* + *BPH17*-PYL, *BPH17* + *BPH21*-PYL, *BPH20* + *BPH32*-PYL, *BPH21* + *BPH25*-PYL, *BPH21* + *BPH17-ptb*-PYL, *BPH25* + *BPH17-ptb*-PYL and *BPH32* + *BPH17-ptb*-PYL, ranged from 50.0% to 68.0%, while the ADM of *BPH20* + *BPH21*-PYL was 33.3%. The ADM rates of PYLs for three genes—*BPH2* + *BPH3* + *BPH17*-PYL (96.0%), *BPH2* + *BPH32* + *BPH17-ptb*-PYL (95.8%) and *BPH20* + *BPH21* + *BPH32*-PYL (92.0%), were higher than those of the corresponding NILs and PYLs for two genes, and were similar to the ADM rates of the donor parents (100%). Furthermore, the ADM rates were negatively correlated with the DSs (Pearson's C = −0.79; $p < 0.001$) and FBRRs (Pearson's C = −0.76; $p < 0.001$).

Table 4. The adult mortality of *Nilaparvata lugens* on near-isogenic lines and pyramided lines carrying brown planthopper resistance genes.

Entry	Adult Mortality (%)	
	Hadano-66	Koshi-2013
BPH2-NIL	68.9 ± 28.5 ^{abcd}	4.0 ± 8.9 ^{b}
BPH3-NIL	30.0 ± 38.0 ^{def}	0.0 ± 0.0 ^{b}
BPH17-NIL	59.0 ± 25.1 ^{abcde}	20.0 ± 14.1 ^{b}
BPH20-NIL	24.0 ± 22.7 ^{def}	4.0 ± 8.9 ^{b}
BPH21-NIL	36.0 ± 37.5 ^{cdef}	12.0 ± 17.9 ^{b}
BPH25-NIL	16.0 ± 15.8 ^{f}	16.0 ± 16.7 ^{b}
BPH26-NIL	50.0 ± 41.4 ^{abcdef}	4.0 ± 8.9 ^{b}
BPH32-NIL	14.0 ± 16.5 ^{f}	12.0 ± 17.9 ^{b}
BPH17-ptb-NIL	22.0 ± 19.9 ^{ef}	20.0 ± 20.0 ^{b}
BPH2+BPH17-PYL	75.0 ± 19.5 ^{abcd}	32.0 ± 17.9 ^{b}
BPH2+BPH25-PYL	87.5 ± 17.9 ^{ab}	12.0 ± 11.0 ^{b}
BPH2+BPH32-PYL	84.0 ± 16.7 ^{abc}	16.0 ± 16.7 ^{b}
BPH2+BPH17-ptb-PYL	84.0 ± 16.7 ^{abc}	16.0 ± 16.7 ^{b}
BPH3+BPH17-PYL	50.0 ± 32.5 ^{abcdef}	24.0 ± 16.7 ^{b}
BPH17+BPH21-PYL	58.3 ± 27.5 ^{abcdef}	16.0 ± 21.9 ^{b}
BPH20+BPH21-PYL	33.3 ± 22.2 ^{cdef}	24.0 ± 16.7 ^{b}
BPH20+BPH32-PYL	62.5 ± 38.5 ^{abcde}	36.0 ± 21.9 ^{b}
BPH21+BPH25-PYL	64.0 ± 26.1 ^{abcde}	24.0 ± 16.7 ^{b}
BPH21+BPH17-ptb-PYL	54.2 ± 35.0 ^{abcdef}	8.0 ± 17.9 ^{b}
BPH25+BPH17-ptb-PYL	68.0 ± 26.8 ^{abcde}	16.0 ± 16.7 ^{b}
BPH32+BPH17-ptb-PYL	62.5 ± 32.9 ^{abcde}	20.0 ± 14.1 ^{b}
BPH2+BPH3+BPH17-PYL	96.0 ± 8.9 ^{ab}	36.0 ± 38.5 ^{b}
BPH2+BPH32+BPH17-ptb-PYL	95.8 ± 8.9 ^{ab}	20.8 ± 20.1 ^{b}
BPH20+BPH21+BPH32-PYL	92.0 ± 11.0 ^{ab}	28.0 ± 26.8 ^{b}
IR71033-121-15	100.0 ± 0.0 ^{a}	44.0 ± 16.7 ^{ab}
PTB33	100.0 ± 0.0 ^{a}	36.0 ± 21.9 ^{b}
Rathu Heenati	100.0 ± 0.0 ^{a}	84.0 ± 35.8 ^{a}
Taichung 65	17.6 ± 16.7 ^{f}	5.0 ± 10.0 ^{b}

Parameter values (means ± standard deviations) followed by the same letter are not significantly different between each *Nilaparvata lugens* population ($p < 0.05$, Tukey–Kramer multiple comparison tests).

2.5. Comparison of ADM Rates for the Koshi-2013 Population on the NILs and PYLs

T65 was susceptible to the Koshi-2013 population with an ADM of 5.0%. Rathu Heenati had the highest ADM among entries (84.0%), which was significantly higher than T65. The ADM rates of the other donor parents, IR71033-121-15 (44.0%) and PTB33 (36.0%), were lower than that of Rathu Heenati. The ADM rates of the NILs were less than or equal to 20.0%. Among the PYLs, the ADM rates of *BPH2* + *BPH17*-PYL (32.0%), *BPH20* + *BPH32*-PYL (36.0%) and *BPH2* + *BPH3* + *BPH17*-PYL (36.0%) were highest. The ADM rates of the six PYLs, *BPH3* + *BPH17*-PYL, *BPH20* + *BPH21*-PYL, *BPH21* + *BPH25*-PYL, *BPH32* + *BPH17-ptb*-PYL, *BPH2* + *BPH32* + *BPH17-ptb*-PYL and *BPH20* + *BPH21* + *BPH32*-PYL ranged from 20% to 28%. The ADM rates of the other PYLs, *BPH2* + *BPH32*-PYL, *BPH2* +

BPH17-ptb-PYL, *BPH17* + *BPH21*-PYL, *BPH25* + *BPH17-ptb*-PYL, *BPH2* + *BPH25*-PYL and *BPH21* + *BPH17-ptb*-PYL ranged from 8.0% to 16.0%.

2.6. Agronomic Characteristics of the NILs and PYLs

Six agronomic traits—days to heading (DTH), panicle length (PL), culm length (CL), flag leaf length (LL), flag leaf width (LW) and panicle number per plant (PN) of the NILs and PYLs are presented in Table 5. The DTHs and PNs of the NILs and PYLs were not significantly different from those of T65. The PLs, CLs, LLs and LWs were similar for NILs and T65, except that the *BPH2*-NIL had longer culms, *BPH25*-NIL had shorter panicles and *BPH3*-NIL had wider flag leaves. The PLs, CLs, LLs and LWs were not significantly different between the PYLs and T65, except that *BPH2* + *BPH17*-PYL, *BPH2* + *BPH25*-PYL, *BPH2* + *BPH32*-PYL, *BPH2* + *BPH17-ptb*-PYL, *BPH21* + *BPH25*-PYL and *BPH2* + *BPH3* + *BPH17*-PYL had longer culms; *BPH21* + *BPH25*-PYL had longer flag leaves; *BPH2* + *BPH32*-PYL, *BPH17* + *BPH21*-PYL and *BPH20* + *BPH32*-PYL had narrower flag leaves; and *BPH3* + *BPH17*-PYL had wider flag leaves.

Table 5. Agronomic traits of near-isogenic lines and pyramided lines for brown planthopper resistance genes.

Entry	Average of Agronomic Trait (AVE ± SD)					
	DTH (day)	CL (cm)	PL (cm)	LL (cm)	LW (cm)	PN
BPH2-NIL	104.0 ± 2.0	120.3 ± 3.2 *	23.8 ± 0.7	36.8 ± 4.5	1.1 ± 0.0	16.6 ± 3.2
BPH3-NIL	100.8 ± 1.3	104.3 ± 4.8	18.9 ± 1.0	30.2 ± 3.3	1.3 ± 0.0 *	16.2 ± 1.5
BPH17-NIL	98.6 ± 1.3	103.2 ± 7.5	23.4 ± 2.2	33.8 ± 8.0	1.2 ± 0.1	15.2 ± 3.1
BPH20-NIL	100.8 ± 1.6	94.2 ± 5.6	20.5 ± 2.0	29.7 ± 4.5	1.1 ± 0.0	14.8 ± 4.1
BPH21-NIL	100.8 ± 1.5	104.2 ± 2.9	20.9 ± 1.2	32.6 ± 3.0	1.0 ± 0.1	13.8 ± 2.6
BPH25-NIL	100.4 ± 0.9	100.0 ± 1.9	18.1 ± 1.9 **	26.7 ± 2.8	1.1 ± 0.1	17.6 ± 1.1
BPH26-NIL	98.4 ± 0.9	102.1 ± 0.9	22.0 ± 1.6	27.3 ± 3.2	1.1 ± 0.0	13.4 ± 1.8
BPH32-NIL	100.2 ± 1.8	97.4 ± 1.5	20.9 ± 0.9	29.3 ± 1.0	1.2 ± 0.0	15.2 ± 1.6
BPH17-ptb-NIL	99.0 ± 0.0	97.4 ± 2.3	20.7 ± 1.0	29.6 ± 1.9	1.1 ± 0.0	14.6 ± 1.7
BPH2+BPH17-PYL	104.6 ± 1.3	115.1 ± 7.9 *	19.8 ± 1.6	27.3 ± 2.6	1.2 ± 0.0	13.8 ± 2.0
BPH2+BPH25-PYL	103.0 ± 1.4	116.2 ± 1.7 *	21.0 ± 0.9	30.2 ± 4.7	1.1 ± 0.1	14.2 ± 1.6
BPH2+BPH32-PYL	104.8 ± 1.3	111.4 ± 8.0 *	19.6 ± 1.6	27.8 ± 5.5	1.0 ± 0.0 **	17.8 ± 3.7
BPH2+BPH17-ptb-PYL	100.8 ± 0.8	110.7 ± 2.3 **	23.0 ± 1.1	26.9 ± 5.0	1.2 ± 0.1	15.2 ± 2.7
BPH3+BPH17-PYL	102.4 ± 1.3	103.3 ± 0.5	18.8 ± 0.9	29.7 ± 3.3	1.5 ± 0.1 *	17.4 ± 2.8
BPH17+BPH21-PYL	98.0 ± 0.0	91.5 ± 1.8	21.9 ± 1.7	30.4 ± 3.7	1.0 ± 0.1 *	14.6 ± 1.5
BPH20+BPH21-PYL	102.8 ± 1.1	77.9 ± 3.3	23.2 ± 2.4	34.7 ± 6.1	1.2 ± 0.0	18.2 ± 2.3
BPH20+BPH32-PYL	105.2 ± 0.4	88.3 ± 1.8	19.3 ± 1.7	34.3 ± 5.8	0.9 ± 0.1 *	16.0 ± 5.2
BPH21+BPH25-PYL	102.8 ± 1.3	110.5 ± 4.9 **	23.0 ± 1.9	41.7 ± 3.5 *	1.1 ± 0.1	18.0 ± 3.6
BPH21+BPH17-ptb-PYL	102.8 ± 2.5	107.5 ± 4.6	22.4 ± 1.6	33.6 ± 4.5	1.1 ± 0.1	16.0 ± 2.6
BPH25+BPH17-ptb-PYL	99.6 ± 1.9	99.6 ± 2.4	19.9 ± 2.0	28.9 ± 3.7	1.2 ± 0.1	15.8 ± 4.8
BPH32+BPH17-ptb-PYL	99.0 ± 0.0	107.6 ± 3.5	20.4 ± 2.2	26.7 ± 3.4	1.1 ± 0.0	18.0 ± 1.6
BPH2+BPH3+BPH17-PYL	103.4 ± 2.9	119.0 ± 2.6 *	22.4 ± 0.7	31.4 ± 2.2	1.1 ± 0.1	16.0 ± 3.4
BPH2+BPH32+BPH17-ptb-PYL	100.6 ± 1.1	102.4 ± 4.4	19.1 ± 0.3	26.2 ± 3.2	1.0 ± 0.1	16.6 ± 5.2
BPH20+BPH21+BPH32-PYL	101.2 ± 0.4	91.7 ± 3.1	19.3 ± 1.4	26.6 ± 4.5	1.0 ± 0.1	17.8 ± 2.8
Taichung 65	99.2 ± 0.4	91.4 ± 2.7	21.2 ± 0.7	29.7 ± 4.2	1.1 ± 0.1	14.6 ± 1.5

DTH: days to heading, CL: culm length, PL: panicle length, LL: flag leaf length, LW: flag leaf width, PN: panicle number per plant. * $p < 0.01$; ** $p < 0.05$ (Dunnett's multiple comparison tests against Taichung 65).

3. Discussion

The seven NILs we developed carried BPH resistance genes on the short arm of chromosome 4 (*BPH17*-NIL, *BPH20*-NIL and *BPH17-ptb*-NIL), on the short arm of chromosome 6 (*BPH3*-NIL and *BPH32*-NIL) and on the long arm of chromosome 12 (*BPH2*-NIL and *BPH21*-NIL). One of the resistance genes on chromosome 12, *BPH2*, was originally identified from ASD7 which was used as a donor parent for many modern resistant varieties (e.g., IR36, IR42 and so on) [31,67]. *BPH2* from ASD7 is identical to *BPH26* in DNA sequence and resistance level [10]. *BPH2* from ASD7 was resistant against the Hatano-66 population (synonym of Hadano-66) but susceptible to Nishigoshi-05, a BPH population collected in Koshi, Kumamoto Prefecture in 2005 [51]. PTB33 was reported to carry one dominant and one recessive gene [40] that were confirmed to be *BPH2* and *BPH3* using conventional genetic analysis [41]. However, there was no report of the exact location of *BPH2* from PTB33. In our study, *BPH2*-NIL had similar resistance patterns to *BPH2* in ASD7: *BPH2*-NIL was highly resistant (ADM of 68.9%) against the Hadano-66 population but less effective (ADM of 4.0%) against the recently collected population, Koshi-2013. Moreover, *BPH2*-NIL (PTB33) and *BPH26*-NIL had similar resistance levels against both Hadano-66 and Koshi-2013, suggesting that PTB33, ADR52 and ASD7 might harbor the same resistance gene. Further sequence analysis for *BPH2* from PTB33 is necessary to understand its genetic basis. Another gene on chromosome 12, *BPH21*, was originally identified from IR71033-121-15, an introgression line derived from *O. minuta* and estimated to be located between two markers, S12094A and B122, on the long arm of chromosome 12 [17]. Recently, *BPH21* has been reported to be allelic to *BPH26* [6] and *BPH18* [68] based on amino acid sequences. Both *BPH18* and *BPH26* were isolated and located at 22.9 Mbp on chromosome 12 [9,10]. Therefore, we estimated that the location of *BPH21* was around 22.9 Mbp on chromosome 12, and the region carrying *BPH21* from IR71033-121-15 was selected using RM1246 (19.2 Mbp) and RM28493 (23.3 Mbp) in this study.

The *BPH17* locus on chromosome 4S from Rathu Heenati has been reported by Sun et al. (2005) [16]. *BPH17* was mapped between two markers, RHD9 (6.2 Mbp) and RHC10 (7.0 Mbp), on chromosome 4S and isolated by Liu et al. (2014) [8]. The amino acid sequence and chromosomal location of *BPH17* from Rathu Heenati were the same as those of *BPH17-ptb* from PTB33 [8]. In this study, resistance of *BPH17*-NIL and *BPH17-ptb*-NIL against the Hadano-66 population differed; however, both NILs had similar effects on the Koshi-2013 population. The different resistant levels might be because the loci were derived from different accessions or varieties of rice. Therefore, the amino acid sequences of PTB33 and Rathu Heenati used in this study on the *BPH17* locus should be determined for future research. Additionally, *BPH20* was detected between two markers, B42 (8.7 Mbp) and B44 (8.9 Mbp) on chromosome 4 [17]. Two NILs for *BPH17* and *BPH20* on the genetic background of 9311 varieties developed by Xiao et al (2016) [51] showed different resistance levels against a BPH population from China [68]. In our study, the resistance levels of *BPH17* and *BPH20* were different in both MSST and antibiosis tests against the Hadano-66 population and against the Koshi-2013 population, which corresponds well with previous research by Xiao et al. (2016) [51]. Therefore, the genes on chromosome 4S of IR71033-121-15, PTB33 and Rathu Heenati might be different. To confirm this, further sequence analyses are needed for the three loci *BPH17*, *BPH17-ptb* and *BPH20*.

Among six genes/QTLs that have been identified on the short arm of chromosome 6 of *O. sativa* and its wild relatives [3,12], *BPH3* and *BPH32* have been widely introduced to elite rice cultivars to improve BPH resistance and were related to durable and broad-spectrum resistance in PTB33 and Rathu Heenati [67,69]. In previous research, *BPH3* was mapped onto chromosome 6 between two markers, RM19291 (1.2 Mbp) and RM8072 (1.4 Mbp) [69]. *BPH32* from PTB33 was identified at the same location as *BPH3* from Rathu Heenati, but the amino acid sequence of *BPH3* was not identical to that of *BPH32* [12]. In our study, the resistance levels of the *BPH3*-NIL were slightly different from those of the *BPH32*-NIL, suggesting that *BPH3* might be different from *BPH32*. A comparison of amino acid sequences between *BPH3* and *BPH32* would be necessary to confirm whether these resistance genes are different.

Among the developed NILs, the *BPH3*-NIL, *BPH17*-NIL, *BPH17-ptb*-NIL and *BPH32*-NIL had around 97.0% of their chromosomal segments from the recurrent parent. This proportion coincides with the theoretical ratio for substituting chromosomal segments from recurrent parents by backcrossing four times. The other NILs had fewer chromosomal segments from T65 than the theoretical rate. The substituted chromosomal segments from the donor parents might be related to undesirable traits such as the suppression of the associated BPH resistance gene. Additionally, due to the low density of available polymorphic SSR markers between T65 and donor DNA around the target genes, the BPH resistance genes on the NILs were selected using two flanking markers that were relatively far apart. Furthermore, *BPH17*, *BPH20* and *BPH17-ptb* on the NILs were selected by flanking markers with longer intervals because of the low density of polymorphic markers between donor parents and T65 around the chromosome 4S region. The intervals between each of the flanking marker pairs for *BPH17* and *BPH17-ptb* were 3.7 Mbp, and that for *BPH20* was 5.7 Mbp. Similarly, the interval for each of the two flanking markers for *BPH2* and *BPH21* on chromosome 12 was 4.1 Mbp because the exact locations of genes had not been identified before we started to develop the NILs by MAS. Therefore, many of the NILs had relatively long chromosome segments derived from the donor parents and there is a possibility that the remaining chromosomal segments from donor DNA around the target genes included linkage drag associated with susceptibility. In further research, ensuring that flanking markers are tightly linked to target genes will avoid linkage drag from donors through MAS and backcrossing.

An improvement of rice resistance levels against BPH is necessary since many genes have become less effective against BPH across Asia [45]. In this study, we developed 15 PYLs carrying two or three genes for BPH resistance. The PYLs tended to increase resistance against the two BPH populations, Hadano-66 and Koshi-2013. Among the 15 PYLs, 12 and nine PYLs had higher ADM rates than corresponding NILs against Hadano-66 and Koshi-2013, respectively; ten PYLs had lower FBRRs compared to corresponding NILs in the MSST against the Hadano-66 population. For example, *BPH2* + *BPH32*-PYL (84.0%) and *BPH2* + *BPH32* + *BPH17-ptb*-PYL (95.8%) had higher resistance levels than those of the *BPH2*-NIL (68.9%), *BPH32*-PYL (14.0%) and *BPH17-ptb*-PYL (22.0%) in antibiosis tests against the Hadano-66 population. The ADM rates of *BPH2* + *BPH17*-PYL (32.0%) and *BPH2* + *BPH3* + *BPH17*-PYL (36.0%) were higher than those of *BPH2*-NIL (4.0%), *BPH3*-NIL (0%) and *BPH17*-NIL (20.0%) against the Koshi-2013 population. Additionally, the FBRR of *BPH2* + *BPH17* (42.3%) was lower than for *BPH2*-NIL (67.6%) and *BPH17*-NIL (58.8%). However, the effectiveness of the PYLs was not consistently higher than that of the corresponding NILs. The effect of PYLs was influenced by specific interactions between gene loci, the specific BPH populations and the screening methods. For example, the resistance levels of *BPH3* + *BPH17*-PYL (50.0%) and *BPH17* + *BPH21*-PYL (58.3%) were not higher than that of *BPH17*-NIL (59.0%) in antibiosis tests (or ADM rates) against the Hadano-66 population. *BPH2* + *BPH25*-PYL (87.5%) showed higher ADM against the Hadano-66 population in comparison to *BPH2*-NIL (68.9%) and *BPH25*-NIL (16.0%), while the ADM rate of *BPH2* + *BPH25*-PYL (12.0%) was lower than that of *BPH25*-NIL (16.0%) against the Koshi-2013 population. A similar tendency has been reported for gene combinations between *BPH1* and *BPH2* [70]; *BPH18* and *BPH32*; *BPH20* and *BPH32*; and *BPH2, BPH18* and *BPH32* [57]. That the resistance levels of most of the PYLs were not significantly higher than those of the corresponding NILs might be related to the relatively small number of replications used in the bioassays (five replications for the antibiosis and three replications for the MSST).

In a previous study, virulence of a BPH population collected during 2005 in Japan had increased compared with the virulence of a population collected in 1966 [51]. Through antibiosis tests, we evaluated BPH resistance against the populations collected in 1966 (Hadano-66) and in 2013 (Koshi-2013). Both represented BPH arriving as migrants to Japan. The Hadano-66 population was virulent to T65 (with no resistance gene) but avirulent to all plants with resistance genes, including Mudgo (*BPH1*), ASD7 (*BPH2*), Rathu Heenai (*BPH3* and *BPH17*), Babawee (*BPH4*), Chin Saba (*BPH8*), Balamawee (*BPH9*) and two NILs, *BPH25*-NIL and *BPH26*-NIL [51,71]. In the present study, most of the NILs, all of the PYLs and the donor parents were still effective against the Hadano-66 population. In contrast, all of the NILs and most of the PYLs were susceptible to the Koshi-2013 population, suggesting that BPH recently arriving to Japan from China has greater virulence than was evident about 50 years ago (i.e., 1966). Among the PYLs, two PYLs, *BPH20* + *BPH32*-PYL and *BPH2* + *BPH3* + *BPH17*-PYL, had relatively high resistance, suggesting that PYLs with combinations of these genes are likely to provide good resistance against the current BPH populations that arrive to Japan (Koshi-2013). Finding new sources of resistance genes will be necessary to further improve resistance against contemporary BPH populations as they gain virulence.

In comparison to the corresponding NILs and PYLs, the resistance levels of PTB33, Rathu Heenati and IR71033-121-15 were higher. This suggests that PTB33, Rathu Heenati and IR71033-121-15 might also contain other BPH resistance gene(s). The other genetic factor(s) for BPH resistance can be revealed by analyzing the segregating populations derived from crosses between the developed PYLs and their donor parents in future studies. Additionally, Rathu Heenati had *QBPH4.1* (5.8-7.8 Mbp) and *QBPH4.2* (15.2-17.2 Mbp) on chromosome 4S rather than *BPH3* and *BPH17* [13]. Therefore, the NILs and PYLs carrying *QBPH4.1* and *QBPH4.2* should be developed and evaluated in further analyses. On the other hand, the lower resistance levels of the NILs and PYLs might be related to the relatively high ratio of substituted chromosomal segments from donors in the NILs (from 3.8 to 55.0 Mbp) and PYLs. There is a possibility that the retained donor chromosomal segments in the genetic background of the NILs and PYLs might be linked to the suppression of BPH resistance. To gain further knowledge of BPH resistance controlled by multiple genes, it will be essential to reduce the donor parent chromosomal segments on the NILs and PYLs by further backcrossing and MAS.

4. Materials and Methods

4.1. Plant Materials

To develop NILs with BPH resistance genes, a *japonica* rice variety, T65, that is susceptible to BPH, was used as a recurrent parent, and three rice varieties resistant to BPH were donor parents. The donor lines were IR71033-121-15, PTB33 and Rathu Heenati. IR71033-121-15 contains two BPH resistance genes, *BPH20* and *BPH21*, from the wild rice species *O. minuta* (Accession number: IRGC101141) [17]. PTB33 (Accession number: IRGC19325) that originated from India contains *BPH2*, *BPH17-ptb* and *BPH32*. Rathu Heenati (Acc. no. IRGC 11730), that originated from Sri Lanka, carries *BPH3* and *BPH17* [16,39]. T65 was crossed with these donor parents and F_1 plants were backcrossed four times with T65 to generate BC_4F_1 plants (Figure 3). At each generation of backcrossing, plants carrying BPH resistance genes from the donor parents were selected by MAS using flanking SSR markers of the target BPH resistance genes (Table 1). The selected BC_4F_1 plants were self-pollinated to produce BC_4F_3, BC_4F_4 and BC_4F_5 plants with BPH resistance genes. Finally, seven NILs with either *BPH2*, *BPH3*, *BPH17*, *BPH20*, *BPH21*, *BPH32* or *BPH17-ptb* were developed. The NILs were used to survey the genetic background and evaluate BPH resistance levels as well as agronomic traits. Two additional NILs, *BPH25*-NIL and *BPH26*-NIL were used in the development of the PYLs [54].

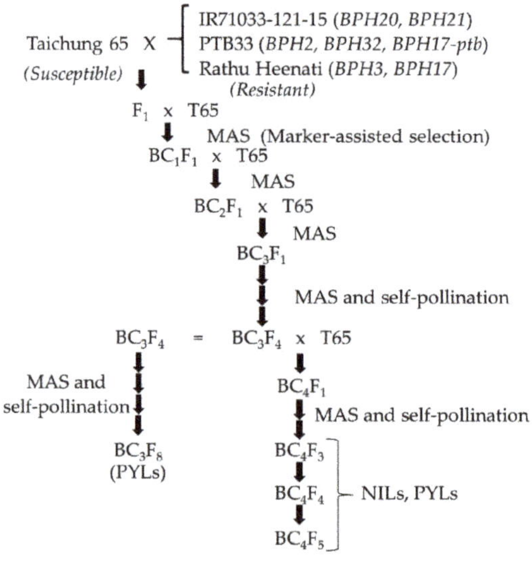

Figure 3. Breeding scheme for the development of near-isogenic lines and pyramided lines containing brown planthopper resistance genes from donor parents, IR71033-121-15, PTB33 and Rathu Heenati.

4.2. The Development of PYLs with BPH Resistance Genes

All the PYLs for two or three BPH resistance genes were developed using the NILs descended from the BC_4F_1 generation, except *BPH20* + *BPH21*-PYL and *BPH32* + *BPH17-ptb*-PYL that were descended from the BC_3F_1 generation. The F_1 plants derived from crosses between NILs were self-pollinated to produce F_2 plants. From 96 F_2 plants, plants that were homozygous for two or three BPH resistance genes were selected by MAS. Several plants from 96 F_3 plants with similar agronomic traits to T65 were selected as final PYLs. The following PYLs carrying two or three BPH resistance genes were evaluated for BPH resistance and agronomic traits: *BPH2* + *BPH17*-PYL, *BPH2* + *BPH25*-PYL, *BPH2* + *BPH32*-PYL, *BPH2* + *BPH17-ptb*-PYL, *BPH3* + *BPH17*-PYL, *BPH17* + *BPH21*-PYL, *BPH20* + *BPH21*-PYL, *BPH20* + *BPH32*-PYL, *BPH21* + *BPH25*-PYL, *BPH21* + *BPH17-ptb*-PYL, *BPH25* + *BPH17-ptb*-PYL, *BPH32* + *BPH17-ptb*-PYL, *BPH2* + *BPH3* + *BPH17*-PYL, *BPH2* + *BPH32* + *BPH17-ptb*-PYL and *BPH20* + *BPH21* + *BPH32*-PYL.

4.3. The MAS for BPH Resistance Genes

To conduct MAS, approximately 2 cm of leaves from two–week old seedlings were collected and dried in a freeze drier for 48 h, and total DNA was extracted using the potassium acetate method [72]. The genotypes of SSR markers on plants in each generation were determined by polymerase chain reaction (PCR) and electrophoresis. The PCR amplification mix (8 µL) contained 3 µL of 1X GoTaq® Green Master Mix (pH 8.5), 0.25 µM of primer and 4 µL of 20 times-diluted DNA. Each PCR amplification included one cycle at 96 °C for 5 min, 35 cycles at 96 °C for 30 s, 55 °C for 30 s and 72 °C for 30 s, followed by one extension cycle at 25 °C for 1 min. PCR products were analyzed by electrophoresis at 200 V using 4% agarose gel with 0.5 µg/mL ethidium bromide in 0.5X TBE buffer for 1 h and photographed under ultraviolet light. During MAS for resistance genes on chromosome 4S, the plants with *BPH17* and *BPH17-ptb*, were selected using two markers, RM8213 and MS10, and the plants with *BPH20* were selected using MS10 and RM5900 (Table 1). The plants with *BPH3* and *BPH32* on the short arm of chromosome 6 were selected using two flanking markers, RM508 and RM588. The plants carrying *BPH2* and *BPH21* located on the long arm of chromosome 12 were screened using RM1246 and RM28493. The plants with *BPH25* were selected using S00310 and MSSR1, and the plants with *BPH26* were selected using RM309, RM28438, InD14, RM28466, RM28481 and MSSR2.

4.4. The Genetic Background Survey of the NILs

In the genetic background survey of the NILs, the bulk DNA from five plants was used. A total of 384 SSR markers distributed on 12 rice chromosomes were used during polymorphism tests with T65 and the donor parents [62]. Among the 384 SSR markers, 254 SSR markers with polymorphisms between IR71033-121-15 and T65 were utilized to identify substituted chromosomal segments from IR71033-121-15 on *BPH20*-NIL and *BPH21*-NIL. Additionally, 244 of 384 SSR markers with polymorphisms between PTB33 and T65 were used to detect substituted chromosomal segments from PTB33 on *BPH2*-NIL, *BPH32*-NIL and *BPH17-ptb*-NIL. To identify substituted chromosomal segments from Rathu Heenati on *BPH3*-NIL and *BPH17*-NIL, 204 of 384 SSR markers with polymorphisms between Rathu Heenati and T65 were used. The whole genome compositions of the developed NILs were graphically displayed following the concept of the graphical genotype proposed by Young and Tanksley (1989) using GGT software version 2.0 [73].

4.5. The BPH Populations and the Characterization of BPH Resistance

Two BPH populations from Japan (Hadano-66 and Koshi-2013) were used to evaluate the NILs and PYLs for their resistance. Hadano-66 was collected in Hadano City, Kanagawa Prefecture, Japan in 1966 [51], and Koshi-2013 was collected in Koshi City, Kumamoto Prefecture, Japan in 2013. Both BPH strains were maintained on the susceptible *japonica* rice variety, Reiho, at 25 °C with 16 h/8 h of light/dark at Kyushu Okinawa Agricultural Research Center of the National Agriculture and Food Research Organization in Japan.

To evaluate resistance, an adaptation of the modified seedbox screening test (MSST) [45,74] was applied at 25 °C using the Hadano-66 strain. To conduct the test, 30 seeds of each of the NILs, PYLs and parent lines were sown to single rows in a plastic tray (23.0 × 30.0 × 2.5 cm) with 2.5 cm between successive rows of seedlings. Two sets of trays—one tray infested by BPH and the other without infestation (the control tray), were used to measure the effects of BPH on plant biomass. One row of Rathu Heenati was added as a resistant control, while three rows of T65 were sown at the center and the two edges as a susceptible control. At seven days after sowing (DAS), the plants in the trays were thinned to 20 plants per row. One tray was infested by the second and third instar nymphs at a density of around 20 BPHs per plant. The experiment was replicated three times. When all the plants of T65 were completely desiccated due to BPH feeding, the DSs of all lines were graded following the standard evaluation system for rice of the International Rice Research Institute [75]. The plants from each row in the two trays were cut above the soil surface and weighed. The fresh biomass reduction rate (FBRR) was calculated using the following formula:

$$\begin{array}{c}\text{Fresh biomass reduction rate}\\ (\text{FBRR})(\%)\end{array} = \left[1 - \frac{\text{Infested plant weight (g)}}{\text{Non-infested plant weight (g)}}\right] \times 100. \quad (1)$$

4.6. *Antibiosis Tests*

Antibiosis tests were conducted at 25 °C following the method described by Myint et al. (2009) [51]. Five plants of each NIL, PYL and parent line were individually sown in 200 mL plastic cups. At four weeks after sowing, the plants were trimmed to 15 cm height and covered with a plastic cage with insect screen windows for ventilation. Each cage was infested with five thin-abdomen brachypterous female BPHs. At five days after infestation, the ADM was recorded (i.e., the number of dead females).

4.7. *Characterization of NILs and PYLs for Agronomic Traits*

The NILs and PYLs were grown in a paddy field at Saga University (Saga, Japan) in 2018 and characterized for their agronomic traits compared to those of T65. Seedlings were transplanted at 28 DAS as one plant per hill, with 20 cm between hills and 25 cm between rows. Each entry was planted as at least three rows (12 plants per a row). Six agronomic traits: DTH, CL, PL, LL, LW and PN were measured for five plants in the same row. DTH was the days from sowing until 50% of panicles flowered. CL was measured from the soil surface to the panicle neck. PL is the length from tip to panicle neck of the longest panicle. The flag leaf width and length were measured from the largest and longest flag leaf of each sampled plant. Panicle number is the number of reproductive panicles of each plant at maturity.

4.8. *Statistical Analysis*

Mean values of BPH resistance (DS, FBRR and ADM) for the NILs and PYLs and agronomic traits were compared using one-way ANOVA. Dunnett's test and Tukey Kramer's test were conducted for multiple comparisons of BPH resistance and agronomic traits, respectively, using R software version 3.5.2.

Author Contributions: P.S.V., D.S.B., F.G.H., H.Y. and D.F. designed the research. S.S.-M. and M.M. provided insects for conducting the research. C.D.N., H.V., D.Z. and D.F. performed the research. C.D.N., H.V. and D.F. developed the plant materials. C.D.N. and D.F. wrote the paper.

Funding: This work was supported by JSPS KAKENHI, grant numbers JP15H04438 and 17K07606.

Acknowledgments: We thank staff of the Insect Pest Management Research Group, Kyushu Okinawa Agricultural Research Center, for rearing and preparing insect populations, and Atsushi Yoshimura, Yoshiyuki Yamagata and the staff belonging to the Plant Breeding Laboratory at Kyushu University for growing and maintaining the plant materials. Furthermore, we thank Kshirod K. Jena for aiding in the development of materials and Elmer Sanchez for developing materials and taking care of the plants. We also wish to thank the Government of Vietnam for the doctoral fellowship granted to C.D.N.

Conflicts of Interest: The authors declare no conflict of interest.

References

1. Bottrell, D.G.; Schoenly, K.G. Resurrecting the ghost of green revolutions past: The brown planthopper as a recurring threat to high-yielding rice production in tropical Asia. *J. Asia Pac. Entomol.* **2012**, *15*, 122–140. [CrossRef]
2. Wei, J.; Jia, D.; Mao, Q.; Zhang, X.; Chen, Q.; Wu, W.; Chen, H.; Wei, T. Complex interactions between insect-borne rice viruses and their vectors. *Curr. Opin. Virol.* **2018**, *33*, 18–23. [CrossRef] [PubMed]
3. Fujita, D.; Kohli, A.; Horgan, F.G. Rice resistance to planthoppers and leafhoppers. *CRC Crit. Rev. Plant Sci.* **2013**, *32*, 162–191. [CrossRef]
4. Horgan, F.G. Integrating gene deployment and crop management for improved rice resistance to Asian planthoppers. *Crop Prot.* **2018**, *110*, 21–33. [CrossRef]
5. Horgan, F.G. Insect herbivores of rice: Their natural regulation and ecologically based management. In *Rice Production Worldwide*; Chauhan, B.S., Jabran, K., Mahajan, G., Eds.; Springer: Cham, Switzerland, 2017; pp. 279–302.
6. Zhao, Y.; Huang, J.; Wang, Z.; Jing, S.; Wang, Y.; Ouyang, Y.; Cai, B.; Xin, X.-F.; Liu, X.; Zhang, C.; et al. Allelic diversity in an NLR gene *BPH9* enables rice to combat planthopper variation. *Proc. Natl. Acad. Sci. USA* **2016**, *113*, 12850–12855. [CrossRef] [PubMed]
7. Du, B.; Zhang, W.; Liu, B.; Hu, J.; Wei, Z.; Shi, Z.; He, R.; Zhu, L.; Chen, R.; Han, B.; et al. Identification and characterization of *Bph14*, a gene conferring resistance to brown planthopper in rice. *Proc. Natl. Acad. Sci. USA* **2009**, *106*, 22163–22168. [CrossRef] [PubMed]
8. Liu, Y.; Wu, H.; Chen, H.; Liu, Y.; He, J.; Kang, H.; Sun, Z.; Pan, G.; Wang, Q.; Hu, J.; et al. A gene cluster encoding lectin receptor kinases confers broad-spectrum and durable insect resistance in rice. *Nat. Biotechnol.* **2014**, *33*, 301–305. [CrossRef] [PubMed]
9. Ji, H.; Kim, S.R.; Kim, Y.H.; Suh, J.P.; Park, H.M.; Sreenivasulu, N.; Misra, G.; Kim, S.M.; Hechanova, S.L.; Kim, H.; et al. Map-based cloning and characterization of the *BPH18* gene from wild rice conferring resistance to brown planthopper (BPH) insect pest. *Sci. Rep.* **2016**, *6*, 34376. [CrossRef] [PubMed]
10. Tamura, Y.; Hattori, M.; Yoshioka, H.; Yoshioka, M.; Takahashi, A.; Wu, J.; Sentoku, N.; Yasui, H. Map-based cloning and characterization of a brown planthopper resistance gene *BPH26* from *Oryza sativa* L. ssp. *indica* cultivar ADR52. *Sci. Rep.* **2014**, *4*, 5872. [PubMed]
11. Wang, Y.; Cao, L.; Zhang, Y.; Cao, C.; Liu, F.; Huang, F.; Qiu, Y.; Li, R.; Luo, X. Map-based cloning and characterization of *BPH29*, a B3 domain-containing recessive gene conferring brown planthopper resistance in rice. *J. Exp. Bot.* **2015**, *66*, 6035–6045. [CrossRef] [PubMed]
12. Ren, J.; Gao, F.; Wu, X.; Lu, X.; Zeng, L.; Lv, J.; Su, X.; Luo, H.; Ren, G. *Bph32*, a novel gene encoding an unknown SCR domain-containing protein, confers resistance against the brown planthopper in rice. *Sci. Rep.* **2016**, *6*, 37645. [CrossRef] [PubMed]
13. Kamolsukyeunyong, W.; Ruengphayak, S.; Chumwong, P.; Kusumawati, L.; Chaichoompu, E.; Jamboonsri, W.; Saensuk, C.; Phoonsiri, K.; Toojinda, T.; Vanavichit, A. Identification of spontaneous mutation for broad-spectrum brown planthopper resistance in a large, long-term fast neutron mutagenized rice population. *Rice* **2019**, *12*, 16. [PubMed]
14. Qiu, Y.; Guo, J.; Jing, S.; Zhu, L.; He, G. Development and characterization of *japonica* rice lines carrying the brown planthopper-resistance genes *BPH12* and *BPH6*. *Theor. Appl. Genet.* **2012**, *124*, 485–494. [CrossRef] [PubMed]
15. Yang, H.; You, A.; Yang, Z.; Zhang, F.; He, R.; Zhu, L.; He, G. High-resolution genetic mapping at the *Bph15* locus for brown planthopper resistance in rice (*Oryza sativa* L.). *Theor. Appl. Genet.* **2004**, *110*, 182–191. [CrossRef] [PubMed]
16. Sun, L.; Su, C.; Wang, C.; Zhai, H.; Wan, J. Mapping of a major resistance gene to the brown planthopper in the rice cultivar Rathu Heenati. *Breed. Sci.* **2005**, *55*, 391–396. [CrossRef]

17. Rahman, M.L.; Jiang, W.; Chu, S.H.; Qiao, Y.; Ham, T.H.; Woo, M.O.; Lee, J.; Khanam, M.S.; Chin, J.H.; Jeung, J.U.; et al. High-resolution mapping of two rice brown planthopper resistance genes, *Bph20(t)* and *Bph21(t)*, originating from *Oryza minuta*. *Theor. Appl. Genet.* **2009**, *119*, 1237–1246. [CrossRef] [PubMed]
18. Hu, J.; Xiao, C.; Cheng, M.; Gao, G.; Zhang, Q.; He, Y. Fine mapping and pyramiding of brown planthopper resistance genes *QBph3* and *Qbph4* in an introgression line from wild rice *O. officinalis*. *Mol. Breed.* **2015**, *35*, 3. [CrossRef]
19. Hu, J.; Xiao, C.; Cheng, M.; Gao, G.; Zhang, Q.; He, Y. A new finely mapped *Oryza australiensis*-derived QTL in rice confers resistance to brown planthopper. *Gene* **2015**, *561*, 132–137. [CrossRef] [PubMed]
20. Mohanty, S.K.; Panda, R.S.; Mohapatra, S.L.; Nanda, A.; Behera, L.; Jena, M.; Sahu, R.K.; Sahu, S.C.; Mohapatra, T. Identification of novel quantitative trait loci associated with brown planthopper resistance in the rice landrace Salkathi. *Euphytica* **2017**, *213*, 38. [CrossRef]
21. Qiu, Y.; Guo, J.; Jing, S.; Zhu, L.; He, G. High-resolution mapping of the brown planthopper resistance gene *Bph6* in rice and characterizing its resistance in the 9311 and Nipponbare near isogenic backgrounds. *Theor. Appl. Genet.* **2010**, *121*, 1601–1611. [CrossRef] [PubMed]
22. Hirabayashi, H.; Kaji, R.; Angeles, E.R.; Ogawa, T.; Brar, D.S.; Khush, G.S. RFLP analysis of a new gene for resistance to brown planthopper derived from *O. officinalis* on rice chromosome 4. *Breed. Res.* **1999**, *48* (Suppl. 1), 82. (In Japanese)
23. Li, R.; Li, L.; Wei, S.; Wei, Y.; Chen, Y.; Bai, D.; Yang, L.; Huang, F.; Lu, W.; Zhang, X.; et al. The evaluation and utilization of new genes for brown planthopper resistance in common wild rice (*Oryza rufipogon* Griff.). *Mol. Entomol.* **2010**, *1*, 1.
24. Huang, D.; Qiu, Y.; Zhang, Y.; Huang, F.; Meng, J.; Wei, S.; Li, R.; Chen, B. Fine mapping and characterization of *BPH27*, a brown planthopper resistance gene from wild rice (*Oryza rufipogon* Griff.). *Theor. Appl. Genet.* **2013**, *126*, 219–229. [CrossRef] [PubMed]
25. He, J.; Liu, Y.; Liu, Y.; Jiang, L.; Wu, H.; Kang, H.; Liu, S.; Chen, L.; Liu, X.; Cheng, X.; et al. High-resolution mapping of brown planthopper (BPH) resistance gene *Bph27(t)* in rice (*Oryza sativa* L.). *Mol. Breed.* **2013**, *31*, 549–557. [CrossRef]
26. Lakshminarayana, A.; Khush, G.S. New genes for resistance to the brown planthopper in rice. *Crop Sci.* **1977**, *17*, 96–100. [CrossRef]
27. Jairin, J.; Phengrat, K.; Teangdeerith, S.; Vanavichit, A.; Toojinda, T. Mapping of a broad-spectrum brown planthopper resistance gene, *Bph3*, on rice chromosome 6. *Mol. Breed.* **2007**, *19*, 35–44. [CrossRef]
28. Kawaguchi, M.; Murata, K.; Ishii, T.; Takumi, S.; Mori, N.; Nakamura, C. Assignment of a brown planthopper (*Nilaparvata lugens* Stål) resistance gene *bph4* to the rice chromosome 6. *Breed. Sci.* **2001**, *51*, 13–18. [CrossRef]
29. Myint, K.K.M.; Fujita, D.; Matsumura, M.; Sonoda, T.; Yoshimura, A.; Yasui, H. Mapping and pyramiding of two major genes for resistance to the brown planthopper (*Nilaparvata lugens* [Stål]) in the rice cultivar ADR52. *Theor. Appl. Genet.* **2012**, *124*, 495–504. [CrossRef] [PubMed]
30. Sun, L.; Liu, Y.; Jiang, L.; Su, C.; Wang, C.; Zhai, H.; Wan, J. Identification of quantitative trait loci associated with resistance to brown planthopper in the *indica* rice cultivar Col.5 Thailand. *Hereditas* **2007**, *144*, 48–52. [CrossRef] [PubMed]
31. Athwal, D.S.; Pathak, M.D.; Bacalangco, E.H.; Pura, C.D. Genetics of resistance to brown planthoppers and green leafhoppers in *Oryza sativa* L. *Crop Sci.* **1971**, *11*, 747–750. [CrossRef]
32. Jeon, Y.H.; Ahn, S.N.; Choi, H.C.; Hahn, T.R.; Moon, H.P. Identification of a RAPD marker linked to a brown planthopper resistance gene in rice. *Euphytica* **1999**, *107*, 23–28. [CrossRef]
33. Sun, L.H.; Wang, C.M.; Su, C.C.; Liu, Y.Q.; Zhai, H.Q.; Wan, J.M. Mapping and marker-assisted selection of a brown planthopper resistance gene *bph2* in rice (*Oryza sativa* L.). *Acta Genet. Sin.* **2006**, *33*, 717–723. [CrossRef]
34. Kabir, M.A.; Khush, G.S. Genetic analysis of resistance to brown planthopper in rice (*Oryza sativa* L.). *Plant Breed.* **1988**, *100*, 54–58.
35. Qiu, Y.; Guo, J.; Jing, S.; Zhu, L.; He, G. Fine mapping of the rice brown planthopper resistance gene *BPH7* and characterization of its resistance in the 93-11 background. *Euphytica* **2014**, *198*, 369–379. [CrossRef]
36. Murata, K.; Fujiwara, M.; Murai, H.; Takumi, S.; Mori, N.; Nakamura, C. *Bph9*, a dominant brown planthopper resistance gene, is located on the long arm of rice chromosome 12. *Rice Genet. Newslett.* **2000**, *17*, 84–86.

37. Ishii, T.; Brar, D.S.; Multani, D.S.; Khush, G.S. Molecular tagging of genes for brown planthopper resistance and earliness introgressed from *Oryza australiensis* into cultivated rice, *O. sativa*. *Genome* **1994**, *37*, 217–221. [CrossRef] [PubMed]
38. Jena, K.K.; Jeung, J.U.; Lee, J.H.; Choi, H.C.; Brar, D.S. High-resolution mapping of a new brown planthopper (BPH) resistance gene, *Bph18(t)*, and marker-assisted selection for BPH resistance in rice (*Oryza sativa* L.). *Theor. Appl. Genet.* **2006**, *112*, 288–297. [CrossRef] [PubMed]
39. Jairin, J.; Teangdeerith, S.; Leelagud, P.; Phengrat, K.; Vanavichit, A.; Toojinda, T. Detection of brown planthopper resistance genes from different rice mapping populations in the same genomic location. *ScienceAsia* **2007**, *33*, 347–352. [CrossRef]
40. Sidhu, G.S.; Khush, G.S. Genetic analysis of brown planthopper resistance in twenty varieties of rice, *Oryza sativa* L. *Theor. Appl. Genet.* **1978**, *53*, 199–203. [CrossRef] [PubMed]
41. Angeles, E.R.; Khush, G.S.; Heinrichs, E.A. Inheritance of resistance to planthoppers and leafhoppers in rice. In Proceedings of the International Rice Genetics Symposium, Manila, Philippines, 27–31 May 1985; International Rice Research Institute: Manila, Philippines, 1986; pp. 537–549.
42. Saxena, R.C.; Barrion, A.A. Biotypes of the brown planthopper *Nilaparvata lugens* (Stål) and strategies in deployment of host plant resistance. *Insect Sci. Appl.* **1985**, *6*, 271–289. [CrossRef]
43. Ketipearachchi, Y.; Kaneda, C.; Nakamura, C. Adaptation of the brown planthopper (BPH), *Nilaparvata lugens* (Stål) (Homoptera: Delphacidae), to BPH resistant rice cultivars carrying *bph8* or *Bph9*. *Appl. Entomol. Zool.* **1998**, *33*, 497–505. [CrossRef]
44. Pathak, M.D.; Khush, G.S. Studies of varietal resistance in rice to the brown planthopper at the International Rice Research Institute. In *Brown Planthopper—Threat to Rice Production in Asia*; International Rice Research Institute: Los Baños, Philippines, 1979; pp. 285–301.
45. Horgan, F.G.; Ramal, A.F.; Bentur, J.S.; Kumar, R.; Bhanu, K.V.; Sarao, P.S.; Iswanto, E.H.; Van Chien, H.; Phyu, M.H.; Bernal, C.C.; et al. Virulence of brown planthopper (*Nilaparvata lugens*) populations from South and South East Asia against resistant rice varieties. *Crop Prot.* **2015**, *78*, 222–231. [CrossRef]
46. Claridge, M.F.; Hollander, J.D. Virulence to rice cultivars and selection for virulence in populations of the brown planthopper *Nilaparvata lugens*. *Entomol. Exp. Appl.* **1982**, *32*, 213–221. [CrossRef]
47. Alam, S.N.; Cohen, M.B. Detection and analysis of QTLs for resistance to the brown planthopper, *Nilaparvata lugens*, in a doubled-haploid rice population. *Theor. Appl. Genet.* **1998**, *97*, 1370–1379. [CrossRef]
48. Peñalver Cruz, A.; Arida, A.; Heong, K.L.; Horgan, F.G. Aspects of brown planthopper adaptation to resistant rice varieties with the *Bph3* gene. *Entomol. Exp. Appl.* **2011**, *141*, 245–257. [CrossRef]
49. Ferrater, J.B.; Naredo, A.I.; Almazan, M.L.P.; de Jong, P.W.; Dicke, M.; Horgan, F.G. Varied responses by yeast-like symbionts during virulence adaptation in a monophagous phloem-feeding insect. *Arthropod-Plant Interact.* **2015**, *9*, 215–224. [CrossRef]
50. Horgan, F.G.; Ferrater, J.B. Benefits and potential trade-offs associated with yeast-like symbionts during virulence adaptation in a phloem-feeding planthopper. *Entomol. Exp. Appl.* **2017**, *163*, 112–125. [CrossRef]
51. Myint, K.K.M.; Yasui, H.; Takagi, M.; Matsumura, M. Virulence of long-term laboratory populations of the brown planthopper, *Nilaparvata lugens* (Stål), and whitebacked planthopper, *Sogatella furcifera* (Horváth) (Homoptera: Delphacidae), on rice differential varieties. *Appl. Entomol. Zool.* **2009**, *44*, 149–153. [CrossRef]
52. Qiu, Y.; Guo, J.; Jing, S.; Tang, M.; Zhu, L.; He, G. Identification of antibiosis and tolerance in rice varieties carrying brown planthopper resistance genes. *Entomol. Exp. Appl.* **2011**, *141*, 224–231. [CrossRef]
53. Ali, M.P.; Alghamdi, S.; Begum, M.; Anwar Uddin, A.B.M.; Alam, M.Z.; Huang, D. Screening of rice genotypes for resistance to the brown planthopper, *Nilaparvata lugens* Stål. *Cereal Res. Commun.* **2012**, *40*, 502–508. [CrossRef]
54. Yara, A.; Phi, C.N.; Matsumura, M.; Yoshimura, A.; Yasui, H. Development of near-isogenic lines for *BPH25(t)* and *BPH26(t)*, which confer resistance to the brown planthopper, *Nilaparvata lugens* (Stål.) in indica rice 'ADR52'. *Breed. Sci.* **2010**, *60*, 639–647. [CrossRef]
55. Liu, Y.; Chen, L.; Liu, Y.; Dai, H.; He, J.; Kang, H.; Pan, G.; Huang, J.; Qiu, Z.; Wang, Q.; et al. Marker assisted pyramiding of two brown planthopper resistance genes, *Bph3* and *Bph27 (t)*, into elite rice Cultivars. *Rice* **2016**, *9*, 27. [CrossRef] [PubMed]
56. Xiao, Y.; Li, J.; Yu, J.; Meng, Q.; Deng, X.; Yi, Z.; Xiao, G. Improvement of bacterial blight and brown planthopper resistance in an elite restorer line Huazhan of *Oryza*. *F. Crop. Res.* **2016**, *186*, 47–57. [CrossRef]

57. Jena, K.K.; Hechanova, S.L.; Verdeprado, H.; Prahalada, G.D.; Kim, S.R. Development of 25 near-isogenic lines (NILs) with ten BPH resistance genes in rice (*Oryza sativa* L.): Production, resistance spectrum, and molecular analysis. *Theor. Appl. Genet.* **2017**, *130*, 2345–2360. [CrossRef] [PubMed]
58. Wang, H.; Shi, S.; Guo, Q.; Nie, L.; Du, B.; Chen, R.; Zhu, L.; He, G. High-resolution mapping of a gene conferring strong antibiosis to brown planthopper and developing resistant near-isogenic lines in 9311 background. *Mol. Breed.* **2018**, *38*, 107. [CrossRef]
59. Horgan, F.G.; Almazan, M.L.P.; Vu, Q.; Ramal, A.F.; Bernal, C.C.; Yasui, H.; Fujita, D. Unanticipated benefits and potential ecological costs associated with pyramiding leafhopper resistance loci in rice. *Crop Prot.* **2019**, *115*, 47–58. [CrossRef] [PubMed]
60. Hu, J.; Li, X.; Wu, C.; Yang, C.; Hua, H.; Gao, G.; Xiao, J.; He, Y. Pyramiding and evaluation of the brown planthopper resistance genes *Bph14* and *Bph15* in hybrid rice. *Mol. Breed.* **2012**, *29*, 61–69. [CrossRef]
61. Fujita, D.; Myint, K.K.M.; Matsumura, M.; Yasui, H. The genetics of host-plant resistance to rice planthopper and leafhopper. In *Planthoppers: New Threats to the Sustainability of Intensive Rice Production Systems in Asia*; Heong, K.L., Hard, B., Eds.; International Rice Research Institute: Los Baños, Philippines, 2009; pp. 389–399.
62. McCouch, S.R.; Teytelman, L.; Xu, Y.; Lobos, K.; Clare, K.; Walton, M.; Fu, B.; Maghirang, R.; Li, Z.; Xing, Y.; et al. Development and mapping of 2240 new SSR markers for rice (*Oryza sativa* L.). *DNA Res.* **2002**, *9*, 199–207. [CrossRef] [PubMed]
63. International Rice Genome Sequencing Project (IRGSP). The map-based sequence of the rice genome. *Nature* **2005**, *436*, 793–800.
64. Temnykh, S.; DeClerck, G.; Lukashova, A.; Lipovich, L.; Cartinhour, S.; McCouch, S. Computational and experimental analysis of microsatellites in rice (*Oryza sativa* L.): Frequency, length variation, transposon associations, and genetic marker potential. *Genome Res.* **2001**, *11*, 1441–1452. [CrossRef] [PubMed]
65. Temnykh, S.; Park, W.D.; Ayres, N.; Cartinhour, S.; Hauck, N.; Lipovich, L.; Cho, Y.G.; Ishii, T.; McCouch, S.R. Mapping and genome organization of microsatellite sequences in rice (*Oryza sativa* L.). *Theor. Appl. Genet.* **2000**, *100*, 697–712. [CrossRef]
66. The Rice Annotation Project (RAP). Available online: https://rapdb.dna.affrc.go.jp/index.html (accessed on 8 November 2019).
67. Khush, G.S.; Virk, P.S. *IR Varieties and Their Impact*; International Rice Research Institute: Los Baños, Philippines, 2005.
68. Xiao, C.; Hu, J.; Ao, Y.T.; Cheng, M.X.; Gao, G.J.; Zhang, Q.L.; He, G.C.; He, Y.Q. Development and evaluation of near-isogenic lines for brown planthopper resistance in rice cv. 9311. *Sci. Rep.* **2016**, *6*, 38159. [CrossRef] [PubMed]
69. Jairin, J.; Teangdeerith, S.; Leelagud, P.; Phengrat, K.; Vanavichit, A.; Toojinda, T. Physical mapping of *Bph3*, a brown planthopper resistance locus in rice. *Maejo Int. J. Sci. Technol.* **2007**, *1*, 166–177.
70. Sharma, P.N.; Torii, A.; Takumi, S.; Mori, N.; Nakamura, C. Marker-assisted pyramiding of brown planthopper (*Nilaparvata lugens* Stål) resistance genes *Bph1* and *Bph2* on rice chromosome 12. *Hereditas* **2004**, *140*, 61–69. [CrossRef] [PubMed]
71. Myint, K.K.M.; Matsumura, M.; Takagi, M.; Yasui, H. Demographic parameters of long-term laboratory strains of the brown planthopper, *Nilaparvata lugens* Stål, (Homoptera: Delphacidae) on resistance genes, *bph20*(t) and *Bph21(t)* in rice. *J. Fac. Agric. Kyushu Univ.* **2009**, *54*, 159–164.
72. Dellaporta, S.L.; Wood, J.; Hicks, J.B. A plant DNA minipreparation: Version II. *Plant Mol. Biol. Rep.* **1983**, *1*, 19–21. [CrossRef]
73. Young, N.D.; Tanksley, S.D. Restriction fragment length polymorphism maps and the concept of graphical genotypes. *Theor. Appl. Genet.* **1989**, *77*, 95–101. [CrossRef] [PubMed]
74. Naeemullah, M.; Sharma, P.N.; Tufail, M.; Mori, N.; Matsumura, M.; Takeda, M.; Nakamura, C. Characterization of brown planthopper strains based on their differential responses to introgressed resistance genes and on mitochondrial DNA polymorphism. *Appl. Entomol. Zool.* **2009**, *44*, 475–483. [CrossRef]
75. *Standard Evaluation System for Rice (SES)*, 5th ed.; International Rice Research Institute: Manila, Philippines, 2014.

plants

Article

Two SNP Mutations Turned off Seed Shattering in Rice

Yu Zhang †, Jiawu Zhou †, Ying Yang †, Walid Hassan Elgamal ‡, Peng Xu, Jing Li, Yasser Z. El-Refaee ‡, Suding Hao and Dayun Tao *

Yunnan Key Laboratory for Rice Genetic Improvement, Food Crops Research Institute, Yunnan Academy of Agricultural Sciences (YAAS), Kunming 650200, China; zhangyu_rice@163.com (Y.Z.); zhjiawu@aliyun.com (J.Z.); yaasyang@126.com (Y.Y.); elgamal.rrtc@gmail.com (W.H.E.); xupeng@xtbg.ac.cn (P.X.); lijinglab@163.com (J.L.); elrefaeey@yahoo.co.in (Y.Z.E.-R.); haosuding@126.com (S.H.)

* Correspondence: taody12@aliyun.com

† These authors contributed equally to this work.

‡ Present address: Rice Research Department, Field Crops Research Institute, Agricultural Research Center, 33717 Sakha, Kafr El-Sheikh, Egypt.

Received: 29 August 2019; Accepted: 5 November 2019; Published: 6 November 2019

Abstract: Seed shattering is an important agronomic trait in rice domestication. In this study, using a near-isogenic line (NIL-*hs1*) from *Oryza barthii*, we found a hybrid seed shattering phenomenon between the NIL-*hs1* and its recurrent parent, a *japonica* variety Yundao 1. The heterozygotes at *hybrid shattering 1* (*HS1*) exhibited the shattering phenotype, whereas the homozygotes from both parents conferred the non-shattering. The causal *HS1* gene for hybrid shattering was located in the region between SSR marker RM17604 and RM8220 on chromosome 4. Sequence verification indicated that *HS1* was identical to *SH4*, and *HS1* controlled the hybrid shattering due to harboring the ancestral haplotype, the G allele at G237T site and C allele at C760T site from each parent. Comparative analysis at *SH4* showed that all the accessions containing ancestral haplotype, including 78 wild relatives of rice and 8 African cultivated rice, had the shattering phenotype, whereas all the accessions with either of the homozygous domestic haplotypes at one of the two sites, including 17 wild relatives of rice, 111 African cultivated rice and 65 Asian cultivated rice, showed the non-shattering phenotype. Dominant complementation of the G allele at G237T site and the C allele at C760T site in *HS1* led to a hybrid shattering phenotype. These results help to shed light on the nature of seed shattering in rice during domestication and improve the moderate shattering varieties adapted to mechanized harvest.

Keywords: Seed shattering; *O. barthii*; *O. sativa*; *HS1*; haplotype

1. Introduction

During rice domestication, seed shattering is one of the most greatly changed traits for seed dispersal. Easy shattering leads to the loss of production [1], and more attention is paid on selection for non-shattering but threshable rice in modern rice breeding [2]. Seed shattering is caused by the formation and degradation of the abscission zone (AZ), which, constituted by a band of small cells, is responsive to signals promoting abscission [3].

Recently, several genes responsible for seed shattering were identified in rice. *qSH1* encodes a BELL homeobox protein. An SNP mutation in the regulatory region of *qSH1* could inhibit its expression, which resulted in defective abscission layer development [4]. The allelic genes of *SH4* [5], *SHA1* [6] and *GL4* [7] encoding a trihelix transcriptional factor, all controlled the seed shattering, however, haplotypes were divergent because of two different SNP mutations. A "G237T" mutation in *SH4* and *SHA1* was responsible for the loss of seed shattering in Asian cultivated rice [5,6], whereas "C760T" transition in *GL4* conferred non-shattering seeds in African cultivated rice [7]. The *SHAT1* gene, which encoded an

APETALA2 transcription factor, was responsible for seed shattering through specifying abscission zone development in rice. The expression of *SHAT1* was positively regulated by the transcription factor *SH4*, which was required for the AZ identification during the early spikelet developmental stage, and *qSH1* functions downstream of *SHAT1* and *SH4*, promoting the AZ differentiation by maintaining the expression of *SHAT1* and *SH4* [8]. *SH5*, which is highly homologous to *qSH1*, also controlled seed shattering by regulating lignin deposition in the pedicel region [9]. *OsGRF4* could increase the expression of two cytokinin dehydrogenase precursor genes resulting in the high cytokinin level, which led to reduced seed shattering [10]. *OSH15* together with *SH5* induced seed shattering by repressing lignin biosynthesis genes [11]; *ObSH3* in *Oryza barthii* encoded a YABBY transcription factor, which was also required for the development of the seed abscission layer [12]. In addition, some other minor genes and allelic interaction at major locus might be involved in the seed shattering domestication as rice underwent a prolonged domestication process, with continuing selection for reduced shattering [13,14].

O. barthii is one of the relatives distributed in West Africa, sharing the same AA genome as Asian cultivated rice. Most of *O. barthii* accessions exhibited the seed shattering. The previous report indicated that *GL4* in *O. barthii* was involved in the non-shattering selection during the African cultivated rice domestication [7], but the nature of seed shattering was still not clear. Here, we report that a novel locus, named *hybrid shattering 1* (*HS1*), controlled the seed shattering in the hybrid between *O. barthii* and *O. sativa*. A near-isogenic line (NIL-*hs1*) from *O. barthii*, and its recurrent parent, a *japonica* variety Yundao 1, showed the non-shattering phenotype. Interestingly, the hybrid between two parents showed the seed shattering, similar to the ancestral wild rice. Whether it was shattering or not was dependent on the different haplotypes of two SNPs at *HS1*. This result could help us understand the complex molecular mechanism of seed shattering.

2. Results

2.1. *HS1 Controlled the Hybrid Seed Shattering in Rice*

We developed a NIL-*hs1* carrying genome fragment from *O. barthii* on chromosome 4 in the Yundao 1 genetic background. Surprisingly, Yundao 1 and the NIL-*hs1* showed the non-shattering seed, while F_1 hybrid exhibited seed shattering (Figure 1A, Table S1). In order to distinguish the differences in abscission layer structure between F_1 hybrid and its parents, longitudinal sections at the seeds base were observed using fluorescent microscopy. The results showed that Yundao 1 displayed the deficiency in abscission zone development near the vascular bundle (Figure 1B), the NIL-*hs1* showed no abscission layer on the palea side and the partial abscission layer on the lemma side between the seed pedicel and the spikelet, respectively (Figure 1D). Conversely, the F_1 hybrid had a continuous abscission zone between the vascular bundle and the epidermis (Figure 1C). These results indicated that seed shattering in the F_1 hybrid resulted from the complete and continuous abscission layer, whereas the loss of seed shattering in Yundao 1 and NIL-*hs1* was caused by the irregular development of the abscission zone.

2.2. *Dominant Complementation of G Allele at G237T Site and C Allele at C760T Site in HS1 Led to Hybrid Seed Shattering Phenotype*

In order to understand whether *HS1* acted as a single Mendelian factor or not, the seed shattering rate was investigated in BC_4F_1 and BC_4F_2 populations derived from the cross between Yundao 1 and NIL-*hs1*. All the BC_4F_1 individuals showed seed shattering. The seed shattering rate in the BC_4F_2 population was segregated into non-shattering and shattering classes in a 246:205 ratio, which fitted the 1:1 ratio ($\chi^2 = 1.870$, $P = 0.172$) (Figure S1). These results indicated that the seed shattering in F_1 hybrid was controlled by a single gene. We designated it as *HS1*.

A population of 790 BC_4F_2 plants was generated for mapping the *HS1*. Eight polymorphic SSR markers in the introgressed region on chromosome 4 were used for genotyping the 790 individuals in the

BC_4F_2 population. *HS1* was mapped into a 0.4 cM region flanked by RM17604 and RM8220, at genetic distances of 0.3 and 0.1 cM, respectively, and co-segregated with RM17616 (Figure 2). The homozygotes from both parents at RM17616 showed a non-shattering phenotype, whereas the heterozygotes at RM17616 showed a shattering phenotype. Based on the GRAMENE public database (http://www.gramene.org), the physical distance between RM17604 and RM8220 was about 434.6 kb. The mapping region of *HS1* was similar to the location of *SH4/GL4* (LOC_Os04g57530/ORGLA04G0254300) identified from the Asian rice and the African rice [4–6]. In order to confirm whether the *HS1* was allelic to *SH4/GL4* or not, the sequence analysis of *SH4* in Yundao 1 and NIL-*hs1* was performed. A total of 13 SNPs, 5 indels in the 2.3 kb of aligned sequenced DNA were identified (Figure 3), which resulted in 1 amino acid insertion, 4 amino acid substitutions, 6 amino acid deletions and pre-stop codon in NIL-*hs1*, respectively (Figure 3). Of these, two base substitutions of G237T and C760T (C760T referred that the C to T SNP mutation in *HS1* was at nucleotide position 760 in *O. barthii*, which was the same as C769T mutation in Yundao 1) resulted in the mutation of Asn79 to Lys79 and Gln258 to a stop codon, respectively. It was reported that the G allele at G237T site and the C allele at C760T site were responsible for the seed shattering during the Asian cultivated rice and the African cultivated rice domestication, respectively [4–6]. Thus, we postulated that *HS1* was identical to *SH4* and *GL4*. The G allele at G237T site and the T allele at C760T in the Yundao 1 background, and the T allele at G237T site and the C allele at C760T in the NIL-*hs1* background all conferred the non-shattering phenotype, whereas the combination of the G allele at G237T site and the C allele at C760T site exhibited the shattering phenotype. Dominant complementation of the G allele at G237T site and the C allele at C760T site in *SH4* led to the hybrid shattering phenotype. Moreover, *SH4* in NIL-*hs1* had a unique deletion of the 227th amino acid residue isoleucine (Ile), compared with that in other AA genome species in the genus *Oryza*.

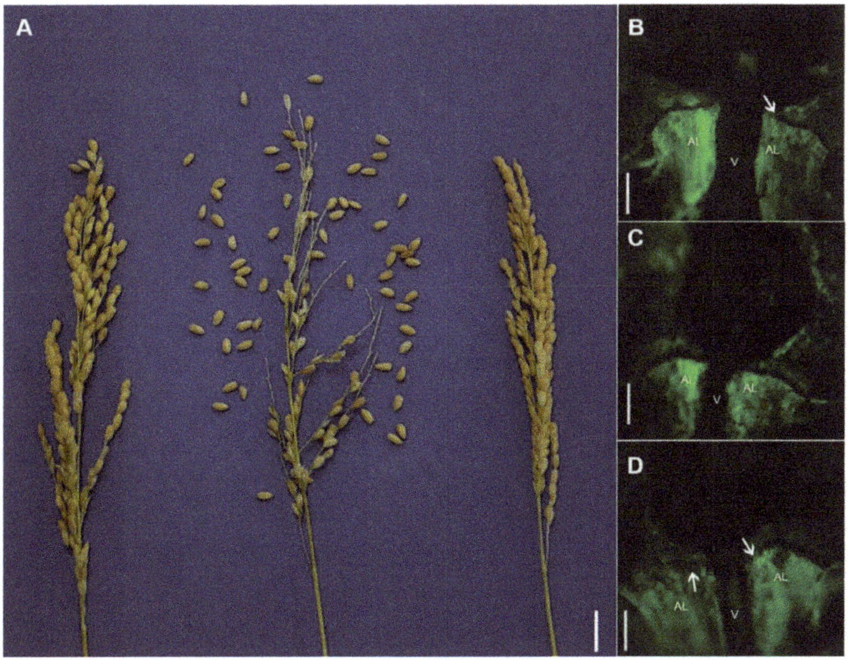

Figure 1. (**A**) The seed shattering rate of Yundao 1 (left), F_1 hybrid (middle) and near-isogenic line (NIL-*hs1*) (right). Scale bars = 0.5 cm. (**B–D**) Fluorescence images of a longitudinal section of the spikelet

and pedicel junction in Yundao 1, F_1 hybrid and NIL-*hs1*, respectively. (**B**) Yundao 1 showed an incomplete in abscission zone. (**C**) F_1 hybrid with a complete abscission layer. (**D**) NIL-*hs1* exhibited a deficiency in abscission layer on the palea side and partial abscission layer on the lemma side. AL: Abscission layer, V: Vascular bundle. White arrow indicates a deficiency in abscission zone. Scale bars = 10 μm.

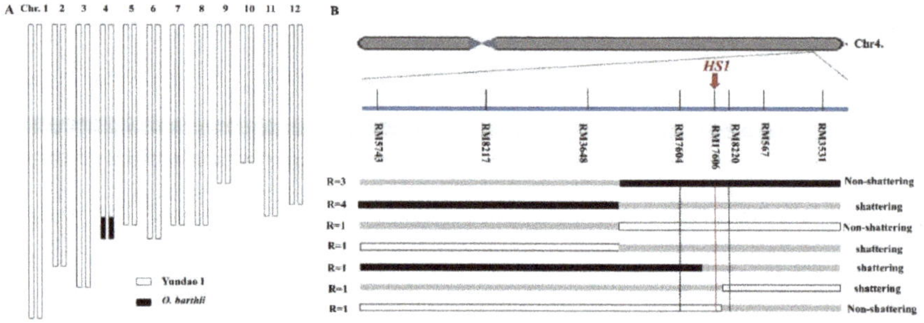

Figure 2. (**A**) Graphical genotypes show that an *O. barthii* chromosomal segment was introgressed into the NIL-*hs1* genome on chromosome 4. (**B**) Genetic mapping of *hybrid shattering 1* (*HS1*) on Chromosome 4, white bar: homozygous Yundao 1; grey bar: heterozygous; black bar: homozygous NIL-*hs1*. "R" means the number of recombinants.

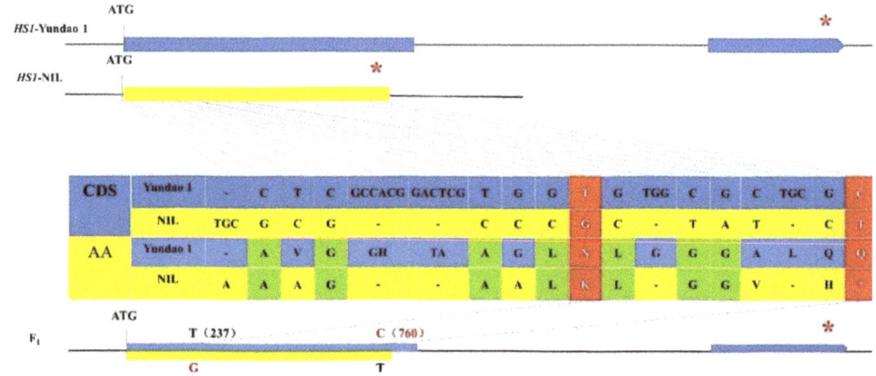

Figure 3. The difference in the coding sequence and amino acid of *HS1* between Yundao 1 and NIL-*hs1*. Synonymous mutations and functional mutations were shown in green and red, respectively. The asterisk indicates the stop codon.

2.3. *Two SNP Mutations Turned off Seed Shattering in Rice*

In order to confirm the function of the two SNPs, we reanalyzed the gene sequence of *SH4* in 95 wild accessions of rice, 119 *O. glaberrima* and 65 *O. sativa* using previously published data. All the accessions harboring both the G allele at G237T site and the C allele at C760T site exhibited the shattering phenotype, including 2 *O. longistaminata* accessions, 22 of 28 *O. barthii* accessions, 8 of 119 *O. glaberrima* accessions, 2 *O. glumaepatula* accessions, 20 of 25 *O. nivara* accessions and 30 of 36 *O. rufipogon* accessions. All the accessions (varieties) with either the G allele at G237T site or the C allele at C760T site showed the non-shattering, including 6 of 28 *O. barthii* accessions, 111 of 119 *O. glaberrima* accessions, 5 of 25 *O. nivara* accessions, 6 of 36 *O. rufipogon* accessions and 65 Asian cultivated varieties. These results indicated that the G allele at G237T site and the C allele at C760T site were ancestral alleles in African rice domestication and Asian rice domestication, respectively, whereas the T allele at both sites that resulted from selection pressure were mutation alleles. The ancestral haplotype (the G allele at G237T site and the C allele at C760T site) induced a shattering phenotype in rice, whereas domestic haplotypes (the G allele at G237T site and the T allele at C760T site, the T allele at G237T site and the C allele at C760T site) all exhibited the loss or reduction of seed shattering (Table 1), which was consistent with our experimental results that the hybrid harboring ancestral haplotype showed a shattering phenotype; however, Yundao 1 and NIL-*hs1* carrying homozygous domestic haplotypes exhibited a non-shattering phenotype.

Table 1. The haplotypes of the *SH4* at G237T and C760T sites in the AA genome species of genus *Oryza*.

Species	G237T	C760T	No. of Accessions (Varieties)	Phenotype
O. longistaminata	G	C	2	Shattering
O. barthii	G	C	22	Shattering
	G	T	6	Non-shattering
O. glaberrima	G	C	8	Shattering
	G	T	111	Non-shattering
O. glumaepatula	G	C	2	Shattering
O. meridionalis	G	C	2	Shattering
O. nivara	G	C	20	Shattering
	T	C	5	Non-shattering
O. rufipogon	G	C	30	Shattering
	T	C	6	Non-shattering
O. sativa, temperate japonica	T	C	30	Non-shattering
O. sativa, tropical japonica	T	C	5	Non-shattering
O. sativa, indica	T	C	25	Non-shattering
O. sativa, aus	T	C	5	Non-shattering

3. Discussion

What causes the differences in seed shattering in different species is totally an open question. Loss or reduction of seed shattering represents a key transition to domestication in rice [15,16]. In this study, it was a serendipitous finding that the hybrid F_1 derived from the cross between the non-shattering Yundao 1 and NIL-*hs1* that displayed the seed shattering phenotype. And we reported that a novel locus *HS1* controlled the hybrid shattering between *O. sativa* and *O. barthii*. Yundao 1 and NIL-*hs1* showed the irregular and partially developed abscission layer, whereas the F_1 hybrid exhibited a continuous abscission layer between seed pedicel and spikelet. *HS1* and *SH4* were mapped into a similar region on chromosome 4 [5], interestingly, *SH4* functioned in the seed shattering on the homozygous background, whereas *HS1* conferred the seed shattering on the heterozygous background. What resulted in this difference? It was suggested that an allelic interaction at a single locus or an epistatic interaction at two independent loci from the non-shattering species *O. sativa* and *O. barthii* determined the phenotype. Interestingly, *HS1* and *SH4* were mapped into a similar region on Chromosome 4 and sequencing analysis confirmed that *HS1*, allelic to *SH4*, carrying ancestral haplotype (the G allele at G237T site and the C allele at C760T site) contributed to shattering in the F_1 hybrid. Thus, an allelic interaction at *SH4* was responsible for the hybrid seed shattering. NIL-*hs1* and most of *O. glaberrima* accessions shared the same *SH4* haplotype, and we also found that the hybrid between the *O. glaberrima* and *O. sativa* also displayed the seed shattering phenotype. As we know, it is difficult to obtain the hybrid progeny between *O. glaberrima* and *O. sativa*, one of the reasons is that interspecific hybrid sterility between *O. glaberrima* and *O. sativa* prevents the formation of hybrid offspring, and another reason is that strong seed shattering in the hybrid increases the difficulty of the crossing. Previous studies reported that one single-nucleotide polymorphism, G237T or C760T, controlled seed shattering in rice independently, because one SNP was fixed and not polymorphic in the wild and cultivated accessions [5–7]. In this study, the relationship of the different haplotypes of *SH4* and the shattering phenotype was analyzed from the whole gene sequence viewpoint. Moreover, two base mutations of G237T and C760T at *SH4* occurred in Trihelix DNA binding domain, indicating that this domain played an important role in seed shattering and either of the nucleotide acid mutations had no effect on the function of the DNA binding domain. The transcription factor *SH4* controlled the AZ identification by positively regulating the expression of *SHAT1*, and *qSH1* could promote the AZ differentiation by maintaining the expression of *SHAT1* and *SH4* [8], suggesting that two amino acid substitutions (Asn79 to Lys79 and Gln258 to a stop codon) in *SH4* might affect the interaction with *SHAT1* and *qSH1*, resulting in the loss of seed shattering. In addition, with the similar genetic background, the shattering degree of Yundao 1 was easier than that of NIL-*hs1*, there were two possibilities: (1) The haplotype of T237 and C760 combination conferred the easier shattering than that of G237 and T760 combination; (2) other genes in the introgressed region could decrease the seed shattering rate by regulating the expression of *SH4*. These results would provide new clues into the molecular basis of seed shattering in rice, and breeders can take the advantage of different haplotype combinations adapted to the moderate shattering degree so as to meet the need for mechanized harvest.

Asian cultivated rice (*O. sativa* L.) was domesticated from wild species *O. rufipogon* thousands of years ago [17,18], whereas *O. glaberrima* Steud. was an African species of rice that was domesticated from the wild progenitor *O. barthii* about 3000 years ago [15,16]. In this study, 79% of *O. barthii* accessions harbored the G allele at G237T site and the C allele at C760T site in *SH4*, but 93% of *O. glaberrima* accessions carried the G allele at G237T site and the T allele at C760T site. It is suggested that the wild-type G at the G237T site was fixed, while the mutated T allele at the C760T site was selected during the African cultivated domestication. The function of the G allele at G237T site (Lysine residue) may be critical for the growth and development of the African wild relative of rice and African cultivated rice so that it was fixed during the gradual domestication process, while the T allele at G237T site contributing to the small grain size was selected, which may be an adaptation to the extreme environment in West Africa, such as drought, soil acidity, iron and aluminum toxicity [7,19,20]. Two haplotypes of *SH4* in *O. rufipogon* existed, whereas the cultivated rice only had one domestic haplotype, these results were also observed in *O. barthii* and *O. glaberrima*. Compared with African cultivated rice, domestic allele in Asian cultivated rice was the G237T mutation, while the C allele at C760T mutation was fixed, indicating that the C alleles at the C760T site might be involved in the selection of a large grain size [7]. This might be one of the reasons that the yield of Asian cultivated rice was higher than that of African cultivated rice. Taken together, the seed shattering characteristics were selected in African cultivated rice and Asian cultivated rice, respectively, which were consistent with the theory that *O. glaberrima* and *O. sativa* were domesticated in parallel [12,18].

4. Materials and methods

4.1. Plant Materials

An *O. barthii* accession, Acc.104284, introduced from the International Rice Research Institute (IRRI), as a donor and male parent, was crossed with a temperate *japonica* variety of *O. sativa*, Yundao 1, from Yunnan province, P. R. China. The male gametes in hybrid F_1 between *O. barthii* and *O. sativa* were fully sterile, and the female gametes in hybrid F_1 were partially fertile, thus, hybrid F_1 as the female parent was consecutively backcrossed with Yundao 1 as the male parent. BC_3 plants were self-pollinated for 13 generations to produce the BC_3F_{14} introgression lines. Four hundred and twenty-six polymorphic SSR markers evenly distributed on 12 chromosomes of rice were used to evaluate the substituted fragments from *O. barthii*. The results indicated that only 14.8 cM segments on chromosome 4 were substituted by the *O. barthii* genome. The individuals harboring the homozygous genome fragment from *O. barthii* were selected as NIL-*hs1*. NIL-*hs1* was crossed with the recurrent parent Yundao 1, and then was self-fertilized to produce the BC_4F_2 population. We found the BC_4F_1 plants all showed seed shattering, while seed shattering and non-shattering were both observed in the BC_4F_2 population. Seven hundred and ninety individuals were used to mapping *HS1* for seed shattering between *O. sativa* and *O. barthii*.

All plant materials were grown in the paddy field at the Experiment Station, YAAS, located in Jinghong, Yunnan Province, P. R. China.

4.2. Evaluation of Seed Shattering Rate

The spikelets of each plant were bagged at the stage of heading. Then, the panicles were collected at the stage of maturity, and panicles freely fell 2 m to the plastic box (60 cm × 90 cm). All the grains shredded prior to the test were counted as shattering grains. The shattering rate was calculated by a percentage of shattered seeds to the total seeds. The shattering rate below 5% or above 50%, was defined as the non-shattering type and the shattering type, respectively.

4.3. Microscopy

Seeds including pedicels were collected at the stage of maturity, and slices made by mature and dry seeds were stained with 1% acridine orange. Abscission layer at seed base was observed by Fluorescent microscopy (OLYMPUS BX53).

4.4. DNA Extraction and SSR Analysis

The experimental procedure for DNA extraction was performed as previously described [21]; rice SSR markers were selected from the Gramene database (http://www.gramene.org) or previously published SSR markers in rice [22]. PCR was performed as follows: a total volume of 10 μL containing 10 ng template DNA, 1 × buffer, 0.2 μM of each primer, 50 μM of dNTPs and 0.5 units of Taq polymerase (Tiangen Company, Beijing, China). The reaction mixture was incubated at 94 °C for an initial 4 min, followed by 30 cycles of 94 °C 30 s, 55 °C 30 s and 72 °C 30 s, and a final extension step of 5 min at 72 °C. PCR products were separated on 8% non-denaturing polyacrylamide gel and detected using the silver staining method.

4.5. Linkage Analysis

A linkage map was constructed on the basis of genetic linkage between the genotype of SSR markers and seed shattering phenotype in the BC_4F_2 population.

4.6. Sequencing

In order to compare the sequence difference of *SH4* between Yundao 1 and NIL-*hs1*, the primer (SH4-F: CCGAACACCAAACGCCTCAG, SH4-R: CCGTACTCCCAATACTCGCAGA) was designed on the 5′UTR and 3′UTR region of *SH4* gene for amplifying the target sequence. PCR mixture (25 uL) contained 0.4 mM of each dNTP, 0.3 uM of each primer, 0.5 units of Taq polymerase (KOD FX DNA polymerase, Toyobo, Japan) and template DNA 100 ng in the GeneAMP PCR system 9700 (Applied Biosystems, Foster City, CA, USA). The PCR program was 94 °C for 2 min, followed by 30 cycles at 98 °C for 10 s, 55 °C for 30 s and 68 °C for 2 min. PCR products were separated in 1% agarose gels.

4.7. Haplotype Analysis of the SH4 Gene

To analyze the haplotype of the *SH4* gene, the nucleotide sequence of *SH4* in the AA genome was downloaded from the GenBank database, wild rice genome project and public data [5,7,15,23–25], 2 SNPs of G237T and C760T were analyzed in 279 rice accessions.

Supplementary Materials: The following are available online at http://www.mdpi.com/2223-7747/8/11/475/s1, Figure S1. Frequency distributions of seed shattering rate and the sequence alignment of *HS1*. Figure S2. The information on haplotypes in *SH4* of 279 accessions in AA genome *Oryza*. Table S1. The seed shattering rate of Yundao 1, NIL-*hs1* and F1 hybrid.

Author Contributions: Y.Z., J.Z. and D.T. planned and designed the research. Y.Y., W.H.E., P.X., J.L., Y.Z.E.-R. and S.H. performed the laboratory experiments. Y.Z., J.Z. and Y.Y. analyzed the data together. Y.Z., J.Z. and D.T. finished the manuscript. All authors reviewed and approved the final manuscript.

Funding: This research was supported by the National Natural Science Foundation of China (Grant Nos. U1502265, 31660380, 31201196), Yunnan Provincial Science and Technology Department, China (Grant Nos. 2015HB079, 2018FG001-086) and the Yunnan Provincial Government (YNWR-QNBJ-2018-359).

Conflicts of Interest: The authors declare no conflict of interest.

Abbreviations

NIL: Near isogenic line; AL: Abscission layer, V: Vascular bundle.

References

1. Ji, H.; Kim, S.-R.; Kim, Y.-H.; Kim, H.; Eun, M.-Y.; Jin, I.-D.; Cha, Y.-S.; Yun, D.-W.; Ahn, B.-O.; Lee, M.C.; et al. Inactivation of the CTD phosphatase-like gene OsCPL1 enhances the development of the abscission layer and seed shattering in rice. *Plant J.* **2010**, *61*, 96–106. [CrossRef]
2. Avik, R.; Debarati, C. Shattering or not shattering: That is the question in domestication of rice (*Oryza sativa* L.). *Genet. Resour. Crop Evol.* **2018**, *65*, 391–395.
3. Oba, S.; Sumi, N.; Fujimoto, F.; Yasue, T. Association between Grain Shattering Habit and Formation of Abscission Layer Controlled by Grain Shattering gene sh-2 in Rice (*Oryza sativa* L.). *Jpn. J. Crop Sci.* **1995**, *64*, 607–615. [CrossRef]
4. Konishi, S.; Izawa, T.; Lin, S.Y.; Ebana, K.; Fukuta, Y.; Sasaki, T.; Yano, M.; Huang, X.; Hansen, N.; Tsuji, N. An SNP Caused Loss of Seed Shattering During Rice Domestication. *Science* **2006**, *312*, 1392–1396. [CrossRef] [PubMed]
5. Li, C.; Zhou, A.; Sang, T. Rice domestication by reducing shattering. *Science* **2008**, *311*, 1936–1939. [CrossRef] [PubMed]
6. Lin, Z.; Griffith, M.E.; Li, X.; Zhu, Z.; Tan, L.; Fu, Y.; Zhang, W.; Wang, X.; Xie, D.; Sun, C. Origin of seed shattering in rice (Oryza sativa L.). *Planta* **2007**, *226*, 11–20. [CrossRef] [PubMed]
7. Wu, W.; Liu, X.; Wang, M.; Meyer, R.S.; Luo, X.; Ndjiondjop, M.-N.; Tan, L.; Zhang, J.; Wu, J.; Cai, H.; et al. A single-nucleotide polymorphism causes smaller grain size and loss of seed shattering during African rice domestication. *Nat. Plants* **2017**, *3*, 17064. [CrossRef] [PubMed]
8. Zhou, Y.; Lu, D.; Li, C.; Luo, J.; Zhu, B.-F.; Zhu, J.; Shangguan, Y.; Wang, Z.; Sang, T.; Zhou, B.; et al. Genetic control of seed shattering in rice by the APETALA2 transcription factor shattering abortion1. *Plant Cell* **2012**, *24*, 1034–1048. [CrossRef]
9. Yoon, J.; Cho, L.-H.; Kim, S.L.; Choi, H.; Koh, H.-J.; An, G. The BEL1-type homeobox gene SH5 induces seed shattering by enhancing abscission-zone development and inhibiting lignin biosynthesis. *Plant J.* **2014**, *79*, 717–728. [CrossRef]
10. Sun, P.; Zhang, W.; Wang, Y.; He, Q.; Shu, F.; Liu, H.; Wang, J.; Wang, J.; Yuan, L.; Deng, H. OsGRF4 controls grain shape, panicle length and seed shattering in rice. *J. Integr. Plant Biol.* **2016**, *58*, 836–847. [CrossRef]
11. Yoon, J.; Cho, L.-H.; Antt, H.W.; Koh, H.-J.; An, G. KNOX Protein OSH15 Induces Grain Shattering by Repressing Lignin Biosynthesis Genes. *Plant Physiol.* **2017**, *174*, 312–325. [CrossRef] [PubMed]
12. Lv, S.; Wu, W.; Wang, M.; Meyer, R.S.; Ndjiondjop, M.-N.; Tan, L.; Zhou, H.; Zhang, J.; Fu, Y.; Cai, H.; et al. Genetic control of seed shattering during African rice domestication. *Nat. Plants* **2018**, *4*, 331–337. [CrossRef] [PubMed]
13. Zheng, Y.; Crawford, G.W.; Jiang, L.; Chen, X. Rice Domestication Revealed by Reduced Shattering of Archaeological rice from the Lower Yangtze valley. *Sci. Rep.* **2016**, *6*, 28136. [CrossRef] [PubMed]
14. Ishikawa, R.; Thanh, P.T.; Nimura, N.; Htun, T.M.; Yamasaki, M.; Ishii, T. Allelic interaction at seed-shattering loci in the genetic backgrounds of wild and cultivated rice species. *Genes Genet. Syst.* **2010**, *85*, 265–271. [CrossRef]
15. Wang, M.; Yu, Y.; Haberer, G.; Marri, P.R.; Fan, C.; Goicoechea, J.L.; Zuccolo, A.; Song, X.; Kudrna, D.; Ammiraju, J.S.S.; et al. The genome sequence of African rice (*Oryza glaberrima*) and evidence for independent domestication. *Nat. Genet.* **2014**, *46*, 982–988. [CrossRef]
16. Huang, X.; Kurata, N.; Wei, X.; Wang, Z.-X.; Wang, A.; Zhao, Q.; Zhao, Y.; Liu, K.; Lu, H.; Li, W.; et al. A map of rice genome variation reveals the origin of cultivated rice. *Nature* **2012**, *490*, 497–501. [CrossRef]
17. Khush, G.S. Origin, dispersal, cultivation and variation of rice. *Oryza Mol. Plant* **1997**, *35*, 25–34. [CrossRef]
18. Brar, D.; Khush, G. Alien introgression in rice. *Plant Mol. Biol.* **1997**, *35*, 35–47. [CrossRef]

19. Sanchez, P.; Wing, R.; Brar, D. *Genetics and Genomics of Rice*; Zhang, Q., Wing, R.A., Eds.; Springer: New York, NY, USA, 2013; pp. 9–25.
20. Vaughan, D.A.; Lu, B.-R.; Tomooka, N. The evolving story of rice evolution. *Plant Sci.* **2008**, *174*, 394–408. [CrossRef]
21. Edwards, K.; Johnstone, C.; Thompson, C. A simple and rapid method for the preparation of plant genomic DNA for PCR analysis. *Nucleic Acids Res.* **1991**, *19*, 1349. [CrossRef]
22. McCouch, S.; Teytelman, L.; Xu, Y.; Lobos, K.; Clare, K.; Walton, M.; Fu, B.; Maghirang, R.; Li, Z.; Xing, Y.; et al. Development and mapping of 2240 new SSR markers for rice (*Oryza sativa* L.) (supplement). *DNA Res.* **2002**, *9*, 257–279. [CrossRef] [PubMed]
23. Zhao, Q.; Feng, Q.; Lu, H.; Li, Y.; Wang, A.; Tian, Q.; Zhan, Q.; Lu, Y.; Zhang, L.; Huang, T.; et al. Pan-genome analysis highlights the extent of genomic variation in cultivated and wild rice. *Nat. Genet.* **2018**, *50*, 278–284. [CrossRef] [PubMed]
24. Huang, X.; Zhao, Y.; Wei, X.; Li, C.; Wang, A.; Zhao, Q.; Li, W.; Guo, Y.; Deng, L.; Zhu, C.; et al. Genome-wide association study of flowering time and grain yield traits in a worldwide collection of rice germplasm. *Nat. Genet.* **2011**, *44*, 32–39. [CrossRef] [PubMed]
25. Thurber, C.S.; Reagon, M.; Gross, B.L.; Olsen, K.M.; Jia, Y.; Caicedo, A.L. Molecular evolution of shattering loci in U.S. weedy rice. *Mol. Ecol.* **2010**, *19*, 3271–3284. [CrossRef]

plants

MDPI

Article

Allelic Differentiation at the *E1/Ghd7* Locus Has Allowed Expansion of Rice Cultivation Area

Hiroki Saito [1,2,*], Yutaka Okumoto [1], Takuji Tsukiyama [1,3], Chong Xu [1,4], Masayoshi Teraishi [1] and Takatoshi Tanisaka [1,4]

1 Graduate School of Agriculture, Kyoto University, Kyoto, Kyoto 606-8502, Japan; okumoto.yutaka.4w@kyoto-u.ac.jp (Y.O.); tsukiyama@nara.kindai.ac.jp (T.T.); xuchong@kiui.ac.jp (C.X.); temple@kais.kyoto-u.ac.jp (M.T.); t_tanisa@kiui.ac.jp (T.T.)
2 Tropical Agriculture Research Front, Japan International Research Center of Agricultural Science, Ishigaki, Okinawa 907-0002, Japan
3 Faculty of Agriculture, Kindai University, Nara, Nara 631-8505, Japan
4 School of Agriculture, Kibi International University, Minami-Awaji 656-0484, Japan
* Correspondence: hirokisaito@affrc.go.jp; Tel.: +81-980-82-2396

Received: 9 September 2019; Accepted: 25 November 2019; Published: 28 November 2019

Abstract: The photoperiod-insensitivity allele *e1* is known to be essential for the extremely low photoperiod sensitivity of rice, and thereby enabled rice cultivation in high latitudes (42–53° north (N)). The *E1* locus regulating photoperiod-sensitivity was identified on chromosome 7 using a cross between T65 and its near-isogenic line T65w. Sequence analyses confirmed that the *E1* and the *Ghd7* are the same locus, and haplotype analysis showed that the *e1/ghd7-0a* is a pioneer allele that enabled rice production in Hokkaido (42–45° N). Further, we detected two novel alleles, *e1-ret/ghd7-0ret* and *E1-r/Ghd7-r*, each harboring mutations in the promoter region. These mutant alleles alter the respective expression profiles, leading to marked alteration of flowering time. Moreover, *e1-ret/ghd7-0ret*, as well as *e1/ghd7-0a*, was found to have contributed to the establishment of Hokkaido varieties through the marked reduction effect on photoperiod sensitivity, whereas *E1-r/Ghd7-r* showed a higher expression than the *E1/Ghd7* due to the nucleotide substitutions in the *cis* elements. The haplotype analysis showed that two photoperiod-insensitivity alleles *e1/ghd7-0a* and *e1-ret/ghd7-0ret*, originated independently from two sources. These results indicate that naturally occurring allelic variation at the *E1/Ghd7* locus allowed expansion of the rice cultivation area through diversification and fine-tuning of flowering time.

Keywords: rice; flowering time; photoperiod sensitivity; allelic variation; fine-tuning

1. Introduction

Rice is a major cereal extensively cultivated in a wide range of latitudes from 55° N to 35° S. Because rice is formerly a facilitative short-day (SD) plant well adapted to warm climate, photoperiodic control of flowering time is a key factor in the regional and seasonal adaptability of rice varieties [1]. In high latitudes (>ca. 40° N), rice cultivation had been impracticable due to the short summer and long-day (LD) more than 15 h during the summer, until early flowering varieties with extremely weak photoperiod sensitivity were raised [2–4]. It was during 1900 to 1930 that such varieties were first released and planted in the northernmost rice cultivation area, Hokkaido, in Japan (42–45° N) [5]. The varieties raised for Hokkaido also enabled rice cultivation even in Hei Long Jiang province (43–53° N) of China [6].

Also in low latitudes (ca. 20° S–20° N), a recent rice breeding program aims to produce varieties with weak photoperiod sensitivity (PS), though a long basic vegetative growth period is necessary at the same time, because such a combination of the two traits for heading will permit almost constant and adequate vegetative growth periods under SD (less than 13.5 h) [7–9]. In addition, in middle latitudes (30–40° N), there is a close relation between the photoperiod sensitivity of varieties and the latitude of their cultivation area [3,4]. Thus, understanding of the genetic factors responsible for photoperiod sensitivity, as well as basic vegetative growth, will be essential for not only guaranteeing stable rice production but also allowing further expansion of rice cultivation area.

Genetic studies on rice flowering (heading) time started in 1915 [10]. Since then, many flowering time loci were reported: among them, *E1* [11–14], *Photosensitivity 1* (*Se1*) [15], and *Earliness 1* (*Ef1*) [16], have been intensively studied about their genetic characteristics, such as allelic variation, response to photoperiod, geographical distribution, and interaction with other loci. The geographical studies showed that these three loci play especially important roles in regional adaptabilities of Japanese and Taiwanese *japonica* rice varieties and *japonica*/*indica* cross varieties in Korea [4–6,15,17–23].

The Committee on Gene Symbolization, Nomenclature and Linkage Groups of the Rice Genetics Cooperative made a rule that the gene symbols which have been commonly used by many workers in the past should be retained [24], and recommended to categorize flowering time genes into three types, earliness and lateness (gene symbol: *E*), photoperiod sensitivity (gene symbol: *Se*), and basic vegetative growth (gene symbol: *Ef*). With the advance of quantitative trait locus (QTL) analysis, however, it has become difficult to categorize newly found QTLs into three types because they are detected only from flowering time data. Since then, the gene symbols, *E*, *Se*, and *Ef*, did not come to be retained. Recent molecular genetic analyses identified three key flowering time loci, *Heading date 1* (*Hd1*) [25], *Early heading date 1* (*Ehd1*) [26], and *Grain number, plant height, and heading date 7* (*Ghd7*) [27], all of which were named regardless of the rule, and subsequent studies on these three loci provided new numerous molecular-based knowledges of rice flowering. Similarly, about the *E1*, *Se1* and *Ef1* loci, the information useful for rice breeding has been accumulated with enormous numbers until now. Therefore, it is significant to clarify the relationships of the three loci, *E1*, *Se1* and *Ef1*, to the loci named regardless of the rule. To date, the *Se1* and *Ef1* loci proved to be identical with the *Hd1* [21,28] and the *Ehd1* loci [7], respectively.

The *E1* was first identified as a late flowering time locus: the functional allele *E1* is completely dominant over the nonfunctional allele *e1* [11,12]. This locus was also involved in plant height. Later, this locus proved to control PS [13], and its functional allele *E1* was shown essentially important in rice varieties for temperate areas in Japan (30–40° N) [17,18] because of firmly inhibiting the panicle primordial differentiation under LD until it becomes SD conditions, and thereby ensuring normal vegetative growth and stable yields. In contrast, a photoperiod-insensitivity allele *e1* was found to be essential for the varieties commercially cultivated in Hokkaido (42–45° N), because of its marked reducing effect on photoperiod sensitivity: use of *e1* enabled rice cultivation in high latitudes where LD conditions continue during the summer [5,21]. The *E1* locus was found to be located on chromosome 7, linked to the *rfs* (rolled fine strip) and *slg* (slender glume) loci with recombination values of 16.3% and 9.1%, respectively [29]. This locus has been well investigated for its effects on photoperiod sensitivity and regional adaptabilities of rice plants [5,11–19], but little is known about the relationship with the loci which were identified by molecular genetic analysis and named regardless of the rule. Recently, the *Ghd7* locus was precisely mapped on chromosome 7, and this locus exert major effects on not only heading date but also number of grains per panicle and plant height [27]. In addition, subsequent molecular analyses of the *Ghd7* locus demonstrated that a loss-of-function allele of *Ghd7* is essential for the extremely early flowering of Hokkaido varieties [30–32]. These reports make a conjecture that the *E1* and *Ghd7* are the same locus.

In the present study, we first analyzed the effects of three alleles at the *E1* locus on photoperiod sensitivity using the Taiwanese *japonica* rice variety "Taichung 65 (T65)" harboring *E1* [33], its isogenic line T65m harboring *e1* [16,33–35], and T65w that harbors a chromosome segment of *O. rufipogon* Griff. including the *E1* locus in the genetic background of T65 [36]. Subsequently, we attempted to determine the precise chromosomal location of the *E1* locus using the progenies from T65 × T65w, and then conducted sequence analysis to learn the sequences of the three alleles, also to investigate the relationship between the two loci, *E1* and *Ghd7*. We finally applied a haplotype analysis of the chromosomal region surrounding the *E1/Ghd7* locus to 44 Hokkaido and 50 Japanese-core-collection varieties in order to prove correctness of the findings by Okumoto et al. (1996) [5] and Ichitani et al. (1998) [21] that *e1* is the key allele for establishing the varieties for the northernmost rice cultivation area, and its history and origin.

2. Results

2.1. Photoperiod Sensitivities of T65, T65w, and T65m

Days to heading (DH) of T65, T65m, and T65w under a SD were 84.7, 81.4, and 82.0 respectively, while those under a LD were 95.6, 90.4, and 118.0, respectively (Figure 1). Thus, the photoperiod sensitivities of T65, T65m, and T65w were estimated at 11.1, 9.0, and 36.0, respectively. Since T65m is an isogenic line of T65 for the *E1* locus, the weaker PS of T65m was attributable to the photoperiod-insensitivity allele e1 at the *E1* locus. T65w showed far stronger photoperiod sensitivity than T65 and T65m. The genotypic difference between T65w and T65 is only in the chromosome region including the *E1* locus, where only T65w harbors a chromosome segment induced from *O. rufipogon* Griff. Since any other photoperiod sensitivity genes have not yet been reported in this region, we conclude that the chromosome segment introduced from *O. rufipogon* Griff. in T65w certainly harbors a strong photoperiod-sensitivity allele, probably at the *E1* locus.

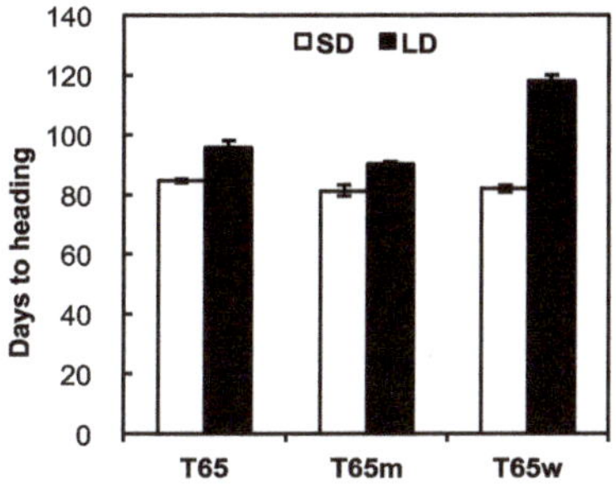

Figure 1. Days to heading of T65, T65m, and T65w under short-day (SD, white bar) and long-day (LD, black bar) conditions.

2.2. Chromosomal Location of the E1 Locus

The F_2 population from the cross between T65 and T65w, comprising 205 plants, showed a continuous distribution of DH within the parental ranges (Figure 2a). We conducted a progeny test using 38 F_3 lines, which were derived from randomly selected F_2 plants. In the test, all the F_3 lines were clearly classified into three groups. The ratio of [T65-type]:[segregating-type]:[T65w-type] lines was 13:14:11, which fitted the 1:2:1 ratio expected for one-locus segregation ($\chi^2 = 8.904$, $P > 0.05$) (Table S1). In contrast, the F_2 population from the cross between T65w and T65m showed a bimodal distribution of DH within the parental ranges, with a clear breakpoint dividing the population into early (T65m-type) and late (T65w-type) groups (Figure 2b). The ratio of early type (34 plants): late type (91 plants) fitted the 1:3 ratio expected for one-locus segregation ($\chi^2 = 0.570$, $P > 0.05$). In the progeny test, all the 40 F_3 lines were clearly classified into three groups. The ratio of [T65m type]:[segregating type]:[T65w type] lines fitted the 1:2:1 ratio expected for one-locus segregation ($\chi^2 = 0.150$, $P > 0.05$) (Table S2). T65m is an isogenic line of T65 harboring a recessive allele *e1* at the *E1* locus. We accordingly inferred that T65w harbors a novel allele at the *E1* locus, whose heading-date delaying effect was stronger than *E1* in T65. We designated this allele *E1-r* (a novel photoperiod-sensitivity allele at the *E1* locus).

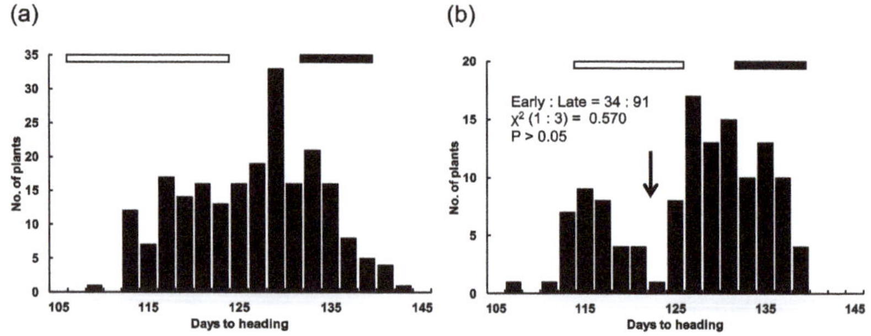

Figure 2. Distributions of days to heading in two F_2 populations from crosses between (**a**) T65 × T65w and (**b**) T65w × T65m. The black bar indicates the range of days to heading of T65w. The white bar indicates the ranges of days to heading of (**a**) T65 and (**b**) T65m. The arrow indicates the breakpoint between early and late heading groups.

Using 546 F_3 plants from the cross between T65 and T65w, we tried to identify the chromosomal location of the *E1* locus. The result showed that the *E1* locus was present in the region with a physical distance of 4.11 Mb between two simple sequence repeat (SSR) markers, RM1253 and RM3635, on chromosome 7 (Figure 3). Subsequently, we attempted to narrow down the candidate region of the *E1* locus, using 1263 F_4 progenies derived from several F_3 recombinants between RM1253 and RM3635; consequently, the chromosomal location of the *E1* locus was narrowed down to the region with a physical length of approximately 228.1-kb between RM5436 and RM21341 (Figure 3). In this region, 11 genes are reported in Rice Annotation Program Database [37]. Among them, we proposed that Os07g0261200, which was reported as *Ghd7*, a repressor of flowering time under LD conditions [27], was likely to be a candidate of *E1*.

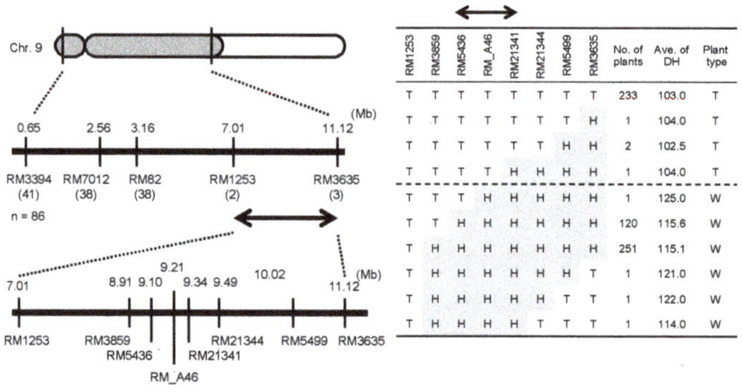

Figure 3. Map-based cloning and graphical genotypes of the candidate region of the E1 locus. "T" and "H" at each marker indicate T65 homozygous and heterozygous, respectively. "T" and "W" at plant type indicate T65-type (early heading) and T65w-type (late heading), respectively.

Sequence analyses showed that the sequences of the alleles at the *Ghd7* locus in T65 and T65m were completely consistent with a functional allele *Ghd7-2* [38] and a nonfunctional allele *ghd7-0a* [27], respectively (Figure 4a,b). Since the genotypic difference in flowering time between T65 and T65m is only at the *E1* locus, this suggests that *E1* and *Ghd7* are the same locus (hereafter we tentatively designate *E1* (=*Ghd7*) as *E1/Ghd7*), and that T65m flowered earlier than T65 because the former harbors a loss-of-function allele *e1/ghd7-0a*. In contrast, the allele of T65w at the *E1* locus harbored four nonsynonymous substitutions and two nucleotide substitutions in the promoter region (Figure 4a,b). Among the substitutions, two in the promoter region were in the transcriptional signal motifs (cis elements): low temperature response element (LTRE) core actor (located at −284) and the TATA box (located at −564). Thus, the two nucleotide substitutions were considered to modify the expression of *E1/Ghd7*. Subsequent expression analysis of *E1/Ghd7* showed that the expression of T65w was higher than that of T65 (Figure 4c). This suggests that the late flowering of T65w is caused by high expression of *E1/Ghd7* due to the nucleotide substitutions in the cis elements.

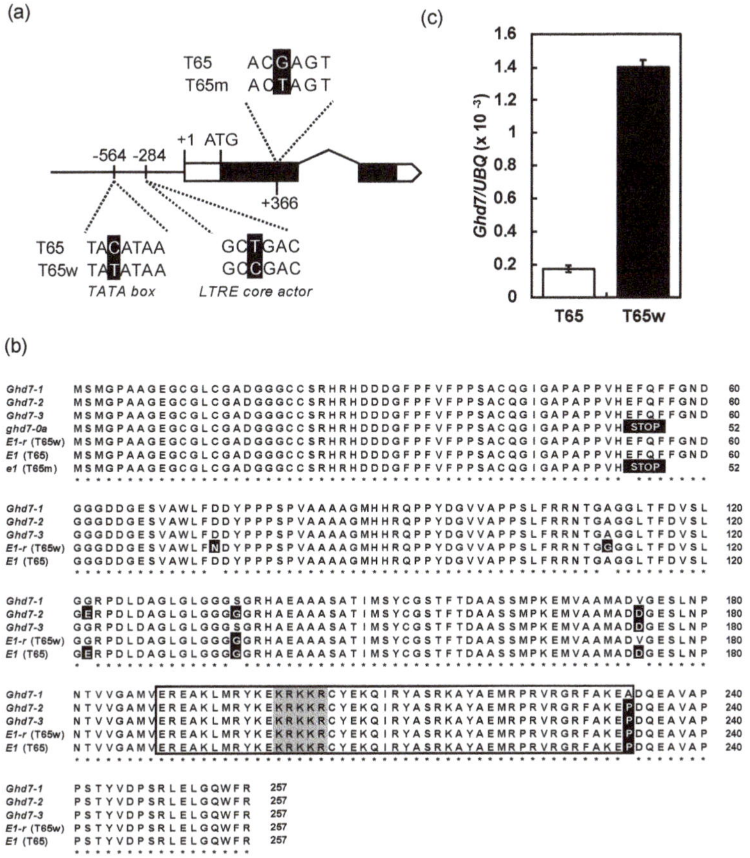

Figure 4. (**a**) Schematic diagrams of the alleles at the *E1/Ghd7* focused on nucleotide substitutions among T65 T65m, and T65w. (**b**) Alignments of amino acid sequences of the alleles at the *E1/Ghd7* locus. The white and black characters with black and gray cells indicate amino acid substitutions and CONSTANS, CO-like, and TOC1 (CCT)-motif, respectively. The box indicates the CCT-motif region. *Ghd7-1*, *Ghd7-2* and *Ghd7-3* were functional alleles [29]. (**c**) Comparison of the expression level of the allele at the *E1/Ghd7* locus between T65 and T65w.

2.3. A Novel Nonfunctional Allele at the E1/Ghd7 Locus

Okumoto et al. (1996) [5] showed that nine Hokkaido varieties tested all harbored a nonfunctional (photoperiod-insensitivity) allele *e1* at the *E1* locus thorough a conventional genetic analysis, and assumed that this allele has played an essential role in the establishment of rice varieties for the Hokkaido district. To confirm this assertion, we analyzed the presence of the nucleotide substitution from GAG (Glu) to TAG (stop codon) in exon 1 at the *E1/Ghd7* locus (Figure 4a) of 44 Hokkaido varieties using a cleaved amplified polymorphic sequence (CAPS) marker. The result showed that 37 varieties harbored the *e1/ghd7-0a* allele, and 7 varieties did not (Table S3). This single nucleotide substitution was not observed in EG5 (Aikoku), which is one of the tester lines for the *E1*, *E2* and *E3* loci involved in the flowering time, and which harbors *e1* allele at the *E1* locus [11–13]. Sequence analysis for the EG5 revealed that a Ty1-copia like retrotransposon (TE) was inserted in the promoter region of the *E1/Ghd7* allele (Figure 5a). The seven varieties, which did not harbor the *e1/ghd7-0a* allele, also harbored the same TE insertion. We named this novel nonfunctional allele *e1-ret/ghd7-0ret*. The expression of the *e1-ret/ghd7-0ret* allele was far lower than the *E1/Ghd7-2* allele in the Japanese variety "Nipponbare" with the reference genome (Figure 5b), indicating that *e1-ret/ghd7-0ret* confers extremely weak photoperiod sensitivity by losing the normal function of the promoter.

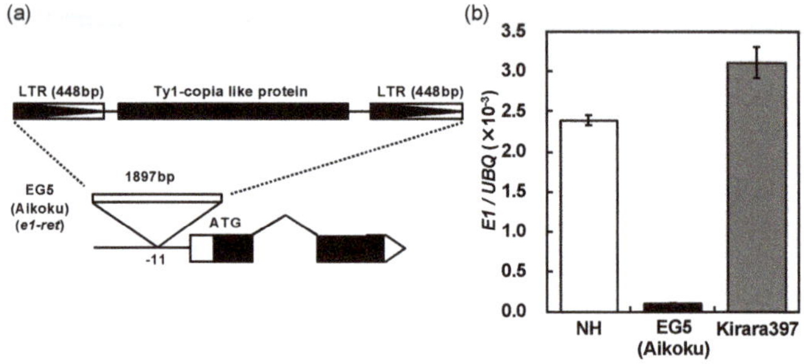

Figure 5. (**a**) Schematic diagram of the allele at the *E1/Ghd7* locus in "EG5 (Aikoku)". (**b**) Comparison of the expression level of the allele at the *E1/Ghd7* locus among three Japanes varieties "Nipponbare" (NH), "EG5 (Aikoku)" and "Kirara397". "EG5 (Aikoku)" and "Kirara397" are a tester line for the *E1*, *E2* and *E3* locus and an elite Hokkaido variety, respectively.

2.4. Haplotype Patterns of the Chromosomal Region Surrounding E1/Ghd7 Locus

We surveyed DNA polymorphisms between EG5 (Aikoku) (*e1-ret/ghd7-0ret*) and Kirara397 (*e1/ghd7-0a*) around *E1/Ghd7* locus. Subsequently, we found three polymorphisms (two SNPs and a 20-bp deletion) other than the nucleotide substitution from GAG (Glu) to TAG (stop codon) and the TE insertion. To know the origins of two nonfunctional alleles *e1/ghd7-0a* and *e1-ret/ghd7-0ret*, we investigated the haplotypes of Hokkaido and Japanese-core collection varieties using five markers surrounding the *E1/Ghd7* locus (three single nucleotide polymorphisms (SNPs), a 20-bp deletion, and a TE insertion). The Japanese-core-collection varieties were classified into at least four haplotypes, Hap2, Hap3, Hap4, and Hap5 (Figure 6 and Table S4). In contrast, Hokkaido varieties were classified into two distinct haplotypes, Hap1 and Hap3. This suggests that Hap1 was derived from Hap2 via nucleotide substitution from GAG (Glu) to TAG (stop codon) in exon 1, whereas Hap3 was derived from Hap4 via the TE insertion in the promoter region. These results indicate that two independent mutational events contributed to the occurrence of the two nonfunctional alleles *e1/ghd7-0a* and *e1-ret/ghd7-0ret*. Interestingly, although Hap1 was found only in Hokkaido varieties, Hap3 was found not only in Hokkaido varieties but also in some Japanese varieties, particularly in the Aikoku-related varieties (Figure 6 and Table S4). Further, the varieties of the Hap3 group, except for Hokkaido varieties, flowered about 20 days later than the Hokkaido varieties, implying that such varieties do not adapt to Hokkaido where autumn comes early (Figure 6 and Table S4). These findings indicate that other genetic factor (s) were involved in the early flowering of Hokkaido varieties belonging to the Hap3 group. We accordingly investigated allelic variations in the *Se1/Hd1* and another major photoperiod-sensitivity gene, *Hd5*, which is known to be involved in the PS in the Hokkaido varieties [30–32]. The results showed that varieties with *e1/ghd7-0a* flowered early regardless of harboring a functional allele(s) (photoperiod-sensitivity allele) at the *Se1/Hd1* and/or *Hd5* locus, whereas varieties with *e1-ret/ghd7-0ret* flowered early only when harboring a nonfunctional allele at either of the *Se1/Hd1* or *Hd5* locus (Figure 7). These results indicate that coexistence of *e1-ret/ghd7-0ret* with a photoperiod-insensitivity allele either at the *Se1/Hd1* or at the *Hd5* locus is necessary to promote flowering under LD conditions.

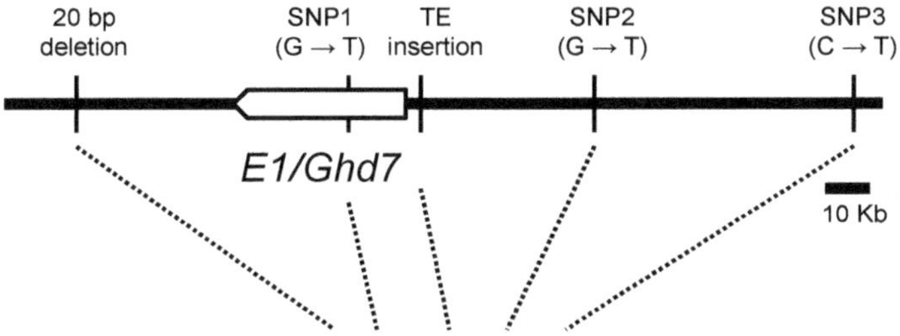

	Haplotype	Variety name	20 bp deletion	SNP1	TE insert	SNP2	SNP3
Hokkaido varieties	Hap 3	Matsumae	-	G	+	G	C
		Tomoemasari	-	G	+	G	C
		Elko	-	G	+	G	C
		Shimahikari	-	G	+	G	C
		Tomoenishiki	-	G	+	G	C
		Fukoku	-	G	+	G	C
		Sorachi	-	G	+	G	C
	Hap 1	Hayakogane	▲20	T	-	T	T
		Ishikari	▲20	T	-	T	T
		Kitahikari	▲20	T	-	T	T
		Bozu	▲20	T	-	T	T
		Iburiwase	▲20	T	-	T	T
		Akage	▲20	T	-	T	T
		Hashiribozu	▲20	T	-	T	T
		Bozu 6	▲20	T	-	T	T
		Wasefukoku	▲20	T	-	T	T
		Megurosakaemochi	▲20	T	-	T	T
		Ishikarishiroke	▲20	T	-	T	T
		Kiyokaze	▲20	T	-	T	T
		Shiokari	▲20	T	-	T	T
		Sasahonami	▲20	T	-	T	T
		Fukuyuki	▲20	T	-	T	T
		Kitaake	▲20	T	-	T	T
		Dohoku 5	▲20	T	-	T	T
		Kitakogane	▲20	T	-	T	T
		Kirara 397	▲20	T	-	T	T
		Hokuto	▲20	T	-	T	T
		Joiku 272	▲20	T	-	T	T
		Kyowa	▲20	T	-	T	T
		Sakigake	▲20	T	-	T	T
		Norin 11	▲20	T	-	T	T
		Joiku 100	▲20	T	-	T	T
		Wasebozu	▲20	T	-	T	T
		Norin 15	▲20	T	-	T	T
		Norin 33	▲20	T	-	T	T
		Shinei	▲20	T	-	T	T
		Akatsuki	▲20	T	-	T	T
		Hayakaze	▲20	T	-	T	T
		Hokkai 126	▲20	T	-	T	T
		Hokkai 242	▲20	T	-	T	T
		Hokkai 244	▲20	T	-	T	T
		Hayamasari	▲20	T	-	T	T
		Tokachikuroge	▲20	T	-	T	T
Japanese rice core collection	Hap 3	Aikoku	-	G	+	G	C
	Hap 2	Ishijiro	▲20	G	-	T	T
		Shinyamadaho 2	▲20	G	-	T	T
		Akamai	▲20	G	-	T	T
		Akamai	▲20	G	-	T	T
		Himenomochi	▲20	G	-	T	T
		Shinsyu	▲20	G	-	T	T
	Hap 4	Hinode	-	G	-	G	C
		Sensyo	-	G	-	G	C
		Yamadabake	-	G	-	G	C
		Kaneko	-	G	-	G	C
		Irumanishiki	-	G	-	G	C
		Okkamodoshi	-	G	-	G	C
		Hirayama	-	G	-	G	C
		Oiran	-	G	-	G	C
		Bozumochi	-	G	-	G	C
		Meguromochi	-	G	-	G	C
		Hassokuho	-	G	-	G	C
		Wataribune	-	G	-	G	C
		Hosokan	-	G	-	G	C
		Akamai	-	G	-	G	C
		Mansaku	-	G	-	G	C
		Joshu	-	G	-	G	C
		Dango	-	G	-	G	C
		Ginbozu	-	G	-	G	C
		Shinrikimochi	-	G	-	G	C
		Shichimenchomochi	-	G	-	G	C
		Moritawase	-	G	-	G	C
		Kameji	-	G	-	G	C
		Omachi	-	G	-	G	C
		Shinriki	-	G	-	G	C
		Kyotoasahi	-	G	-	G	C
		Kabashiko	-	G	-	G	C
		Sekiyama	-	G	-	G	C
		Kemomi	-	G	-	G	C
		Karahoshi	-	G	-	G	C
		Akamai	-	G	-	G	C
		Karahoushi	-	G	-	G	C
		Okabo	-	G	-	G	C
		Hakamuri	-	G	-	G	C
		Aichiasahi	-	G	-	G	C
		Raiden	-	G	-	G	C
		Houmanshindenine	-	G	-	G	C
	Hap5	Nagoyashiro	-	G	-	G	T

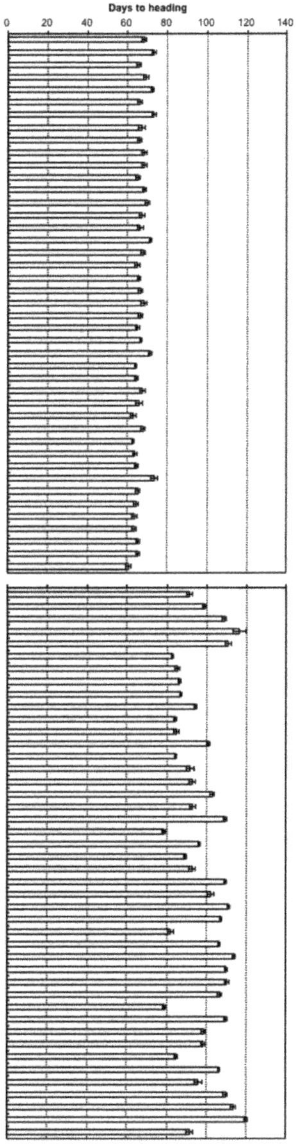

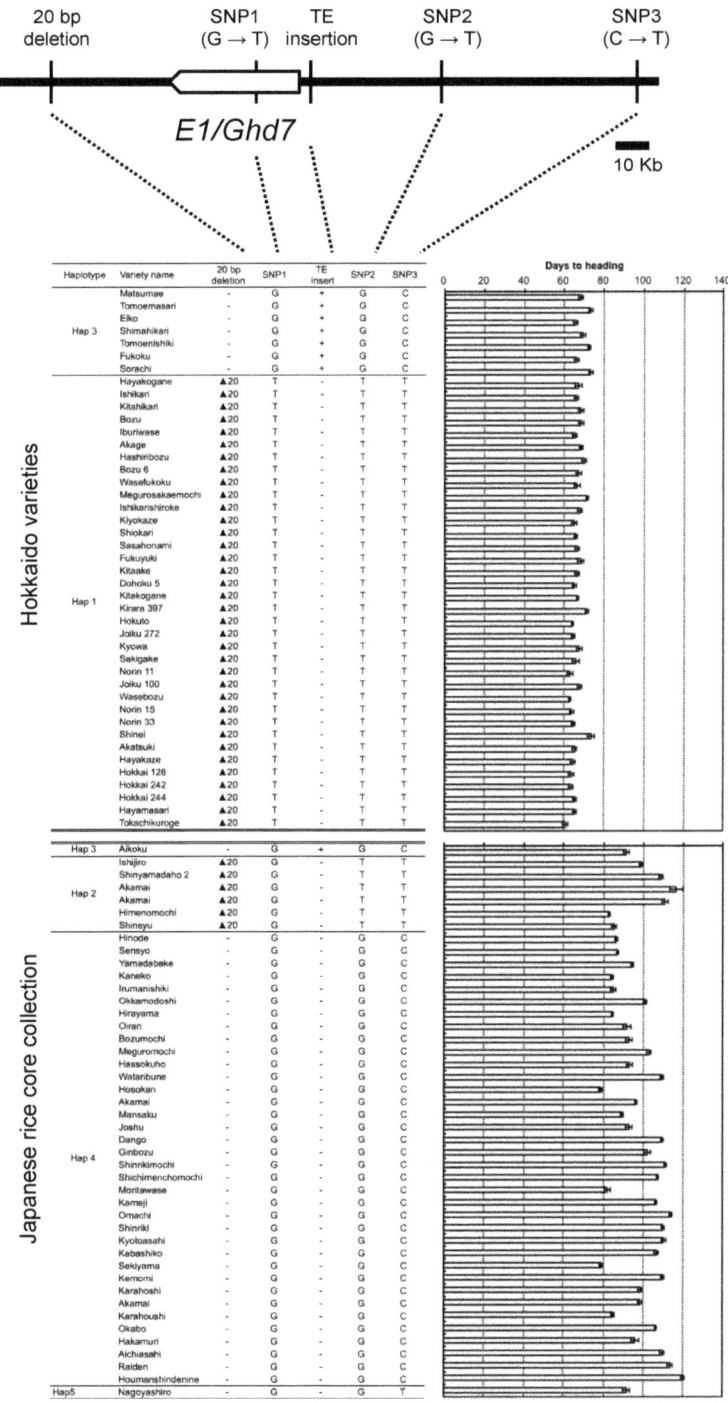

Figure 6. Haplotypes around the *E1/Ghd7* locus and days to heading of Hokkaido and Japanese core collection varieties.

Variety name	Genotype		
	E1	*Se1*	*Hd5*
Matsumae	*e1-ret*	*se1*	*Hd5*
Tomomasari	*e1-ret*	*se1*	*Hd5*
Eiko	*e1-ret*	*se1*	*Hd5*
Tomonishiki	*e1-ret*	*se1*	*Hd5*
Fukoku	*e1-ret*	*se1*	*Hd5*
Sorachi	*e1-ret*	*Se1*	*hd5*
Shimahikari	*e1-ret*	*se1*	*hd5*
Shinnriki	*e1-ret*	*Se1*	*Hd5*
Ouu20	*e1-ret*	*Se1*	*Hd5*
Aikoku 1)	*e1-ret*	*Se1*	*Hd5*
Kirara397	*e1*	*Se1*	*Hd5*
Ishikari	*e1*	*Se1*	*Hd5*
Hayamasari	*e1*	*Se1*	*hd5*
Hokkai244	*e1*	*Se1*	*hd5*
Akage 2)	*e1*	*se1*	*Hd5*
Bozu	*e1*	*se1*	*Hd5*
Bozu6	*e1*	*se1*	*hd5*
Norin11	*e1*	*se1*	*hd5*

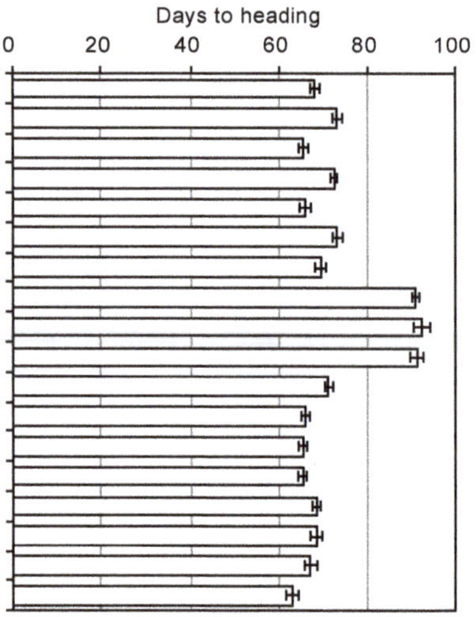

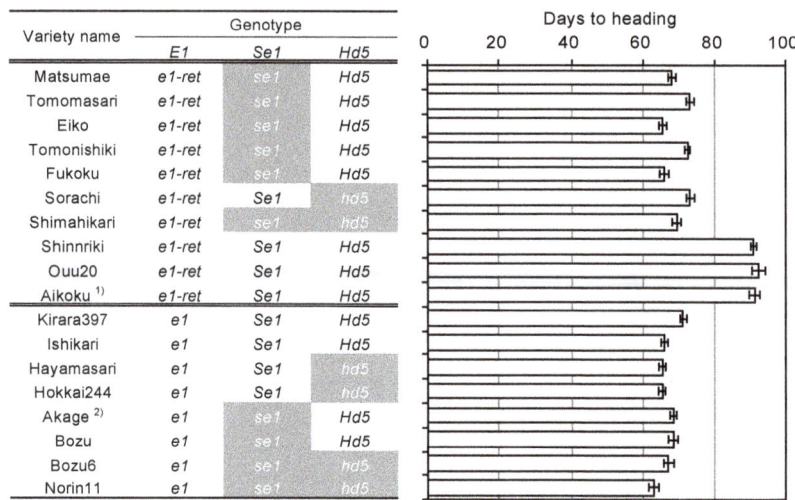

Figure 7. Gene combinations for the *E1*, *Se1*, and *Hd5* loci and days to heading. 1) This "Aikoku" variety belongs to the Japanese core collections. 2) This "Akage" variety belongs to the Hokkaido varieties.

3. Discussion

Since Hoshino (1915) [10], many genes (loci) controlling flowering time have been reported (reviewed in [39–41]). Among them, the *E1* is an important locus closely associated with the regional adaptability of rice varieties: its photoperiod insensitivity allele *e1* enabled rice cultivation even in Hokkaido, one of the northernmost rice cultivation area (42–45° N) [5,13,21]. A dominant photoperiod-sensitivity allele *E1* at the *E1* locus widely deployed among Japanese varieties for all regions other than Hokkaido in Japan [17,18,21]. Ichitani et al. (1998) [21] reported that the *E1* locus was identical to the *Heading date 4* (*Hd4*) locus, which was identified by Quantitative Trait Locus (QTL) analysis of flowering time using progenies from the cross of the *indica* variety Kasalath and the *japonica* variety Nipponbare [25]. Fujino and Sekiguchi (2005) [30] identified two QTLs, *qDTH-7-1* and *qDTH-7-2*, for flowering time using progenies from the cross between two Hokkaido varieties, Hoshinoyume and Nipponbare. They concluded that *qDTH-7-1* is the same locus as the *E1* (*Hd4*). Later, Xue et al. (2008) [27] isolated *the grain number, plant height, and heading date 7* (*Ghd7*) locus on chromosome 7, whose functional allele *Ghd7* encodes a *CONSTANS*, *CO-like*, and *TOC1* (CCT) domain-containing protein that delays flowering under LD conditions. They also reported that a nonfunctional allele *ghd7-0a*, harboring a premature termination in the predicted coding region, deployed among varieties commercially cultivated in Hei Long Jiang province, China (43–53° N). In the present study, we successfully determined the chromosomal location of the *E1* locus within a 228.1-kb physical region on chromosome 7 (Figure 3). According to the rice public database RAP-DB [38], 11 loci (genes) exist in this region. Among the genes, only Os07g0261200 (=*Ghd7*) showed a SNP in exon 1 of a photoperiod insensitive allele *e1* in T65m (Figure 4). This substitution was the same as that of the nonfunctional allele *ghd7-0a* [27]. From these results, we concluded that the *E1*, *Hd4*, *qDTH-7-1* and *Ghd7* are the same locus. Then we finally designate this locus as *E1/Ghd7*. It is noteworthy that we identified a novel photoperiod-insensitivity allele *e1-ret/ghd7-0ret* that harbored a TE insertion in the promoter region of *E1/Ghd7*. We conclude that this TE-insertional mutation causes the loss of function of the *E1/Ghd7* allele.

In addition, we identified a novel strong photoperiod-sensitivity allele *E1-r/Ghd7-r*, which harbors three nonsynonymous substitutions in the coding sequence (CDS) and two SNPs in the promoter region: one is in the LTRE core actor (CCGAC), and the other is in the TATA-box (Figure 5a). The LTRE core actor was identified in the regulatory regions of all cold-induced genes in Arabidopsis [42]. The CCGAC core motif, also known as C-repeat / drought response element (CRT/DRE), is essential for transcriptional activation in response to cold, drought, and/or high-salt treatments [43]. Kim et al. (2002) [44] showed that light signaling mediated by phytochrome activates cold-induced gene expression through CRT/DRE. The expression of *E1-r/Ghd7-r* was higher than that of a PS allele of *E1/Ghd7-2* in Nipponbare (Figure 5b). The expression level of *E1/Ghd7* is regulated by red light signal and correlated well with its LD specific activity [45,46]. These suggest that the mutations in the promoter region modify the *E1-r/Ghd7-r* expression, which delays flowering under LD conditions. Elucidation of the influences of the three amino acid substitutions in the CDS of *E1-r/Ghd7-r* awaits further study.

Haplotype analysis showed that two photoperiod-insensitive alleles, *e1/ghd7-0a* and *e1-ret/ghd7-0ret*, originated independently from two sources (Figure 6). *e1/ghd7-0a* widely deployed among improved and landrace varieties in Hokkaido, including "Akage", which was one of the pioneers of the Hokkaido rice varieties in the late 1800's [47]. This indicates that *e1/ghd7-0a* is a pioneer allele, leading to raising extremely early heading varieties with extremely weak photoperiod sensitivity during 1900–1930 (see Introduction). In contrast, *e1-ret/ghd7-0ret* was detected in "Fukoku", one of the past leading varieties in Hokkaido. "Fukoku" was bred from the cross between the Japanese warm region variety "Nakate-Aikoku", and the Hokkaido variety "Bozu 6". "Nakate-Aikoku" harbors *e1-ret/ghd7-0ret*, whereas "Bozu 6" does not. This indicates that *e1-ret/ghd7-0ret* of "Fukoku" was derived from "Nakate-Aikoku" (Figure S1). Interestingly, although most of the Aikoku-related varieties harbor *e1-ret/ghd7-0ret* (Table S4), none of them flowered as early as "Fukoku". This indicates that some genetic factor(s) other than *e1-ret/ghd7-0ret* is also responsible for the early flowering of "Fukoku". The varieties with *e1-ret/ghd7-0ret* flowered extremely early when harboring a nonfunctional allele either at the *Se1/Hd1* locus or at the *Hd5* locus (Figure 7). Therefore, we conclude that a nonfunctional allele either at the *Se1/Hd1* locus or at the *Hd5* locus is necessary for early flowering of the Hokkaido varieties with *e1-ret/ghd7-0ret*. Fujino et al. (2013) [32] reported that the nonfunctional allele at the *Hd5* locus was a spontaneous mutant gene that occurred in the Hokkaido local landrace "Bozu," and that this allele contributed to the expansion of rice cultivation to the northern area of Hokkaido. In contrast, the nonfunctional allele at the *Se1/Hd1* locus widely deployed among the varieties for the Tohoku-Hokuriku region (37–40° N) [15,18]. In the early rice breeding in Hokkaido, many Aikoku-related varieties, which were chiefly cultivated in the Tohoku-Hokuriku region, Japan, were frequently used as cross-parents to increase the genetic diversity and improve grain quality [47]. This suggests that *e1-ret/ghd7-0ret* was introduced from the Aikoku-related varieties, and the combination with the nonfunctional allele at the *Hd1* or *Hd5* locus made *e1-ret/ghd7-0ret* available for rice breeding programs in Hokkaido. It is also suggested that mutations in the promoter region often make functional differentiations of alleles at the *E1/Ghd7* locus, bringing about flowering time diversification in rice.

Recent molecular genetic studies revealed that there are large natural allelic variations in key loci controlling flowering time, such as *Ghd7*, *Hd1*, *DTH2*, and *DTH7*, which contribute to the diversity of flowering time, and to the regional adaptability by adjusting flowering time to distinct environmental conditions [27,32,48–51]. Zhang et al. (2015) [52] and Zheng et al. (2016) [53] classified *Ghd7* alleles into three (strong function, weak function and non-function) and two (function and non-function) groups based on the sequence polymorphisms in the coding region, respectively. On the other hand, Lu et al. (2012) [38] analyzed 104 varieties (*O. sativa*) and three wild rice accessions (*O. rufipogon*) and found that 76 SNPs and six insertions and deletions within a 3932-bp DNA fragment of *Ghd7*. Among them, the functional C/T mutation in the promoter region was related to plant height probably by altering gene expression. In this study, we identified two novel alleles with functional differentiations in the promoter region of the *E1/Ghd7* allele. The interaction of these alleles with other flowering time genes except *Hd1* and *Hd5* have not yet been elucidated, additional detailed analysis of their effects on flowering time will contribute to fine-tuning of the flowering time well adapting to the climatic conditions in each area.

Photoperiod sensitivity is an important trait responsible for regional and seasonal adaptability of rice varieties. In this study, we detected two novel alleles at the *E1/Ghd7* locus. It is known that many other loci are involved in the photoperiod sensitivity pathway of flowering in rice. Recent studies showed that photoperiod sensitivity loci, *Se1/Hd1*, *OsPRR37* and *Ghd8*, each have a large allelic variation [49,54–57]. Therefore, analyzing the functional and inter-locus interactions should be advanced, which will lead to practice the fine-tuning of flowering time in rice breeding programs. In addition, we elucidated that the *E1* and *Ghd7* are the same locus. Until now, numerous genetic information has been accumulated for each locus (*E1*, *Ghd7*). For further genetic analyses of the *E1/Ghd7* locus, however, all information about *E1* and *Ghd7* should be available.

4. Materials and Methods

4.1. Photoperiod Sensitivities of T65, T65m, and T65

The Taiwanese *japonica* rice variety "Taichung 65 (T65)", and its isogenic line T65m and its near-isogenic line T65w were used. T65 harbors a photoperiod sensitivity allele *E1* at the *E1* locus, while T65m harbors a photoperiod-insensitivity allele *e1*, which is derived from the cross with Bozu5 [16,19,34,35]. T65w is a chromosome substitution line which was developed by introducing the *E1* region of *O. rufipogon* (W107) into the genetic background of T65 [36]. Five seeds were sown on field soil in a 3.6 L pot and covered with granulated soils. Seedlings were thinned to one plant per pot 14 days after sowing, and were grown under two photoperiod conditions, SD (12-h light/12-h dark) and LD (14-h light/10-h dark). Photoperiod treatments were conducted using two growth cabinets without temperature control. In addition to natural daylight (8:00–18:00), supplementary artificial light was used for the 12-h and 14-h light conditions. The degree of photoperiod sensitivity of each line was expressed as a difference of DH between SD and LD. The experiment was conducted by using five plants per line with three replications from 1 May to October 16. Heading date was recorded for each plant when the first panicle emerged from the sheath of the flag leaf.

4.2. Chromosomal Location of the E1 Locus

Two F_2 populations from crosses of T65 × T65w, comprising 205 plants, and T65w × T65m, comprising 125 plants, were subjected to genetic analysis of heading date. They were grown in a paddy field of Kyoto University, Kyoto, Japan (35°01′ N). Seeds were sown on April 25 in 2007, and seedlings were transplanted on May 16 in 2007. The progeny test was conducted with 40 F_3 lines (sowing on April 30 in 2008 and transplanting on May 21 in 2008). Each F_3 line with approximately 25 plants was the progeny of an F_2 plant randomly selected from the F_2 population. To narrow down the candidate region of the *E1* locus, the F_4 progenies derived from several F_3 recombinants between RM1253 and RM3635 were cultivated (sowing on May 1 in 2009 and transplanting on May 22 in 2009).

4.3. Expression Analysis of E1/Ghd7

Plants of T65, T65m, and T65w were grown in a growth cabinet with a temperature controller under a LD condition (14.5-h light, 30 °C/9.5-h dark, 25 °C) at 70% relative humidity. Seedlings were grown on sand in 3.6 L pots (two plants/pot) with additional liquid fertilizer (Kimura's B Culture Solution, Nippon Medical and Chemical Instruments Co., Ltd., Osaka, Japan). Thirty days before flowering, leaves were collected at 4-h intervals during that day. Total RNAs were extracted with Trizol reagent (Life Technologies Inc., Gaithersburg, Maryland, USA) according to the manufacturer's protocols. Total RNA was subjected to DNA digestion by treatment with RNase-free DNase I (Takara Bio Inc.). The transcriptor first-strand cDNA synthesis kit (Roche Applied Science, Indianapolis, Indiana, USA) was used to reverse-transcribe cDNA from 1 μg of RNA using anchored-oligo (dT)18 primers. Real-time PCR analysis was performed by TaqMan PCR using a LightCycler 1.5 (Roche Applied Science) according to the manufacturer's instructions. The primer sets of *Ghd7* and *UBQ* genes and Universal Probe Library probes of each gene were designed with ProbeFinder version 2.45 (Roche; https://www.roche-applied-science.com/). Primer and probe sequences are shown in Table S5. Expression analysis using the standard curve method was performed to determine the expression level of each gene. The relative expression level of each gene was calculated using *UBQ* gene. The RNA gene standards for the seven genes were applied to their plasmids prepared by the pGEM-T Easy Vector System (Promega Corp., Madison, Wisconsin, USA) using PCR amplicons from the total RNA of T65.

4.4. Identification of the Genotype at the E1/Ghd7 Locus

To identify the *E1/Ghd7* genotype, we developed two DNA markers based on an SNP and a TE insertion. CAPS marker analysis was based on a nucleotide substitution from GAG (Glu) to TAG (stop codon) in exon 1, producing an additional restriction enzyme (*Spe* I) site. A pair of PCR primers (Ghd7_CAPS_F: 5′-CCAACTTGCCCTGTCTTCTT-3′, Ghd7_CAPS_R: 5′-AGCTGCTGCAAGCCAGTAAT-3′) was designed to amplify the 950 bp. PCR was performed with a 20 μL reaction mixture containing 2 μL template DNA, 10× PCR buffer, 25 mM $MgCl_2$, 2 mM of each deoxyriboside-triphosphate (dNTP), 0.2 μL *Taq* DNA polymerase (5 U/μL), 4 μL of a 2.5 mM solution of each primer, and 3.2 μL H_2O. PCR conditions were as follows: 94 °C for 5 min, followed by 35 cycles (1 min at 94 °C, 1 min at 60 °C, and 2 min at 72 °C) with a final extension of 7 min at 72 °C. The amplified products were digested with *Spe* I at 37 °C for 6 h. After digestion, the nonfunctional allele produced three fragments (81, 287, and 582 bp), whereas the functional allele produced two fragments (81 and 869 bp). Amplicons and digested amplicons were separated on 1% agarose gel. After electrophoresis, the gel was stained with ethidium bromide, and the DNA fragments were visualized under UV light. Insertion and deletion (INDEL) marker analysis was based on an 1897-bp copia-like TE insertion. A pair of PCR primers (Ghd7_INDEL_F: 5′-CGTTTCAGCAATAGCATTATGG-3′, Ghd7_INDEL_R: 5′-GCGGGTAGTCATCGAACAG-3′) was designed to amplify the 824 bp in the wild type and the 2721 bp in the insertion type. PCR was performed with a 20 μL reaction mixture containing 2 μL template DNA, 10× PCR buffer, 25 mM $MgCl_2$, 2 mM of each dNTP, 0.2 μL *Taq* DNA polymerase (5 U/μL), 4 μL of a 2.5 mM solution of each primer, and 3.2 μL H_2O. PCR conditions were as follows: 94 °C for 5 min, followed by 35 cycles (1 min at 94 °C, 1 min at 60 °C, and 2 min at 72 °C) with a final extension of 7 min at 72 °C. Amplicons were separated on 1% agarose gel. After electrophoresis, the gel was stained with ethidium bromide, and the DNA fragments were visualized under UV light.

4.5. Haplotype Patterns of the Chromosomal Region around the E1/Ghd7 Locus

Three DNA polymorphisms (two single-nucleotide substitutions and a 20 bp deletion) within the 100-kb chromosomal region surrounding *Ghd7* existed among 44 Hokkaido, 50 Japanese-core-collection [58] and 71 Aikoku-related varieties, which are considered to the derivatives of "Aikoku" variety based on their names. The Japanese rice core collection is a limited set of accessions representing, with a minimum set repetitiveness, the genetic diversity among Japanese rice varieties [58]. The two substitutions (Hap_SNP1 and Hap_SNP2) were detected by CAPS marker analyses. Hap_SNP1 harbors a G-to-T substitution, resulting in changing the recognition site of the restriction enzyme *Hpy188* I. Hap_SNP2 harbors a C-to-T substitution, resulting in changing the recognition site of *Hpy188* I. Primer pairs for Hap_SNP1 and Hap_SNP2 were designed to amplify 623 and 295 bp, respectively (Table S6). In addition, two polymorphism surveys of the *copia*-like TE insertion and the SNP in the first exon of *Ghd7* were performed. The CAPS and INDEL marker analyses to detect the insertion and SNP were described above.

Supplementary Materials: The following are available online at http://www.mdpi.com/2223-7747/8/12/550/s1, Figure S1: Pedigree of the varieties, which harbor *e1-ret/ghd7-0ret* allele. Bold with underline indicates that the varieties harbor *e1-ret/ghd7-0ret* allele. Gray indicates varieties whose allele is unknown. Other indicates that the varieties harbor *e1/ghd7-0a* allele., Table S1: Frequency distributions of days to heading in F_3 lines (T65 × T65w), Table S2: Frequency distributions of days to heading in F_3 lines (T65m × T65w), Table S3: Lists of Hokkaido and Japanese core collection varieties used in this study, Table S4: Lists of Aikoku-related varieties used in this study, Table S5: Primer and probe sequences for expression analysis, Table S6 Primer sequences for haplotype analysis.

Author Contributions: Conceptualization, H.S., Y.O. and T.T. (Takatoshi Tanisaka); Methodology, H.S., Y.O. and T.T. (Takatoshi Tanisaka); Formal analysis, H.S., T.T. (Takuji Tsukiyama) and M.T.; Investigation, H.S. and C.X.; Writing—original draft preparation, H.S.; Writing—review and editing, H.S., Y.O. and T.T. (Takatoshi Tanisaka); Project administration, H.S. and T.T. (Takatoshi Tanisaka).

Acknowledgments: We thank the genetic resources center, NARO (National Agriculture and Food Research Organization) for providing the seeds of Hokkaido, Japanese-core-collection and Aikoku-related varieties. T65m was kindly provided from Shigetoshi Sato in Ryukyu University and Kuo Hai Tsai in National Chung-Hsing University. T65w was kindly provided from Yoshio Sano in Hokkaido University. We also would like to thank Syo Asano and Shinsuke Hamada for their supports of our experiments.

Conflicts of Interest: The authors declare no conflict of interest.

References

1. Khush, G.S. Green revolution: The way forward. *Nat. Rev. Genet.* **2001**, *2*, 815–820. [CrossRef] [PubMed]
2. Wada, E. Studies on the response of heading to daylength and temperature in rice plants. : II. Response of varieties and the relation to their geographical distribution in Japan. *Jpn. J. Breed.* **1952**, *2*, 55–62. [CrossRef]
3. Hosoi, T. Studies on meteorological fluctuation in the growth of rice plants. V. Regional references of thermo-sensitivity, photo-sensitivity, basic vegetative growth and factors determining the growth duration of Japanese varieties. *Jpn. J. Breed.* **1981**, *31*, 239–250. [CrossRef]
4. Tanisaka, T.; Inoue, H.; Uozu, S.; Yamagata, H. Basic vegetative growth and photoperiod sensitivity of heading-time mutants induced in rice. *Jpn. J. Breed.* **1992**, *42*, 657–668. [CrossRef]
5. Okumoto, Y.; Ichitani, K.; Inoue, H.; Tanisaka, T. Photoperiod insensitivity gene essential to the varieties grown in the northern limit region of paddy rice (*Oryza sativa* L.) cultivation. *Euphytica.* **1996**, *92*, 63–66. [CrossRef]
6. Zhang, Y. An Economic Analysis on Growth and Technological Progress of Rice Production in Three Northeastern Provinces of China. *J. Rural Problem* **2002**, *38*, 1–12. [CrossRef]
7. Saito, H.; Yuan, Q.; Okumoto, Y.; Doi, K.; Yoshimura, A.; Inoue, H.; Teraishi, M.; Tsukiyama, T.; Tanisaka, T. Multiple alleles at *Early flowering 1* locus making variation in the basic vegetative growth period in rice (*Oryza sativa* L.). *Theor. Appl. Genet.* **2009**, *119*, 315–323. [CrossRef]
8. Saito, H.; Ogiso-Tanaka, E.; Okumoto, Y.; Yoshitake, Y.; Izumi, H.; Yokoo, T.; Matsubara, K.; Hori, K.; Yano, M.; Inoue, H.; et al. *Ef7* encodes an ELF3-like protein and promotes rice flowering by negatively regulating the floral repressor gene *Ghd7* under both short-and long-day conditions. *Plant Cell Physiol.* **2012**, *53*, 717–728. [CrossRef]
9. Yuan, Q.; Saito, H.; Okumoto, Y.; Inoue, H.; Nishida, H.; Tsukiyama, T.; Teraishi, M.; Tanisaka, T. Identification of a novel gene *ef7* conferring an extremely long basic vegetative growth phase in rice. *Theor. Appl. Genet.* **2009**, *119*, 675–684. [CrossRef]
10. Hoshino, Y. On the inheritance of the flowering time in peas and rice. *J. Col. Agric.* **1915**, *6*, 229–288.
11. Syakudo, K.; Kawase, T. Studies on the quantitative inheritance (11): A Rice (*Oryza sativa* L.) (d) Inheritance of the heading duration and the quantitative function of the causal genes in its determination. : (1) On the quantitative function of the genes E_1, E_2 and D_1. *Jpn. J. Breed.* **1953**, *3*, 6–12. [CrossRef]
12. Syakudo, K.; Kawase, T.; Yoshino, K. Studies on the quantitative inheritance (13): A Rice (*Oryza sativa* L.) (d) Inheritance of the heading duration and the quantitative function of the causal genes in its determination. : (2) On the quantitative function of the genes E_3, E_4 and E_5. *Jpn. J. Breed.* **1954**, *4*, 83–91. [CrossRef]
13. Yamagata, H.; Okumoto, Y.; Tanisaka, T. Analysis of genes controlling heading time in Japanese rice. *Rice Genet.* **1986**, *1*, 351–359.
14. Okumoto, Y.; Tanisaka, T.; Yamagata, H. Genotypic difference in response to light interruption in Japanese rice varieties. *Rice Genet.* **1991**, *2*, 778–780.
15. Yokoo, M.; Kikuchi, F.; Nakane, A.; Fujimaki, H. Genetical analysis of heading time by aid of close linkage with blast resistance in rice. *Bull. Nat. Inst. Agric. Sci. Ser.* **1980**, *D31*, 95–126.
16. Tsai, K.H. Studies on earliness genes in rice, with special reference to analysis of isoalleles at the *E* locus. *Jpn. J. Genet.* **1976**, *51*, 115–128. [CrossRef]
17. Okumoto, Y.; Tanisaka, T.; Yamagata, H. Heading-time genes of the rice varieties grown in south-west-warm region in Japan. *Jpn. J. Breed.* **1991**, *41*, 135–152. [CrossRef]

18. Okumoto, T.; Tanisaka, T.; Yamagata, H. Heading-time genes of the rice varieties grown in the Tohoku-Hokuriku region in Japan. *Jpn. J. Breed.* **1992**, *42*, 121–135. [CrossRef]
19. Okumoto, Y.; Yoshimura, A.; Tanisaka, T.; Yamagata, H. Analysis of a rice variety Taichung 65 and its isogenic early-heading lines for late-heading genes *E1*, *E2* and *E3*. *Jpn. J. Breed.* **1992**, *42*, 415–429. [CrossRef]
20. Ichitani, K.; Okumoto, Y.; Tanisaka, T. Photoperiod sensitivity gene of *Se-1* locus found in photoperiod insensitive rice cultivars of the northern limit region of rice cultivation. *Breed. Sci.* **1997**, *47*, 145–152.
21. Ichitani, K.; Okumoto, Y.; Tanisaka, T. Genetic analysis of low photoperiod sensitivity of rice cultivars for the northernmost regions of Japan. *Plant Breed.* **1998**, *117*, 543–547. [CrossRef]
22. Ichitani, K.; Inoue, H.; Nishida, H.; Okumoto, Y.; Tanisaka, T. Interactive effects of two heading-time loci, *Se1* and *Ef1*, on pre-flowering developmental phases in rice (*Oryza sativa* L.). *Euphytica.* **2002**, *12*, 227–234. [CrossRef]
23. Saito, H.; Okumoto, Y.; Teranishi, T.; Yuan, Q.; Nakazaki, T.; Tanisaka, T. Heading time genes responsible for the regional adaptability of 'Tongil-type short-culmed rice cultivars' developed in Korea. *Breed. Sci.* **2007**, *57*, 135–143. [CrossRef]
24. Kinoshita, T. Report of the committee on gene symbolization, nomenclature and linkage groups. *Rice Genet. Let.* **1986**, *3*, 3.
25. Yano, M.; Sasaki, T. Genetic and molecular dissection of quantitative traits in rice. *Plant Mol. Biol.* **1997**, *35*, 145–153. [CrossRef]
26. Doi, K.; Izawa, T.; Fuse, T.; Yamanouchi, U.; Kubo, T.; Shimatani, Z.; Yano, M.; Yoshimura, A. *Ehd1*, a B-type response regulator in rice, confers short-day promotion of flowering and controls *FT*-like gene expression independently of *Hd1*. *Gene Dev.* **2004**, *18*, 926–936. [CrossRef]
27. Xue, W.; Xing, Y.; Weng, X.; Zhao, Y.; Tang, W.; Wang, L.; Zhou, H.; Yu, S.; Xu, C.; Li, X.; et al. Natural variation in *Ghd7* is an important regulator of heading date and yield potential in rice. *Nat. Genet.* **2008**, *40*, 761–767. [CrossRef]
28. Yano, K.; Yamamoto, E.; Aya, K.; Takeuchi, H.; Lo, P.C.; Hu, L.; Yamasaki, M.; Yoshida, S.; Kitano, H.; Matsuoka, M. Genome-wide association study using whole genome sequencing rapidly identifies new genes influencing agronomic traits in rice. *Nat. Genet.* **2016**, *48*, 927–934. [CrossRef]
29. Okumoto, Y.; Tanisaka, T. Trisomic analysis of a strong photoperiod-sensitivity gene *E1* in rice (*Oryza sativa* L.). *Euphytica* **1997**, *95*, 301–307. [CrossRef]
30. Fujino, K.; Sekiguchi, H. Mapping of QTLs conferring extremely early heading in rice (*Oryza sativa* L.). *Theor. Appl. Genet.* **2005**, *111*, 393–398. [CrossRef]
31. Nonoue, Y.; Fujino, K.; Hirayama, Y.; Yamanouchi, U.; Lin, S.Y.; Yano, M. Detection of quantitative trait loci controlling extremely early heading in rice. *Theor. Appl. Genet.* **2008**, *116*, 715–722. [CrossRef] [PubMed]
32. Fujino, K.; Yamanouchi, U.; Yano, M. Roles of the *Hd5* gene controlling heading date for adaptation to the northern limits of rice cultivation. *Theor. Appl. Genet.* **2013**, *126*, 611–618. [CrossRef] [PubMed]
33. Inoue, H.; Nishida, H.; Okumoto, Y.; Tanisaka, T. Identification of an early heading time gene found in the Taiwanese rice cultivar Taichung 65. *Breed. Sci.* **1998**, *48*, 103–108. [CrossRef]
34. Tsai, K.H.; Oka, H. Genetic studies of yielding capacity and adaptability in crop plants. 4. Effects on an earliness gene, m^b in the genetic background of a rice variety, Taichung 65. *Bot. Bull. Acad. Sinica.* **1970**, *11*, 16–26.
35. Itoh, Y.; Sano, Y. Phyllochron dynamics under controlled environments in rice (*Oryza sativa* L.). *Euphytica* **2006**, *150*, 87–95. [CrossRef]
36. Eiguchi, M.; Sano, Y. Evolutionary significance of chromosome 7 in an annual type of wild rice. *Rice Genet. Newsl.* **1995**, *12*, 187.
37. The Rice Annotation Project Database. Available online: http://rapdb.dna.affrc.go.jp (accessed on 29 August 2019).
38. Lu, L.; Yan, W.H.; Shao, D.; Xing, Y.Z. Evolution and association analysis of *Ghd7* in rice. *PLoS ONE* **2012**, *7*, e34021. [CrossRef]
39. Vergara, B.S.; Chang, T.T. *The Flowering Response of the Rice Plant to Photoperiod: A Review of the Literature*, 2nd ed.; IRRI: Manila, Philippine, 1985; pp. 1–65.
40. Izawa, T. Adaptation of flowering-time by natural and artificial selection in Arabidopsis and rice. *J. Exp. Bot.* **2011**, *58*, 3091–3097. [CrossRef]

41. Tsuji, H.; Taoka, K.; Shimamoto, K. Regulation of flowering in rice: Two florigen genes, a complex gene network, and natural variation. *Curr. Opin. Plant Biol.* **2011**, *14*, 45–52. [CrossRef]
42. Baker, S.S.; Wilhelm, K.S.; Thomashow, M.F. The 5′-region of *Arabidopsis thaliana cor15a* has cis-acting elements that confer cold-, drought- and ABA-regulated gene expression. *Plant Mol. Biol.* **1994**, *24*, 701–713. [CrossRef]
43. Yamaguchi-Shinozaki, K.; Shinozaki, K. A novel cis-acting element in an Arabidopsis gene is involved in responsiveness to drought, low-temperature, or high-salt stress. *Plant Cell* **1994**, *6*, 251–264.
44. Kim, H.J.; Kim, Y.K.; Park, J.Y.; Kim, J. Light signaling mediated by phytochrome plays an important role in cold-induced gene expression through the C-repeat/dehydration responsive element (C/DRE) in Arabidopsis thaliana. *Plant J.* **2002**, *29*, 693–704. [CrossRef] [PubMed]
45. Itoh, H.; Izawa, T. The coincidence of critical day length recognition for florigen gene expression and floral transition under long-day conditions in rice. *Mol. Plant* **2013**, *6*, 635–649. [CrossRef] [PubMed]
46. Osugi, A.; Itoh, H.; Ikeda-Kawakatsu, K.; Takano, M.; Izawa, T. Molecular dissection of the roles of phytochrome in photoperiodic flowering in rice. *Plant Physiol.* **2011**, *157*, 1128–1137. [CrossRef] [PubMed]
47. Shinada, H.; Yamamoto, T.; Yamamoto, E.; Hori, K.; Yonemaru, J.; Matsuba, S.; Fujino, K. Historical changes in population structure during rice breeding programs in the northern limits of rice cultivation. *Theor. Appl. Genet.* **2014**, *127*, 995–1004. [CrossRef]
48. Yano, M.; Katayose, Y.; Ashikari, M.; Yamanouchi, U.; Monna, L.; Fuse, T.; Baba, T.; Yamamoto, T.; Umehara, Y.; Nagamura, Y.; et al. *Hd1*, a major photoperiod sensitivity quantitative trait locus in rice, is closely related to the *Arabidopsis* flowering time gene *CONSTANS*. *Plant Cell* **2001**, *12*, 2473–2484. [CrossRef]
49. Fujino, K.; Wu, J.; Sekiguchi, H.; Ito, T.; Izawa, T.; Matsumoto, T. Multiple introgression events surrounding the *Hd1* flowering-time gene in cultivated rice, *Oryza sativa* L. *Mol. Gen. Genom.* **2010**, *284*, 137–146. [CrossRef]
50. Wu, W.; Zheng, X.M.; Lu, G.; Zhong, Z.; Gao, H.; Chen, L.; Wu, C.; Wang, H.J.; Wang, Q.; Zhou, K.; et al. Association of functional nucleotide polymorphisms at *DTH2* with the northward expansion of rice cultivation in Asia. *Proc. Natl. Acad. Sci. USA* **2013**, *110*, 2775–2780. [CrossRef]
51. Gao, H.; Jin, M.; Zheng, X.M.; Chen, J.; Yuan, D.; Xin, Y.; Wang, M.; Huang, D.; Zhang, Z.; Zhou, K.; et al. *Days to heading 7*, a major quantitative locus determining photoperiod sensitivity and regional adaptation in rice. *Proc. Natl. Acad. Sci. USA* **2014**, *111*, 16337–16342. [CrossRef]
52. Zhang, J.; Zhou, X.; Yan, W.; Zhang, Z.; Lu, L.; Han, Z.; Zhao, H.; Liu, H.; Song, P.; Hu, Y.; et al. Combinations of the *Ghd7*, *Ghd8* and *Hd1* genes largely define the ecogeographical adaptation and yield potential of cultivated rice. *New Phytol.* **2015**, *208*, 1056–1066. [CrossRef]
53. Zheng, X.M.; Feng, L.; Wang, J.; Qiao, W.; Zhang, L.; Cheng, Y.; Yang, Q. Nonfunctional alleles of long-day suppressor genes independently regulate flowering time. *J. Int. Plant Biol.* **2016**, *58*, 540–548. [CrossRef] [PubMed]
54. Takahashi, Y.; Teshima, K.M.; Yokoi, S.; Innan, H.; Shimamoto, K. Variations in Hd1 proteins, *Hd3a* promoters, and *Ehd1* expression levels contribute to diversity of flowering time in cultivated rice. *Proc. Natl. Acad. Sci. USA* **2009**, *106*, 4555–4560. [CrossRef] [PubMed]
55. Wei, X.; Xu, J.; Guo, H.; Jiang, L.; Chen, S.; Yu, C.; Zhou, Z.; Hu, P.; Zhai, H.; Wan, J. *DTH8* suppresses flowering in rice, influencing plant height and yield potential simultaneously. *Plant Physiol.* **2010**, *153*, 1747–1758. [CrossRef] [PubMed]
56. Ebana, K.; Shibaya, T.; Wu, J.; Matsubara, K.; Kanamori, H.; Yamane, H.; Yamanouchi, U.; Mizubayashi, T.; Kono, I.; Shomura, A.; et al. Uncovering of major genetic factors generating naturally occurring variation in heading date among Asian rice cultivars. *Theor. Appl. Genet.* **2011**, *122*, 1199–1210. [CrossRef]
57. Koo, B.H.; Yoo, S.C.; Park, J.W.; Kwon, C.T.; Lee, B.D.; An, G.; Zhang, Z.; Li, J.; Li, Z.; Paek, N.C. Natural variation in *OsPRR37* regulates heading date and contributes to rice cultivation at a wide range of latitudes. *Mol. Plant* **2013**, *6*, 1877–1888. [CrossRef]
58. Ebana, K.; Kojima, K.; Fukuoka, S.; Nagamine, T.; Kawase, M. Development of mini core collection of Japanese rice landrace. *Breed Sci.* **2008**, *58*, 281–291. [CrossRef]

plants

MDPI

Article

In Silico Identification of QTL-Based Polymorphic Genes as Salt-Responsive Potential Candidates through Mapping with Two Reference Genomes in Rice

Buddini Abhayawickrama [1], Dikkumburage Gimhani [1], Nisha Kottearachchi [1,*], Venura Herath [2], Dileepa Liyanage [1] and Prasad Senadheera [3]

1 Department of Biotechnology, Faculty of Agriculture and Plantation Management, Wayamba University of Sri Lanka, Makandura, Gonawila 60170, Sri Lanka; bpramudika@wyb.ac.lk (B.A.); dgimhani@wyb.ac.lk (D.G.); dileepasripal@gmail.com (D.L.)

2 Department of Agricultural Biology, Faculty of Agriculture, University of Peradeniya, Peradeniya 20400, Sri Lanka; venura@agri.pdn.ac.lk

3 Department of Botany, Faculty of Natural Sciences, Open University of Sri Lanka, Nawala 11222, Sri Lanka; spsen@ou.ac.lk

* Correspondence: nisha@wyb.ac.lk

Received: 27 December 2019; Accepted: 5 February 2020; Published: 11 February 2020

Abstract: Recent advances in next generation sequencing have created opportunities to directly identify genetic loci and candidate genes for abiotic stress responses in plants. With the objective of identifying candidate genes within the previously identified QTL-hotspots, the whole genomes of two divergent cultivars for salt responses, namely At 354 and Bg 352, were re-sequenced using Illumina Hiseq 2500 100PE platform and mapped to Nipponbare and R498 genomes. The sequencing results revealed approximately 2.4 million SNPs and 0.2 million InDels with reference to Nipponbare while 1.3 million and 0.07 million with reference to R498 in two parents. In total, 32,914 genes were reported across all rice chromosomes of this study. Gene mining within QTL hotspots revealed 1236 genes, out of which 106 genes were related to abiotic stress. In addition, 27 abiotic stress-related genes were identified in non-QTL regions. Altogether, 32 genes were identified as potential genes containing polymorphic non-synonymous SNPs or InDels between two parents. Out of 10 genes detected with InDels, tolerant haplotypes of *Os01g0581400*, *Os10g0107000*, *Os11g0655900*, *Os12g0622500*, and *Os12g0624200* were found in the known salinity tolerant donor varieties. Our findings on different haplotypes would be useful in developing resilient rice varieties for abiotic stress by haplotype-based breeding studies.

Keywords: abiotic stress; rice; salinity; whole genome re-sequencing

1. Introduction

Rice, being the staple food crop of many nations, is considered as a high priority crop for research programs that focused on ensuring food security [1–3]. Rice is mostly cultivated under natural rain-fed systems frequently exposed to various abiotic and biotic stress conditions throughout the world. Development of improved rice varieties for abiotic stress tolerance is the most affordable strategy to increase rice production using marginal and non-arable lands. Among major abiotic stress conditions, salinity, the presence of increased levels of salts, predominantly sodium chloride, is considered the second most limiting factor for rice production next to drought [4]. In every year, nearly two million hectares of irrigated land become uncultivable due to the buildup of salts [5]. In addition, sodic soil which is accumulated with excessive sodium ions cause unfavorable conditions for agriculture by adversely affecting the soil physical properties. Thus, the interaction between soil sodicity and salinity could seriously compromise the rice growth in the field [6,7]. However, due to the genetic complexity of the trait, development of resilient varieties against salinity stress cannot be achieved by a single step strategy. Due to the polygenic nature of the trait, many Quantitative Trait Loci (QTLs) and Quantitative Trait Nucleotides (QTNs) have been reported linking either with salinity tolerance or susceptibility traits distributed throughout the genome in many rice lines [8–13].

Although rice is sensitive to salt, especially at the seedling stage and reproductive stage, vast diversity for this trait across the rice varieties offers a promising tool for improving salt tolerance in rice. Pokkali and Nona bokra are popular traditional salt-tolerant *indica* rice varieties that tolerate up to 80 mM NaCl at the seedling stage and serve as donors for rice salt tolerance [14]. The major strategies for improving salinity tolerance are reducing Na^+ toxicity by limited Na^+ net influx, Na^+ compartmentalization and removal of Na^+ into the apoplast to achieve a good Na^+/K^+ balance in the shoot under saline condition [3]. It is reported that Pokkali, demonstrates both 'Na^+ exclusion' and 'ion balance' mechanisms while Nipponbare, a moderate tolerant *japonica* variety showed only 'ion balance' [14]. Besides, accumulation of compatible osmolytes for osmotic protection, antioxidant regulation and minimalizing the exposure time of cells to ionic imbalance are observed as components of salt tolerance [15–17]. By QTL mapping, genomic locations of such mechanisms are primarily recognized, giving an insight into the understanding of gene-level identification. Fine mapping followed by map-based cloning is the common approach that has been practicing to reveal candidate genes from QTLs [18,19]. For example, *SKC1* gene that encodes HKT-type transporter is one of the salinity tolerant genes identified through dissecting *Saltol* QTL by map-based cloning [20]. Harnessing QTLs and QTNs of salinity tolerance from diverse rice accessions and introgression them to generate salt-tolerant varieties can be achieved by marker-assisted breeding, which is based on genomic sequences. The Next Generation Sequencing (NGS) technique has been successful in generating DNA sequences of organisms revealing genomic variations at a low cost. It is becoming more popular than the use of marker-based polymorphism techniques. There are many studies indicating how NGS facilitates rice improvement by exploration and exploitation of many functional genes that regulate agronomic traits [21–24].

Feltus et al. (2004) [25] have reported that there are 408,898 candidate DNA polymorphisms including single nucleotide polymorphisms (SNPs) and InDels distinguished between *indica* and *japonica*. These SNPs and InDels can be exploited for gene mapping, association studies and DNA marker-assisted breeding. If there is an SNP or InDel in a gene or regulatory sequence of the gene, there will be a chance to affect the function of the gene either adversely or favorably relative to the function of the gene of reference genome by creating either a missense mutation or premature termination or preventing stop signal or shifting the amino acid sequence leading to phenotypic variations. Mishra et al. (2016) [26] have reported that some of SNP haplotypes of HKT family genes were associated with salt tolerance in Indian wild rice germplasms while some other SNP haplotypes were sensitive to salt stress indicating the impact of SNP variations for the phenotype. The popular *SUB1A* allele of ethylene response factor-like gene that carries an SNP mutation conferring submergence tolerance in vegetative stage of rice is another evidence for the contribution of SNP mutation towards favorable agronomic traits [27]. Therefore, mining of SNPs and InDels of candidate genes is useful for detecting possible phenotypic variations which would be important in breeding programs.

The availability of whole genome information, gene expression profiles and *in silico* gene annotation tools have enabled physical identification of candidate genes by aligning genetic map and the putative QTLs. This approach helps to shortlist promising candidate genes of the trait by analyzing SNPs, InDels and structural variations which can later be validated by expression studies and promoter analysis. Instead of costly conventional fine mapping done with large inbred populations that need significant labor and time, QTLs and QTNs targeted annotation of the NGS derived sequences has revealed many candidate genes in various disciplines of plants [2,8,10,28,29].

Many researchers have conducted QTL mapping studies using Simple Sequence Repeat (SSR) markers, but they could not develop genetic linkage maps with more than 300 markers due to lack of polymorphism [30,31]. Therefore, SSR marker-based maps usually generate many gaps that are difficult to be used directly in candidate gene discovery studies. Currently, SNP markers have become more popular as they generate the vast number of polymorphic sites among individuals. For example, Thomson et al. (2017) [32] have reported that usually 1300–2500 SNP polymorphic markers could be generated from a bi-parental population of rice derived from either *indica* × *indica* or *indica* × *japonica*, if Cornell_6K_array_Infinium_Rice (C6AIR) chip containing about 6000 SNPs is used. Therefore, it appears that due to the availability of huge re-sequencing data, high-density SNPs-based maps have been developed [33,34]. Gimhani et al. (2016) [35] were also able to produce SNP-based highly dense and saturated molecular maps with the C6AIR chip covering 1460.81 cM of the rice genome with an average interval of 1.29 cM between marker loci using a Recombinant Inbred Line (RIL) population derived from At 354 and Bg 352. At 354 is a salinity tolerant elite rice *indica* variety with the pedigree of Pokkali and Bg 94-1 and Bg 352 is a salinity susceptible elite rice *indica* variety with the pedigree of Bg 380/Bg 367-4. Both At 354 and Bg 352 are recommended, high yielding, improved rice varieties in Sri Lanka with a relatively short growth duration of 105 days. Gimhani et al. (2016) reported 14 QTL hotspots and 11 solitary QTLs for salt tolerance flanked with SNP markers narrowing down to less than 1 Mb intervals indicating the potential of use in gene mining studies. We noted that the same regions of these QTLs were reported in other studies validating the potentiality of accommodating candidate genes for abiotic stress, mapped using other breeding populations (Supplementary Table S1). Therefore, it is worthwhile for attempting physical identification of the particular regions via NGS-based approaches instead of conventional fine mapping techniques that consume much time. Hence, as an extension of the same study, we sequenced two varieties—At 354 and Bg 352—with reference to *Oryza sativa japonica* group cultivar Nipponbare and *Oryza sativa indica* group cultivar Shuhui498 (R498) and reported revealing of candidate genes underlined by those QTL hotspots. We performed a gene ontology (GO) analysis to functionally characterize the potential candidate genes. We also outlined the variant calling procedure of the At 354 and Bg 352 genomes, the short-listing approach of the candidate genes leading to salinity tolerance and their possible allelic differences.

2. Results

2.1. Whole Genome Sequencing and Comparison with Nipponbare and R498 Reference Genomes

Whole genome sequencing of At 354 and Bg 352 generated 11.5 and 13.5 Gb of raw data, respectively. More than 90% of the data exceeded Q30 Phred quality score for both of the varieties with mean depth coverage of 30X. The GC percentages of At 354 and Bg 352 were found to be 42.75 and 49.03 respectively. The reads of At 354 and Bg 352 were aligned to two reference genomes. *Oryza sativa japonica* group cultivar Nipponbare IRGSP-1.0 with 374,304,577 bp length was used as the reference genome and the mapped lengths of At 354 and Bg 352 were 349,124,521 bp (93.27%) and 348,205,846 bp (93.03%) respectively. Out of total generated reads, 108 × 10^6 reads of At 354 and 96 × 10^6 reads of Bg 352 were aligned to the Nipponbare genome with an average of 27.9X and 24.9X read depth and 94.65% and 70.76% genome wide coverage respectively (Table 1). Also, the reads were aligned to *Oryza sativa indica* group cultivar R498 with a length of 390,322,188 bp and more than 95% of the length of R498 genome was mapped to At 354 (374,732,599 bp) and Bg 352 (373,811,968 bp). Out of total generated reads, 110 × 10^6 reads of At 354 and 96 × 10^6 reads of Bg 352 were aligned to the R498 genome with an average of 27.3X and 24.4X read depth and 96.98% and 72.15% genome wide coverage, respectively (Table 1).

Table 1. Summary of sequencing statistics.

Reference	Ref Length		Mapped Sites	Total Reads	Mapped Reads	Mapped Bases	Mean Depth	GC%	Ts/Tv
Nipponbare	374,304,577	At 354	349,124,521 (93.27%)	114,142,434	108,034,211 (94.65%)	10,446,593,443	27.91	42.75	2.54
		Bg 352	348,205,846 (93.03%)	135,985,268	96,223,079 (70.76%)	9,333,912,611	24.94	49.03	
R498	390,983,850	At 354	374,732,599 (95.84%)	114,127,820	110,684,704 (96.98%)	10,689,439,220	27.34	42.75	2.48
		Bg 352	373,811,968 (95.61%)	135,973,740	98,099,869 (72.15%)	9,548,473,921	24.42	49.03	

2.2. Identification of Variants in At 354 and Bg 352 Genomes

The genome-wide SNPs and InDels on At 354 and Bg 352 were examined with reference to the Nipponbare and R498 genomes. The frequency distributions of total SNPs and InDels of two varieties with respect to Nipponbare and R498 were shown in Figure 1. Identification of variants with comparison to Nipponbare genome showed that a total of 2,734,000 variants (2,478,369 SNPs and 255,631 InDels) in At 354 and a total of 2,726,469 variants (2,477,244 SNPs and 249,225 InDels) in Bg 352. With reference to R498, only 1,122,726 (1,044,783 SNPs and 77,943 InDels) and 1,107,112 (1,038,244 SNPs and 68,868 InDels) of total variants were observed in At 354 and Bg 352 respectively. The highest SNPs density was observed in chromosome 10 (776.5 and 778.4 in At 354 and Bg 352 respectively) in both varieties while the lowest was on chromosome 4 (557.9) and 5 (535.5) in At 354 and Bg 352 respectively with reference to Nipponbare. However, with reference to R498, the highest SNPs density was observed in chromosome 12 (357.8) in At 354 and chromosome 4 (421.2) in Bg 352 while the lowest SNPs density was observed in chromosome 2 (191.5) in At 354 and chromosome 7 (164.4) in Bg 352 respectively. Most of the SNP changes observed were of transition type with a Ts/Tv ratio of 2.54 in both varieties with respect to Nipponbare reference and a Ts/Tv ratio of 2.48 with respect to R498 reference genome. With regards to InDel density, the highest was observed in chromosome 2 and 3 while the lowest was in chromosome 12 and 4 in At 354 and Bg 352 respectively with reference to Nipponbare. With reference to R498, the highest InDel density was shown in chromosome 8 of both varieties while the lowest was shown in chromosome 3 and 10 of At 354 and Bg 352 respectively (Table 2).

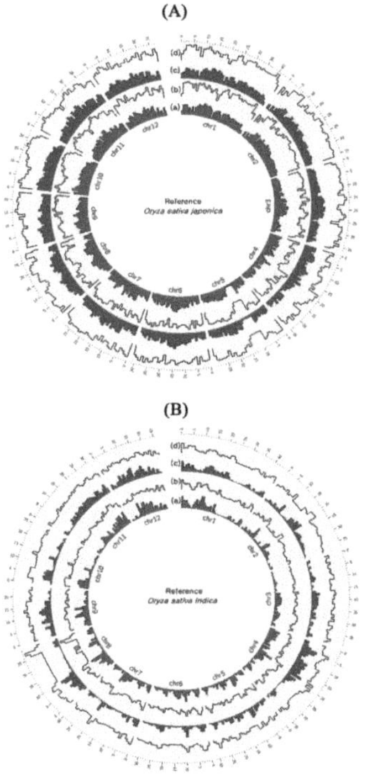

Figure 1. Frequency distribution of single nucleotide polymorphisms (SNPs) and InDels in At 354 and Bg 352. (**A**) with reference to Nipponbare, (**B**) with reference to R498. (a) SNPs–At 354 (b)–InDels At 354 (c) SNPs–Bg 352 (d) InDels–Bg 352.

Table 2. Occurrence and density of SNPs and InDels within the 12 chromosomes in At 354 and Bg 352 genomes after quality filtering.

Ch	No of SNPs		Density (SNPs/100 kb)		No of InDels		Density (InDels /100kb)		Total No of Variants	
	Nipponbare	R498	Nipponbare	R498	Nipponbare	R498	Nipponbare	R498	Nipponbare	R498
At 354										
1	285,179	118,433	659.1	267.0	31,896	7703	73.7	17.4	317,075	126,136
2	242,403	72,315	674.5	191.5	26,628	5104	74.1	13.5	269,031	77,419
3	230,918	76,883	634.1	193.7	25,899	5133	71.1	12.9	256,817	82,016
4	198,076	121,443	557.9	338.8	19,779	6776	55.7	18.9	217,855	128,219
5	167,936	75,130	560.6	240.5	18,590	4704	62.1	15.1	186,526	79,834
6	217,319	89,233	695.4	274.9	22,179	5432	71	16.7	239,498	94,665
7	207,603	79,157	699.1	261.4	20,678	5046	69.6	16.7	228,281	84,203
8	186,119	91,487	654.4	305.4	19,016	11,184	66.9	37.3	205,135	102,671
9	166,630	78,222	724.1	315.9	16,561	6223	72	25.1	183,191	84,445
10	180,213	52,393	776.5	204.8	17,107	4854	73.7	19.0	197,320	57,247
11	218,524	94,898	753	298.6	20,369	8158	70.2	25.7	238,893	103,056
12	177,449	95,189	644.5	357.8	16,929	7626	61.5	28.7	194,378	102,815
Bg 352										
Ch	No of SNPs		Density (SNPs/100kb)		No of InDels		Density (InDels /100kb)		Total No of Variants	
	Nipponbare	R498	Nipponbare	R498	Nipponbare	R498	Nipponbare	R498	Nipponbare	R498
1	282,510	112,753	652.9	254.2	31,008	7567	71.7	17.1	313,518	120,320
2	234,214	87,750	651.7	232.4	25,297	5958	70.4	15.8	259,511	93,708
3	240,771	82,763	661.2	208.5	26,549	5411	72.9	13.6	267,320	88,174
4	214,856	151,002	605.2	421.2	20,144	7562	56.7	21.1	235,000	158,564
5	160,435	77,146	535.5	247.0	17,570	4690	58.6	15.0	178,005	81,836
6	211,331	65,387	676.3	201.4	21,022	4176	67.3	12.9	232,353	69,563
7	215,710	49,767	726.4	164.4	21,225	3449	71.5	11.4	236,935	53,216
8	180,107	75,759	633.2	252.9	17,993	8474	63.3	28.3	198,100	84,233
9	159,509	74,401	693.1	300.5	15,603	6456	67.8	26.1	175,112	80,857
10	180,638	49,286	778.4	192.7	16,604	3137	71.5	12.3	197,242	52,423
11	215,175	108,091	741.4	340.1	19,436	6214	67	19.6	234,611	114,305
12	181,988	104,139	661.0	391.5	16,774	5774	60.9	21.7	198,762	109,913

2.3. QTL-Based SNPs and InDels of Abiotic Stress-Related Genes

QTL-based screening was performed on previously identified salinity stress-related QTL hotspots [35], and we observed slight deviations (0.1 Mb to 3.0 Mb) in the corresponding locations of QTL hotspots with reference to R498 (Figure 2). As expected, a low number of total variants were observed in R498 in each and every QTL examined compared to the Nipponbare. The most abundant variants were found in QTL hotspot 9 of At 354 parent with reference to Nipponbare while the least abundant variants were found in hotspot 10 of Bg 352 parent with reference to R498 (Supplementary Table S2). We found 1236 genes within QTL hotspots and the highest number of genes (215) was found on QTL hotspot 9 while the lowest number (51) was found on QTL hotspot 11. Out of them, 106 genes were associated with abiotic stress. The highest number of stress-related genes (19) was detected within QTL hotspot 2 located on chromosome 2. The lowest number of genes were on hotspots 6 and 12 located on chromosomes 4 and 11, respectively (Supplementary Table S2). In this study, we examined genes located in non-QTL regions to minimize the exclusion of other potential salinity-related genes. Accordingly, we selected 27 genes known for their association with salinity. Therefore, altogether 133 genes were used to examine the allelic differences for salinity.

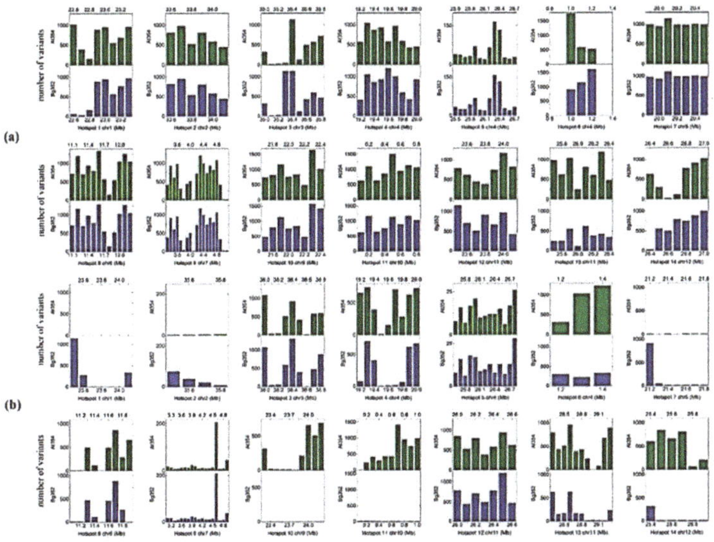

Figure 2. Location of 14 Quantitative Trait Loci (QTLs) on rice chromosomes and the total variants (SNPs and InDels) distribution within the QTLs in 100 kb windows. (**a**) with reference to Nipponbare, (**b**) with reference to R498.

In the above 133 genes, the variants located in exons, introns, 5′ UTR and 3′ UTR regions were analyzed (Supplementary Table S3). Accordingly, *Os01g0581400* which was reported as serine-threonine protein kinase-related domain-containing protein possessed 21 nucleotide variants in the 5′ UTR, 16 variants in the exons and 10 variants in the introns in At 354 comparatively to Nipponbare sequence while Bg 352 possessed only 1 nucleotide variant in 3′ UTR. *Os11g0661600* (similar to peroxidase), *Os11g0669100* (calmodulin binding protein-like family protein), *Os11g0621825* (similar to universal stress protein) were the genes with the highest number of nucleotide variants in exon regions in both of the varieties. *Os11g0621825* (a protein similar to universal stress protein) possessed 110 variants in At 354 and 129 variants in Bg 352 in the intron regions, comparatively to Nipponbare. *Os04g0423400* (similar to *OSIGBa0076I14.3* protein) possessed 10 and 11 nucleotide variants in At 354 and Bg 352 respectively in the 3′ UTR.

2.4. Screening Candidate Genes Based on Polymorphic Nucleotide Variants between Two Parents

This study was aiming at finding polymorphic nucleotide variants between two parents with the potential of salinity tolerance based on allelic differences. Therefore, we screened the genes that showed at least one difference in exons of the nucleotide sequence in one parent comparatively to the other parent from the above 133 gene list. As a result, we found 31 genes located within the QTL regions and three genes located outside the QTLs containing polymorphic variations in the exon region. Each one of them had either one of missense or frame shift or loss of stop codon or early gain of the start codon. Table 3 and Supplementary Table S4 shows the polymorphism type and the location of above 34 candidate genes extracted from the gene sequences of At 354 and Bg 352. Accordingly, we observed 84 variants including 72 SNPs and 12 InDels in 34 genes compared to Nipponbare reference while 73 variants including 63 SNPs and 10 InDels compared to R498 reference. Two InDel variants found in *Os01g0307500* and *Os04g0423400* with reference to Nipponbare were absent with reference to R498.

Os01g0581400 gene of At 354 (GenBank accession number: MK440689) was found with a 12 bp deletion and three missense mutations. In Bg 352, the gene (GenBank accession number: MK440690) encoded the full sequence with 765 amino acids while the sequence of At 354 shifted from 262 position and terminated with 761 amino acid residues due to the 12 bp deletion (Figure 3). GO analysis indicated that *Os01g0581400* was responsible for protein phosphorylation in relation to stress (Supplementary Table S5, [36]).

The gene *Os02g0766700* located within QTL hotspot 2 exhibited two missense mutations in Bg 352 leading to change in amino acid residues from lysine to asparagine and phenylalanine to leucine. According to GO analysis, this gene was reported to provide a regulatory function as a transcription factor in Abscisic Acid (ABA) signaling, water deprivation and salt stress (Supplementary Table S5, [37,38]. Another gene, *Os02g0782500*, located on the same QTL was found with one missense variant in At 354 which changed the glycine to serine.

The gene *Os03g0839200* on QTL hotspot 3 associated with protein detoxification had a 3 bp deletion and a 3 bp insertion at two different locations in Bg 352 (GenBank accession number: MK440692). These two mutations caused a change in amino acid sequence from 490 position in Bg 352 and terminated with 516 amino acids. In At 354, the gene (GenBank accession number: MK440691) indicated encoding the full sequence as of Nipponbare with 516 amino acids (Figure 3). Another gene *Os03g0795900*, a heat stress transcription factor associated with tolerance to environmental stress [39], was found with two missense variants in At 354 sequence, changing serine into alanine and proline into serine.

In Bg 352, the gene *Os04g0117600* (GenBank accession number: MK492739) had a 3 bp insertion and caused a frame shift in amino acid sequence starting from 310 position and terminated with 690 amino acids while At 354 (GenBank accession number: MK492738) showed encoding of full sequence with 689 amino acids. The gene is indicated as tRNA-dihydrouridine synthase-like gene [40]. Go analysis indicated that it could be involved in oxidation reduction biological processes.

Os05g0390500 of At 354 (GenBank accession number: MK492742) exhibited 2 bp insertion which leads to the loss of stop codon and extended the sequence up to 537 amino acid residues. The Bg 352 (GenBank accession number: MK492743) sequence of the same gene indicated encoding for 536 amino acid protein similar to Nipponbare (Figure 3). This gene was located at QTL hotspot 7 from which the salt tolerance was contributed by At 354 parent as indicated by the additive effect of the QTL (Table 3). Although *Os06g0318500* was found with four missense alternative variants in Bg 352 with reference to Nipponbare, the gene was found only with three missense alternative variants with reference to R498, encoding three different amino acid residues in respective positions. GO analysis revealed that this gene functions similar to Sodium/hydrogen exchanger as reported by Panahi et al. (2013) and Reguera et al. (2014) [41,42].

Table 3. Candidate genes identified based on polymorphic InDels in exons regions of two parents.

Gene	QTL Hotspots and Additive Effect [a]	Variation Type	Nipponbare Location	Nipponbare Reference	Nipponbare At 354	Nipponbare Bg 352	R498 Location	R498 Reference	R498 At 354	R498 Bg 352	Amino acid Position
Os01g0581400	01, At 354	SNP	22539348	A	G	A	23425343	G	G	A	I 20 M
		SNP	22539497	A	G	A	23425492	G	G	A	N 70 S
		Indel	22540408	ACTGCGGCGGCGGC	AC	ACTGCGGCGGCGGC	23426414	AC	AC	ACTGCGGCGGCGGC	frame shift
Os03g0839200	03, At 354	SNP	35266328	A	A	G	38332616	A	A	G	D 20 G
		SNP	35266337	T	T	C	38332625	T	T	C	V 23 A
		SNP	35267146	G	G	A	38333434	G	G	A	D 293 N
		Indel	38335687	CGAAG	CGAAG	CG	38335687	CGAAG	CGAAG	CG	frame shift
		Indel	35269310	CC	CC	CCATC	38335755	CC	CC	CCATC	frame shift
Os04g0117600	06, Bg 352	SNP	1047005	C	C	T	1342824	T	C	T	P 16 L
		SNP	1047150	G	A	G	1342968	G	A	G	G 64 D
		SNP	1048077	C	C	T	1343864	T	C	T	T 134 M
		SNP	1048092	A	A	G	1343879	G	A	G	N 139 S
		Indel	1051506	AA	AA	AAGTA	1346152	AAGTA	AA	AAGTA	frame shift
		SNP	1053033	G	C	G	1347696	G	C	G	S 585 T
Os05g0390500	07, At 354	Indel	18933030	AGG	AGGGG	AGG	20113492	AGGGG	AGGGG	AGGGG	frame shift
		SNP	18933023	T	T	G	20113527	G	T	G	S 529 A
Os07g0181000	09, Bg 352	Indel	4266957	CGCCAC	CGCCACAGCCAC	CGCCAC	4219095	CGCCAC	CGCCAC	CGCCAC	frame shift
Os07g0225300 *		Indel	6968059	TGGCGGCG	TGGCGGCGTCGGCGGCG	TGGCGGCG	6890507	TGGCGGCG	TGGCGGCGTCGGCGGCG	TGGCGGCG	frame shift
		SNP	6968496	G	G	A	6890944	A	G	A	G 147 S
		SNP	6968832	A	T	A	6891280	A	T	A	M 259 L
		SNP	6968847	C	G	C	6891295	C	G	C	P 3264 A
Os10g0107000	11, Bg 352	Indel	482854	GGTCGTCG	GGTCGTCGTCG	GGTCGTCG	686457	GGTCGTCG	GGTCGTCGTCG	GGTCGTCG	frame shift
		SNP	483138	C	C	T	686741	T	C	T	S 179 N
		SNP	489170	C	G	G	686974	G	G	G	E 101 D
		Indel	489349	CGCCGCCGGAGCCG	CGCCG	CGCCGCCGGAGCCG	687153	CGCCGCCGGAGCCG	CGCCG	CGCCGCCGGAGCCG	frame shift
		Indel	489388	CTGATGA	CTGA	CTGATGA	687195	CTGATGA	CTGA	CTGATGA	frame shift

Table 3. *Cont.*

			Nipponbare				R498				
Gene	QTL Hotspots and Additive Effect [a]	Variation Type	Location	Reference	At 354	Bg 352	Location	Reference	At 354	Bg 352	Amino acid Position
Os11g0655900	13, At 354	Indel	26270259	CG	CGCCGGAG	CG	29205256	CG	CGCCGGAG	CG	frame shift
		SNP	26270278	C	C	G	29205275	G	G	G	L 71 V
Os12g0622500	14, Bg 352	SNP	26579061	A	A	G	25614979	G	A	G	H 15 R
		Indel	26579981	ATT	ATT	ATTTT	25615899	ATTTT	ATT	ATTTT	frame shift
Os12g0624200	14, Bg 352	Indel	26669266	CCGTCGTCGTCGTCGTC	CCGTCGTCGTCGTCGTC	CCGTCGTCGTCGTC	25694767	CCGTCGTCGTCGTC	CCGTCGTCGTCGTCGTC	CCGTCGTCGTCGTC	frame shift

* Not within QTL hotspots [a] Allele donor was obtained from Additive effect of Gimhani et al. (2016) [35].

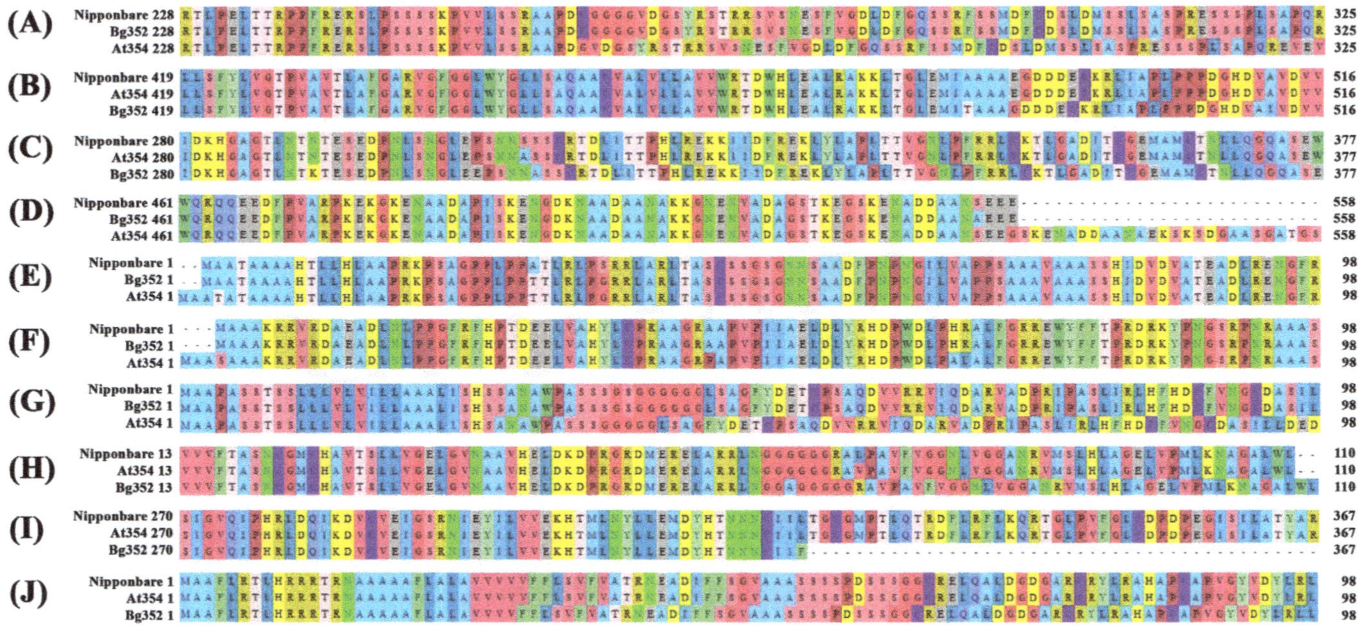

Figure 3. Sequence alignment showing the amino acid sequence coded by putative candidate genes detected with InDel variations. (**a**) *Os01g0581400*, (**b**) *Os03g0839200*, (**c**) *Os04g0117600*, (**d**) *Os05g0390500*, (**e**) *Os07g0181000*, (**f**) *Os07g0225300*, (**g**) *Os10g0107000*, (**h**) *Os11g0655900*, (**i**) *Os12g0622500*, (**j**) *Os12g0624200*.

Os07g0181000 which is associated with kinase activity and ion binding exhibited 6 bp insertion in At 354 (GenBank accession number: MK492744) which resulted in extended amino acid sequence with 580 amino acids while Bg 352 (GenBank accession number: MK492745) had the complete sequence coding for 578 amino acid protein. *Os07g0225300* of At 354 (GenBank accession number: MK492754) showed an 8 bp insertion along with two missense variants. The 8 bp insertion has occurred just before the starting codon thereby leading for gaining of a start codon at three residues before the Nipponbare reference sequence (Figure 3).

The QTL hotspot 11 on chromosome 10 was spotted with five candidate genes in which two of them had frame shifts. In At 354, *Os10g0107000* (GenBank accession number: MK492746) which is responsive to oxidative stress [43] possessed a 3 bp deletion, 3 bp insertion and a 9 bp deletion causing a frame shift in amino acid sequence starting from 28 position and terminated with 326 amino acids while Bg 352 (GenBank accession number: MK492747) encoded the full length of sequence with 329 amino acids. *Os11g0621825,* which codes for a protein similar to universal stress protein [44], was found with two missense mutations in Bg352. The gene *Os11g0655900* which is important for cell redox homeostasis and electron transportation had a 6 bp insertion in At 354 (GenBank accession number: MK492750) causing a frame shift in its amino acid sequence starting from 65 position and terminating at 110 position. Both Bg 352 (GenBank accession number: MK492751) and Nipponabre coded for amino acid sequences with 108 amino acid residues (Table 3).

Out of three candidate genes of QTL hotspot 14 located on chromosome 12, two genes had frame shifts in Bg 352. A 2 bp insertion in *Os12g0622500* of Bg 352 resulted in a 323 amino acid protein due to early gain of stop codon while At 354 had the full sequence coding for 487 amino acids. The gene *Os12g0624200* was found with a 3 bp deletion in Bg 352 and the mutation caused a frame shift starting from the 30 position and terminating at 586 position while At 354 encoded as that of the Nipponbare sequence with 587 amino acids. According to GO analysis, this gene encodes an integral membrane protein that involves transport activity (Supplementary Table S5). In addition, *Os01g0583100, Os01g0591000, Os02g0148100, Os03g0838400, Os03g0839000, Os03g0848400, Os04g0116600, Os04g0430800, Os05g0393800, Os05g0455500, Os09g0559800, Os10g0103800, Os10g0105400, Os10g0109600, Os11g0656000, Os11g669100* and *Os12g0623500* exhibited different missense variants leading to amino acid residue changes in one parent compared to the other parent (Supplementary Table S4).

2.5. Comparative Analysis of InDels in Predicted Candidate Genes with indica Rice Lines

We compared the InDels of the predicted genes in a panel of *indica* rice lines and results revealed that their occurrence varied from approximately 4% to 75%. Of them, the allele of 3bp deletion in *Os10g0107000* was the most abundant InDel while the allele of 9bp insertion in *Os07g0225300* appeared to be a rare allele in the tested population (Figure 4, Supplementary Table S6). We noted that the 12 bp deletion of *Os01g0581400* (GenBank accession number: MK440689) in At 354 was also present in other *indica* varieties such as Nona bokra and Pokkali (Figure 4, Supplementary Figure S1). In At 354, *Os10g0107000* (GenBank accession number: MK492746) possessed a 3 bp deletion, 3 bp insertion and a 9 bp deletion causing a frame shift in amino acid sequence. The same mutations were observed in other salt-tolerant *indica* varieties such as FL478 and Pokkali. *Os11g0655900* had a 6 bp insertion in At 354 (GenBank accession number: MK492750). Interestingly, the same mutation with 6 bp insertion was noted in Pokkali. The 2 bp insertion of *Os12g0622500* in Bg 352 (GenBank accession number: MK492752) was also present in Nona bokra and Pokkali while the 3 bp deletion observed in Bg 352 allele of *Os12g0624200* (GenBank accession number: MK492753) was detected in Nona bokra (Figure 4, Supplementary Figure S1). Accordingly, five genes with InDels found in this study were present in other known salt-tolerant varieties demonstrating evidence for their sequence validation.

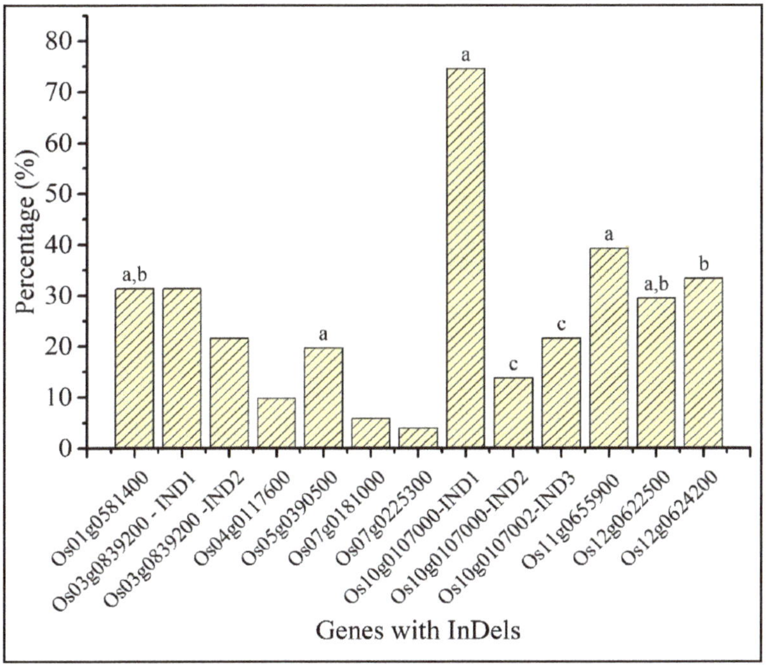

Figure 4. Presence of the InDels in the fifty *indica* rice panel. Inclusion of salt-tolerant donor varieties; Pokkali-IRIS 313-8244, Nona bokra-IRIS 313-7736 and FL478-CX219 are indicated by a, b, and c respectively.

2.6. Analysis of the Promoter Sequences of the Genes with InDels

We examined the *cis*-acting elements on abiotic stress, within promoter regions of candidate genes detected with Indels, to speculate their association with salinity, comparatively to At 354 and Bg 352. Supplementary Table S7 summarizes the particular *cis*-acting elements found within the 1000 bp 5′ upstream of each of the candidate genes. Nine types of abiotic stress-related *cis*-acting elements were found in this study. They are namely, ABRE, CAAT box, DPBF, GAGA, GBOX, IBOX, ROOT, SEF3, and SEF4 which belong to different transcription factor families involved in abiotic stress-related pathways. The highest number of abiotic stress-related *cis* regulatory elements were found in *Os12g0622500* and the lowest number were found in the *Os04g0117600*. The *Os10g0107000* gene had a comparatively notable difference in terms of the type and the number of *cis*-acting elements. In Bg 352, there were 30 *cis*-acting elements in *Os10g0107000* gene while At 354 had only 24 *cis*-acting elements and the DPBF element was absent in At 354.

2.7. PCR-Based InDel Marker for the Detection of Genotypic Polymorphism

Although the accuracy of sequencing is proved, it is still a requirement to confirm the genotypic variations found by *in silico* experiments. Therefore, we selected the longest InDel present among 10 genes, which was the 12 bp deletion in At 354 of *Os01g0581400* allele and designed an InDel marker (PKW) to reveal the polymorphism. The PCR product which was electrophoresed in 3% agarose showed a polymorphic banding pattern matching exactly with the corresponding genotype. Accordingly, At 354, several RILs and International Rice Research Institute (IRRI) germplasm (Pokkali-IRIS 313-8244, Kurulutudu-IRIS 313-8925, H6-IRIS 313-9472 and Puttu Nellu-IRIS 313-9969) were identified as mutated genotype possessing 12 bp deletion in *Os01g0581400* (Figure 5, Supplementary Table S8). The results of this experiment have shown that prediction uncertainty of *in-silico* searchers could be eliminated by combining the task with wet laboratory experiment.

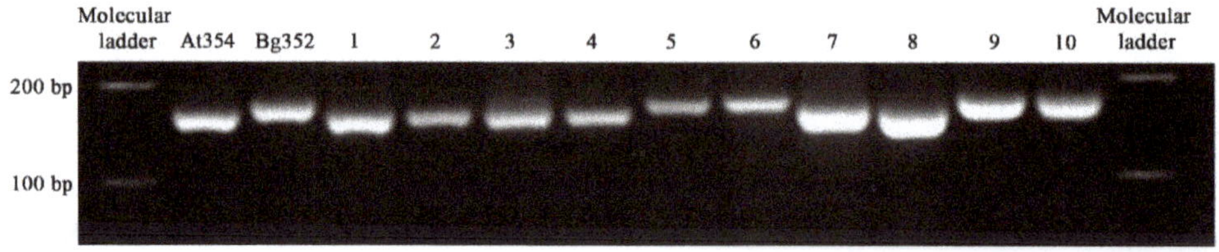

Figure 5. InDel marker for the identification of Polymorphism in *Os01g0581400*. Lane (1). Pokkali-IRIS 313-8244, (2). Kurulutudu-IRIS 313-8925 (3).H6- IRIS 313-9472, (4).Puttu Nellu-IRIS 313-9969, (5).Honderawala-IRIS 313-11382, (6).Herath Banda-IRIS 313-11741, (7 to 10). RILs.3.

3. Discussion

In the present study, the whole genomes of two elite *indica* rice varieties, namely At 354 and Bg 352, were re-sequenced and mapped to both *Oryza sativa* L. cv. Nipponbare reference genome and *Oryza sativa indica* group cultivar R498. As Nipponbare is the current and most comprehensively curated reference genome for the *Oryza sativa*, our analysis was mainly conducted comparatively to Nipponbare while the data were validated using R498 *indica* reference. The near-complete R498 genome is an extra resource for studying genetic variations in rice belonging to *indica* subspecies [45]. Although the majority of reads were mapped to both reference genomes, unmapped read rates of 6% and 29% were observed for At 354 and Bg 352 respectively for Nipponbare and 3% and 27% were observed respectively for R498. The unmapped reads rate of At 354 is comparable with other *indica* rice varieties such as Godawee (8.35%), Swarna (11%) and IR64 (10%) [1,46,47]. Also, Subbaiyan et al. (2012) have observed an average unmapped rate of about 7.5% among 6 *indica* rice inbreds [48]. GC content of At 354 and Bg 352 has been 42.75% and 49.03% respectively, in line with the GC content of monocots that vary within the range of 34% to 49% usually [49]. According to the analysis of chromosome wise variations, a lower number of variants were observed in both varieties with respect to R498 than those of Nipponbare genome. Obviously, it is expected to capture a low number of variants comparatively to R498, because two parents belong to *indica* subspecies. The analysis of chromosome wise variations indicated that IR24, SH527 [50], Godawee [46], Swarna [47] and six elite *indica* rice inbreds [48] contained the highest and the lowest total number of variants on chromosome 1 and 9 respectively as that of At 354 and Bg 352 with respect to Nipponbare. However, the highest and the lowest total number of variants were observed on chromosome 4 and 10 with respect to R498. We calculated the density of occurrence of variants, in order to determine the genomic distribution of SNPs and InDels. The SNP and InDels densities of At 354 and Bg 352 were consistent with other *indica* rice varieties [43–45]. As reported by Tenaillon et al. (2001), a greater SNPs rate could be correlated with a higher level of genomic diversity [51]. In the present study, we observed the ratio of transitions to transversions as 2.54 and 2.48 respectively to Nipponbare and R498 indicating more transition SNPs than transversion SNPs showing transition bias. This incidence has been previously reported in rice genomes revealing 2.0 to 2.5 transitions to transversions ratio [46–48]. In order to maintain RNA stability and conserve the protein structure, transitional mutations have occurred more frequently than transversions during evolution [52].

In this study, abiotic stress-related genes located within previously identified QTL hotspots were analyzed to identify the variants in At 354 and Bg 352. Altogether, we used 14 QTL hotspots flanked less than 1 Mb intervals explaining 12.5–46.7% of phenotypic variation in salinity-related traits [35]. We could find 106 abiotic stress-related genes associated with these QTL hotspots. According to a frequency distribution analysis of genome-wide variations conducted by Jiang et al. (2017) [53], 10 highest SNPs and InDels rich regions were identified in the rice genome. Of them, three regions, chromosome 2

(33–35 Mb), 5 (19–22 Mb) and 6 (10–22 Mb) were exactly matched with the intervals of QTL hotspot 2, 4 and 8 of At 354 × Bg 352, respectively with reference to both Nipponbare and R498. This observation gives evidence to justify the polymorphic nature of the respective QTL regions, indicating the possible existence of allelic variations. Also, we detected the highest number of stress-related genes (19) on the QTL hotspot 2 indicating its potential contribution for salt responsive phenotypic variation. QTL hotspots were previously detected under phytotron conditions of the International Rice Research Institute, Philippines where all possible salinity-related QTLs might not have been expressed [35]. Therefore, in addition to the abiotic stress-related genes within the QTL hotspots, we considered 27 other abiotic stress-related genes which were involved in salt-tolerant pathways. Hence, altogether 133 genes were analyzed and polymorphic variants between At 354 and Bg 352 were observed to identify potential candidate genes. As exons are significantly important due to their function in presenting mRNA and coding the proteins, here we focused mainly on the alternative variants in exons between At 354 and Bg 352.

The SNPs are single-point mutations observed in the genomic DNA of organisms. Some of the SNPs cause the amino acid substitution in the corresponding amino acid sequence (missense mutations) of the genes while others are not (silent mutations) [54]. The missense mutations affect the protein function indirectly through effects on protein folding, stability, flexibility, and aggregation. Modification of the protein to be more flexible or rigid, compared to the respective native structure affects the protein function adversely. If a missense mutation occurred in an active site of a protein structure, it could possibly alter the biological or biochemical reactions and change the kinetics of the reaction and affect the normal protein function [54,55]. There are a number of studies that have shown the functional consequences of SNPs. Wang et al. (1997) [56] have shown that a missense mutation (*adg2-1*) in the ADPG Pyrophosphorylase large subunit gene either affects the stability of the ADGase large subunit protein or its assembly into holoenzyme in *Arabidopsis thaliana*. The *S1-24* mutant in a highly conserved zinc finger domain of *OsCESA7* gene in rice is due to a missense mutation, causing brittle culms, dwarfism and partial sterility. The influence of this mutation is predicted to be in affecting the interactions between different CESA subunits and *OsCESA7* [57]. Tang et al. (2018) [58] have shown that a missense mutation in a plastid ribosomal protein (*RPS4*) in Chinese cabbage has impaired the rRNA processing and affected the ribosomal function. This information indicates that missense mutations occurred due to SNP variations in the gene sequences play an important role in affecting the gene functions in plants. Thus, *in silico* information on SNP variants reported in the present study, could possibly play an important role in regulating the functions of the genes under abiotic stress condition.

The candidate genes with InDels identified in the present study were compared with past research studies in order to speculate their function in relation to stress tolerance. Due to the fact that InDels of *Os01g0307500* and *Os04g0423400* did not appeared with reference to the R498, we did not consider them as true variants.

The gene, *Os01g0581400* identified in the present study, also was reported by Chen et al. (2010) [36] indicating that it contains a juxtamembrane (JM) domain which regulates the proper function of receptor-like kinases (RLKs) by autophosphorylation. The RLKs play an important role in plant responses such as development, hormone perception, defense and response to pathogens [59,60]. *Os01g0581400* had a 12 bp deletion in At 354 (GenBank accession number: MK440689) which truncated the sequence to 761 amino acids while Bg 352 (GenBank accession number: MK440690) showed coding of full sequence with 765 amino acids. This gene was found from QTL hotspot 1 and At 354 had been the respective allele donor for salt tolerance as shown by additive effect (Table 3). GO analysis also indicated that it functions as a protein kinase (GO:0004672) and its involvement in protein phosphorylation (GO:0006468) (Supplementary Table S5). It is interesting to note that At 354 allele which had 12 bp deletion leading to altered amino acid sequence was also present in Nona bokra and Pokkali which are popular salinity tolerant varieties (Supplementary Figure S1). Moreover, we could reveal the two types of alleles detected, in *in silico* analysis using a novel InDel marker confirming their physical presence in diverse varieties (Figure 5). Although the presence of two types of alleles of *Os01g0581400* was proved in the RIL population and other diverse germplasm, we could not interpret the haplotype contribution to salt responsiveness due to genetic complexity of the trait, because none of the salt-tolerant donors has all the desirable alleles for tolerance mechanisms while a salt-susceptible line may also contain few desirable alleles, affecting for unpredictable cumulative effect. The same perception was supported by Islam et al. (2019) [61], who reported the same gene, *Os01g0581400* as a salt responsive candidate gene indicating that possible activation of protein kinase domain-containing protein (*LOC_Os01g39970*) of the particular gene under salt stress by QTL-meta-analysis, a precise estimation technique. However, they were also unable to demonstrate the association between the genes with meta-QTL linked makers and salt tolerance due to complexity of the trait. Therefore, we suggest that the InDel marker developed from the mutation of *Os01g0581400* would be useful to develop near-isogenic lines nullifying the complexity caused by other genes, to investigate the allele contribution for salt tolerance.

The *Os01g0583100,* potentially being another candidate gene located on the same QTL hotspot, possessed one missense mutation (Supplementary Table S4). It was reported as protein phosphatase 2C (*PP2C6)* family member of rice. The *Os01g0583100* is regulated by ABA via ABA-responsive elements located on its promoter (GO:0048364, GO:0004722) (Supplementary Table S5) [62]. Yoshida et al. (2010) [63] have reported the function of PP2C genes in relation to water stress (GO:0009414) and drought conditions while Li et al. (2015) [64] have reported their importance in controlling root architecture and drought tolerance.

The *Os01g0591000* (*OsALDH2C4*) and *Os05g0455500* (*OsALDH18B1*), possessing missense mutations, located on QTL hotspot 1 and 7 respectively (Supplementary Table S4), were found belonging to rice aldehyde dehydrogenase (ALDH) protein superfamily [65]. Kotchoni et al. (2010) [65] have reported that *OsALDH18B1* which is unique for rice, encodes an enzyme for proline synthesis (*P5CS*) (GO:0006561, GO:0004029, GO:0043878) and is important for salt stress adaptation and tolerance. Moreover, the ALDHs are capable of detoxifying the reactive aldehyde molecules which are produced under different abiotic stress conditions and maintain the redox balance in the cells [65,66].

The candidate gene *Os02g0766700* (*OsbZIP23*) located on QTL hotspot 2, is a member of basic leucine zipper (bZIP) transcription factor family in rice, and it contains two missense variants in Bg 352 variety when compared to the At 354. Several gene expression studies have reported its sensitivity to drought, salt and osmotic stress responses [37,67]. Moreover, Xiang et al. (2008) have observed that *OsbZIP23* is highly expressed in leaf tissues and its overexpression may enhance salt tolerance [38]. GO analysis also indicated its involvement in response to water deprivation (GO:0009414) and salt stress (GO:0009651).

The *Os02g0782500* screened with a missense mutation, was also identified previously as an abiotic stress-responsive gene in rice and the function was categorized under small heat stress protein (sHSP) class III by Yi et al. (2013) [68]. Also, Waters et al. (2008) [69] have reported that sHSPs are expressed in other plant response stresses such as drought, salinity, UV, osmotic and oxidative stresses in addition to heat and cold responses (GO:0009408, GO:0009644, GO:0009651, GO:0042542) (Supplementary Table S5). Therefore, the polymorphism between two alleles needs to be further characterized with regards to the function under salinity.

In our study, the salinity susceptible variety, Bg 352 was found with a 3bp insertion, 3bp deletion and three missense variants in *Os03g0839200* (GenBank accession number: MK440692) that shifted the frame of the amino acid from 490 to 516. In accordance with the additive effect, this mutation indicates a possible contribution to salt susceptibility. Neerja et al. (2018) [70] have conducted a research on the transporter genes and found that the *Os03g0839200* has been associated with multidrug and toxic compound extrusion (MATE) efflux family protein (GO:0006855, GO:0015238, GO:0015297), which is an integral component of the membrane involving in salt toxic ion extrusion.

The gene, *Os04g0117600* was found possessing a 3 bp insertion in Bg 352 (GenBank accession number: MK492739) and the mutation caused a frame shift in amino acid sequence starting from 310 position and terminated with 690 amino acids. At 354 predicted the full sequence with 689 amino acids (GenBank accession number: MK492738). It was noted that the gene was located in the telomeric region of chromosome 4. According to the GO analysis, *Os04g0117600* could be involved in tRNA dihydrouridine synthesis (GO:0002943), metal ion binding (GO:0046872) and oxidation reduction processes (GO:0055114).

An InDel variation was found in *Os05g0390500* in QTL hotspot 7 located at 19.8–20.5 Mb of chromosome 5. GO analysis indicated that *Os05g0390500* is responsive to salt stress (GO:0009651) (Supplementary Table S5). *Os06g0318500* gene found with four SNP variations, is one of the five Na^+/H^+ exchanger (NHX) genes present in rice (GO:0009651, GO:0015385, GO:0015386) and several studies have shown that *NHX* genes are capable of regulating the Na^+ and/or K^+ uptake under high salinity conditions faced by plants [41,42,71]. Yang et al. (2016) [72] have reported that *Os07g0181000*, which contained a 6 bp insertion mutation in At 354 allele (GenBank accession number: MK492744), is a photosynthesis-related gene. However, the relevance of the gene to salinity needs to be further investigated.

The *Os10g0107000* on QTL hotspot 11 was found with 3 InDels including a 3 bp insertion, 9 bp and 3 bp deletions in At 354 (GenBank accession number: MK492746) and the same mutations were also observed in other salt-tolerant *indica* varieties, FL478 and Pokkali. *Os10g0107000* which was identified as a class III peroxidase family gene (GO:0004601, GO:0006979) has been upregulated in response to cadmium stress in rice [43]. In addition, Wang et al. (2015) [73] have observed that class III peroxidase genes are differentially expressed in response to abiotic stress in maize and play a significant role in roots.

Another two genes, *Os11g0655900 (OsGRX23)* and *Os11g0656000 (OsGRX24)* belonged to CC-type Glutaredoxin (GRX) family contained an InDel and missense mutations respectively. It was reported that GRXs regulate and participate in the redox-dependent signaling pathways (GO:0045454, GO:0009055, GO:0022900) and provide protection to plants over oxidative stress while being involved in several metabolic pathways [74,75]. Moreover, Garg et al. (2010) [74] have shown that *Os11g0655900* is differentially expressed in rice seedlings under different abiotic stress conditions. The *Os11g0655900* gene possessed a 6 bp insertion in At 354 (GenBank accession number: MK492750) while Bg 352 (GenBank accession number: MK492751) closely aligned with Nipponbare and R498. We noted that the allele contribution for salt tolerance was by At 354 although distortion was occurred by the extension of two additional amino acids. However, reference genomes- (Nipponbare and R498)-based alignments do not indicate which could be the distorted allele, whether At 354 or the reference genomes. Sometimes the At 354 allele could encode the correct version of the amino acid sequence because it gives two additional amino acids. Also, we observed that Pokkali contained the same 6 bp insertion in the particular location indicating that prevalence of the same allele in another salt-tolerant donor variety.

We observed that *Os12g0622500* gene in Bg 352 (GenBank accession number: MK492752) had a 2 bp insertion truncating the sequence to 323 amino acids due to early gain of stop codon while At 354 had the full sequence length with 487 amino acids. We observed that the same mutation was present in Nona bokra giving evidence for the prevalence in another salt-tolerant donor variety. The *Os12g0624200* belonging to the Ca^{2+}/cation antiporter superfamily (GO:0055085) was detected with a 3 bp deletion in Bg 352 (GenBank accession number: MK492753) that shifted the amino acid frame starting from 30th amino acid position with reference to the Nipponbare and R498 genomes. Also, the allele of Bg 352 truncated the sequence to 586 amino acids while At 354 encoded the full sequence with 587 amino acids. Furthermore, studies have shown that *Os12g0624200* is significantly upregulated in response to salinity and dehydration conditions imposed on rice suggesting its involvement in stress tolerance [76,77].

Not only the coding sequences, but also the 5′ upstream regions including the *cis*-acting elements of promoter sequences usually affect the expression of the genes. Therefore, analyzing variations of the *cis*-acting elements gives an insight into the understanding of functional variations of genes. Hence, we analyzed the *cis*-acting elements of the 10 candidate genes with InDels to examine their involvement in stress-related pathways and also to speculate possible causal factors in addition to the InDel variations (Supplementary Table S7). As a whole, the analysis of *cis*-acting elements in the promoter regions of the candidate genes, ABRE, CAAT box, DPBF, GAGA, GBOX, IBOX, ROOT, SEF3, and SEF4 were present approximately an equal number in all genes indicating their involvement in stress-related mechanisms as reported in several studies [78–84].

Recent research advances have shown that environmental stresses, do not only imbalance the ionic and osmotic homeostasis in plants but also weaken photosynthesis, redox reactions, and cellular energy depletion. Therefore, plants harbor a broader, overlapping set of genes that are involved in both biotic and abiotic stress responses and developmental processes, increasing the evidence that plant signaling does not operate as independent or parallel pathways [85]. Golldack et al. (2014) [86] have proposed a model on cross-talk of ABA, gibberellic acid and jasmonate signaling plant responses to abiotic stressors such as drought and salt, linking with other pathways leading to ROS detoxification, lipid signaling and structural adaptation of membranes. Thus, the ABA-related candidate genes found in this study (*Os01g0583100 and Os02g0766700*) could be involved in different inter-connected networks and pathways to function against salinity stress. Similarly, bZIPs, RLKs, PP2Cs, and other candidate genes found in this study would be involved in linking abiotic stresses such as heat, cold, osmotic, oxidative and salinity stress by mediating signaling cross-talks [14,85,87,88].

4. Materials and Methods

4.1. Plant Material and DNA Extraction

At 354 and Bg 352 varieties were selected for whole genome re-sequencing in this study. The rice seeds were grown under controlled conditions and the genomic DNA was extracted from leaf tissues of 2-week-old seedlings of At 354 and Bg 352 using the CTAB method [89].

4.2. Rice Whole Genome Re-Sequencing and Variant Calling

High throughput whole genome re-sequencing was performed using Illumina's paired-end sequencing technology on the Hiseq 2000 platform for two rice varieties. The paired-end libraries were constructed for Bg 352 and At 354 according to the manufacturer's protocol (Illumina Inc., Hayward, CA 94545, USA). After the clonal cluster generation, the DNA was sequenced by Illumina's sequencing by synthesis (SBS) technology. The sequencing data were converted into raw data with 101 bp size reads and obtained BAM and fastq files for further analysis. After the raw reads were gone through the quality control process, the quality-filtered reads were mapped to two reference genomes, *Oryza sativa japonica* group cultivar Nipponbare IRGSP-1.0 (GenBank Assembly Accession: GCA_001433935.1) and *Oryza sativa indica* group cultivar Shuhui498 (R498) (GenBank Assembly Accession: GCA_002151415.1) using Burrows Wheeler Alignment (BWA) program [90] with default parameters. The duplicates in the aligned reads were removed and the alignment results were merged to generate indexed BAM files. Basic statistics including GC%, read depth, coverage and Q20/Q30 were calculated using the alignment results. The mapped reads were used to detect SNPs and InDels. After removing duplicates with Sambamba and identifying variants with SAMTools, information on each variant was gathered and classified by chromosomes. The variants were further filtered using parameters i.e., variant quality score $\geq$100 and zygosity (homozygous). Circos software was used to visualize the frequency distribution of the SNPs and InDels on 12 rice chromosomes of At 354 and Bg 352 with respect to Nipponbare genome and R498 genome.

4.3. Variation Analysis on Abiotic Stress-Related Genes and Prediction of Candidate Genes for Salinity

The QTL hotspot regions which were previously identified by QTL mapping of At 354 × Bg 352 were queried for prospective abiotic stress-related genes in the regions [35]. In addition, known salt tolerance-related genes were also selected. Gramene and NCBI GenBank Database [91,92] were used for identifying the abiotic stress-related genes. IRGSP-1.0 annotations were used in identifying the locations of the genes within QTLs. Respective coordinates of the R498 were obtained by matching the DNA sequences of IRGSP-annotated genes with R498. The SNP and InDel variants of the genes in the QTL regions of At 354 and Bg 352 genomes were classified according to their locations such as exons, introns, 5′ and 3′ untranslated regions (UTR) with respect to the Nipponbare. The SNPs and InDels which were polymorphic between At 354 and Bg 352 in the exons were further examined in the R498 reference. The amino acid changes due to SNPs were observed based on the Short Genetic Variations database (dbSNP) of NCBI [93]. The open reading frames (ORF) for the coding sequences (CDS) of the selected genes (with InDels) were predicted by NCBI ORF finder [94].

4.4. Comparative Analysis of InDels in Predicted Candidate Genes with indica Rice Lines

The nucleotide sequences of predicted genes in a panel of 50 rice cultivars including Sri Lankan rice varieties and popular salt-tolerant donor varieties, FL 478 (CX219), Nona bokra (IRIS 313-7736) and Pokkali (IRIS 313-8244) were retrieved from Rice SNP-Seek Database [95]. The concordance of InDel variations was examined with *indica* rice panel.

4.5. Analysis of the Promoter Sequences of the Genes with InDels

The 1000 bp 5′ upstream region of the selected candidate genes were retrieved for At 354 and Bg 352 varieties as the promoter sequences. The plant *cis*-acting regulatory DNA elements of each gene were obtained from the NEW PLACE database version 30.0 [96]. Out of the total *cis*-acting elements, the abiotic stress-related *cis*-acting elements were filtered using the already available literature [97].

4.6. GO Analysis

The candidate gene sequences of At 354 and Bg 352 were annotated with Blast2GO software using the blastn algorithm and the Cloud Blast database in order to identify the molecular function, biological process and the cellular components [98].

4.7. Data Availability

The gene sequence data for this study can be found at GenBank Repository. (https://www.ncbi.nlm.nih.gov/genbank/).

The VCF files for At 354 and Bg 352 with reference to Nipponbare and R498 genomes has been archived under the following accessions of EVA. (https://www.ebi.ac.uk/ena/data/view/PRJEB35319).

Project: PRJEB35319.

Analyses: ERZ1143791, ERZ1143792, ERZ1143793, ERZ1143794, ERZ1143795, ERZ1143796, ERZ1143797, ERZ1143798, ERZ1143799, ERZ1143800.

5. Conclusions

In this study, we re-sequenced two elite *indica* rice varieties—At 354 (salt-tolerant) and Bg 352 (salt susceptible)—with reference to *japonica* cultivar Nipponbare and *indica* cultivar R498 and detected high genetic variations through SNPs and InDels between two parents, particularly in their chromosomes and QTL regions. We identified a narrow deviation in QTL locations between Nipponbare and R498 references, ranging from 0.1 Mb to 3 Mb. In total, 106 abiotic stress-related genes were identified in QTL regions, most of which had polymorphic nucleotide variants between two parents. Of them, 34 genes were identified for the presence of polymorphic SNPs and InDels between parents with respect to Nipponbare, but only 32 variants were confirmed with the reference material of R498. Altogether 10 genes that contained InDels leading to altered amino acid sequences were identified and their mutated sequences were able to be validated due to their presence in other *indica* varieties. Further studies need to be focused on the functional characterization of particular alleles by expression studies under different salinity levels and exposure times in order to elucidate the contribution of mutations on salt tolerance. The different haplotypes revealed in this study would be useful for genetic improvement of rice through haplotype-based molecular breeding.

Supplementary Materials: The following are available online at http://www.mdpi.com/2223-7747/9/2/233/s1. Supplementary Table S1. Supporting references for QTL hotspots co-located in the same chromosomal regions reported in other breeding studies. Supplementary Table S2. Variants identified in abiotic stress-related genes with reference to Nipponbare genome. Supplementary Table S3. Abiotic stress-related *cis* acting elements revealed by the analysis of the promoter sequences of the genes with InDels. Supplementary Table S4. Candidate genes identified based on polymorphic SNPs in exons regions of two parents. Supplementary Table S5. GO analysis of the selected candidate genes. Supplementary Table S6. Presence of InDels in the selected rice panels. Supplementary Figure S1. *InDels* detected in the candidate genes and their sequence validation. A. *Os01g0581400* B. *Os10g0107000* C. *Os11g0655900* D. *Os12g0622500* E. *Os12g0624200*. Supplementary Table S7 Abiotic stress-related Cis acting elements revealed by the analysis of the promoter sequences of the genes with InDels. Supplementary Table S8 InDel marker information.

Author Contributions: Conceptualization, N.K. and D.G.; methodology, B.A., N.K. and D.G.; formal analysis, B.A. with the cooperation of V.H., D.L. and P.S.; investigation, B.A. and D.L.; writing—original draft preparation, B.A.; writing—review and editing, N.K., D.G., V.H., D.L. and P.S.; supervision, N.K., D.G. and V.H.; funding acquisition, N.K. All authors have read and agreed to the published version of the manuscript.

Funding: This research was funded by NATIONAL RESEARCH COUNCIL, Sri Lanka, grant number NRC 16-016.

Acknowledgments: We acknowledge International Rice Research Institute, Philippines for providing SNP data via Global Rice Science scholarship 2011 program. Also, we thank Rice Research and Development Institute, Batalegoda, Sri Lanka for providing rice seeds. Moreover, we acknowledge National Science Foundation, Sri Lanka for providing a postgraduate research scholarship under NSF/SCH/2019/01.

Conflicts of Interest: The authors declare no conflict of interest.

References

1. Jain, M.; Moharana, K.C.; Shankar, R.; Kumari, R.; Garg, R. Genomewide discovery of DNA polymorphisms in rice cultivars with contrasting drought and salinity stress response and their functional relevance. *Plant Biotechnol. J.* **2014**, *12*, 253–264. [CrossRef] [PubMed]
2. Ramya, M.; Raveendran, M.; Subramaniyam, S.; Jagadeesan, R. In silico analysis of drought tolerant genes in rice. *Int. J.* **2010**, *3*, 36–40.
3. Reddy, I.N.B.L.; Kim, B.K.; Yoon, I.S.; Kim, K.H.; Kwon, T.R. Salt Tolerance in Rice: Focus on Mechanisms and Approaches. *Rice Sci.* **2017**, *24*, 123–144. [CrossRef]
4. Wang, Y.; Zhang, L.; Nafisah, A.; Zhu, L.; Xu, J.; Li, Z. Selection efficiencies for improving drought/salt tolerances and yield using introgression breeding in rice (*Oryza sativa* L.). *Crop J.* **2013**, *1*, 134–142. [CrossRef]
5. Umali, D.L. *Irrigation-Induced Salinity*; The World Bank: Washington, DC, USA, 1993; ISBN 978-0-8213-2508-7.
6. Oster, J.D.; Shainberg, I. Soil responses to sodicity and salinity: Challenges and opportunities. *Aust. J. Soil Res.* **2001**, *39*, 1219–1224. [CrossRef]
7. Srivastava, P.K.; Gupta, M.; Pandey, A.; Pandey, V.; Singh, N.; Tewari, S.K. Effects of sodicity induced changes in soil physical properties on paddy root growth. *Plant Soil Environ.* **2014**, *60*, 165–169.
8. De Leon, T.B.; Linscombe, S.; Subudhi, P.K. Identification and validation of QTLs for seedling salinity tolerance in introgression lines of a salt tolerant rice landrace "Pokkali". *PLoS ONE* **2017**, *12*, e0175361. [CrossRef]
9. Koyama, M.L.; Levesley, A.; Koebner, R.M.; Flowers, T.J.; Yeo, A.R. Quantitative trait loci for component physiological traits determining salt tolerance in rice. *Plant Physiol.* **2001**, *125*, 406–422. [CrossRef]
10. Naveed, S.A.; Zhang, F.; Zhang, J.; Zheng, T.Q.; Meng, L.J.; Pang, Y.L.; Xu, J.L.; Li, Z.K. Identification of QTN and candidate genes for Salinity Tolerance at the Germination and Seedling Stages in Rice by Genome-Wide Association Analyses. *Sci. Rep.* **2018**, *8*, 1–11. [CrossRef]
11. Pang, Y.; Chen, K.; Wang, X.; Wang, W.; Xu, J.; Ali, J.; Li, Z. Simultaneous Improvement and Genetic Dissection of Salt Tolerance of Rice (*Oryza sativa* L.) by Designed QTL Pyramiding. *Front. Plant Sci.* **2017**, *8*, 1–11. [CrossRef]
12. Tiwari, S.; Sl, K.; Kumar, V.; Singh, B.; Rao, A.R.; Sv, A.M.; Rai, V.; Singh, A.K.; Singh, N.K. Mapping QTLs for Salt Tolerance in Rice (*Oryza sativa* L.) by Bulked Segregant Analysis of Recombinant Inbred Lines Using 50K SNP Chip. *PLoS ONE* **2016**, *11*, e0153610. [CrossRef] [PubMed]

13. Wang, Z.; Chen, Z.; Cheng, J.; Lai, Y.; Wang, J.; Bao, Y.; Huang, J.; Zhang, H. QTL Analysis of Na+ and K+ Concentrations in Roots and Shoots under Different Levels of NaCl Stress in Rice (*Oryza sativa* L.). *PLoS ONE* **2012**, *7*, e51202. [CrossRef] [PubMed]
14. Hossain, M.R.; Bassel, G.W.; Pritchard, J.; Sharma, G.P.; Ford-Lloyd, B.V. Trait Specific Expression Profiling of Salt Stress Responsive Genes in Diverse Rice Genotypes as Determined by Modified Significance Analysis of Microarrays. *Front. Plant Sci.* **2016**, *7*, 567. [CrossRef] [PubMed]
15. Lima, M.; Da, G.D.S.; Lopes, N.F.; Zimmer, P.D.; Meneghello, G.E.; Ferrari, C.; Mendes, C.R. Detection of genes providing salinity-tolerance in rice. *Acta Sci. Biol. Sci.* **2014**, *36*, 79–85. [CrossRef]
16. Gupta, B.; Huang, B. Mechanism of Salinity Tolerance in Plants: Physiological, Biochemical, and Molecular Characterization. *Int. J. Genom.* **2014**, *2014*, 701596. [CrossRef]
17. Liang, W.; Ma, X.; Wan, P.; Liu, L. Plant salt-tolerance mechanism: A review. *Biochem. Biophys. Res. Commun.* **2018**, *495*, 286–291. [CrossRef]
18. Wang, G.-L.; Song, W.; Ruan, D.; Sideris, S.; Ronald, P. The cloned gene, Xa21, confers resistance to multiple Xanthomonas oryzae pv. oryzae Isolates in transgenic plants. *Mol. Plant. Microbe. Interact.* **1997**, *9*, 850–855. [CrossRef]
19. Jahan, N.; Zhang, Y.; Lv, Y.; Song, M.; Zhao, C.; Hu, H.; Cui, Y.; Wang, Z.; Yang, S.; Zhang, A.; et al. QTL analysis for rice salinity tolerance and fine mapping of a candidate locus qSL7 for shoot length under salt stress. In *Plant Growth Regulation*; Springer: New York, NY, USA, 2019.
20. Thomson, M.J.; de Ocampo, M.; Egdane, J.; Rahman, M.A.; Sajise, A.G.; Adorada, D.L.; Tumimbang-Raiz, E.; Blumwald, E.; Seraj, Z.I.; Singh, R.K.; et al. Characterizing the Saltol quantitative trait locus for salinity tolerance in rice. *Rice* **2010**, *3*, 148–160. [CrossRef]
21. Lim, J.-H.; Yang, H.-J.; Jung, K.-H.; Yoo, S.-C.; Paek, N.-C. Quantitative Trait Locus Mapping and Candidate Gene Analysis for Plant Architecture Traits Using Whole Genome Re-Sequencing in Rice. *Mol. Cells* **2014**, *37*, 149–160. [CrossRef]
22. Matsumoto, T.; Wu, J.; Itoh, T.; Numa, H.; Antonio, B.; Sasaki, T. The Nipponbare genome and the next-generation of rice genomics research in Japan. *Rice* **2016**, *9*, 33. [CrossRef]
23. Nguyen, T.D.; Moon, S.; Nguyen, V.N.T.; Gho, Y.; Chandran, A.K.N.; Soh, M.S.; Song, J.T.; An, G.; Oh, S.A.; Park, S.K.; et al. Genome-wide identification and analysis of rice genes preferentially expressed in pollen at an early developmental stage. *Plant Mol. Biol.* **2016**, *92*, 71–88. [CrossRef] [PubMed]
24. Zhou, X.; Bai, X.; Xing, Y. A Rice Genetic Improvement Boom by Next Generation Sequencing. *Curr. Issues Mol. Biol.* **2018**, 109–126. [CrossRef] [PubMed]
25. Feltus, F.A.; Wan, J.; Schulze, S.R.; Estill, J.C.; Jiang, N.; Paterson, A.H. An SNP resource for rice genetics and breeding based on subspecies Indica and Japonica genome alignments. *Genome Res.* **2004**, *14*, 1812–1819. [CrossRef] [PubMed]
26. Mishra, S.; Singh, B.; Panda, K.; Singh, B.P.; Singh, N.; Misra, P.; Rai, V.; Singh, N.K. Association of SNP Haplotypes of HKT Family Genes with Salt Tolerance in Indian Wild Rice Germplasm. *Rice* **2016**, *9*, 15. [CrossRef]
27. Xu, K.; Xu, X.; Fukao, T.; Canlas, P.; Maghirang-Rodriguez, R.; Heuer, S.; Ismail, A.M.; Bailey-Serres, J.; Ronald, P.C.; Mackill, D.J. Sub1A is an ethylene-response-factor-like gene that confers submergence tolerance to rice. *Nature* **2006**, *442*, 705–708. [CrossRef]
28. Lee, Y.W.; Gould, B.A.; Stinchcombe, J.R. Identifying the genes underlying quantitative traits: A rationale for the QTN programme. *AoB Plants* **2014**, *6*, 1–14. [CrossRef]
29. Ma, Y.; Qin, F.; Tran, L.S.P. Contribution of genomics to gene discovery in plant abiotic stress responses. *Mol. Plant* **2012**, *5*, 1176–1178. [CrossRef]
30. Temnykh, S.; Declerck, G.; Lukashova, A.; Lipovich, L.; Cartinhour, S.; Mccouch, S. Computational and Experimental Analysis of Microsatellites in Rice (*Oryza sativa* L.): Frequency, Length Variation, Transposon Associations, and Genetic Marker Potential. *Genome Res.* **2001**, *11*, 1441–1452. [CrossRef] [PubMed]
31. Xing, Y.; Tan, Y.; Hua, J.; Sun, X.; Xu, C.; Zhang, Q. Characterization of the main effects, epistatic effects and their environmental interactions of QTLs on the genetic basis of yield traits in rice. *Theor. Appl. Genet.* **2002**, *105*, 248–257. [CrossRef]
32. Thomson, M.J.; Singh, N.; Dwiyanti, M.S.; Wang, D.R.; Wright, M.H.; Perez, F.A.; Declerck, G.; Chin, J.H.; Malitic-layaoen, G.A.; Juanillas, V.M.; et al. Large-scale deployment of a rice 6 K SNP array for genetics and breeding applications. *Rice* **2017**, *10*, 40. [CrossRef]

33. McCouch, S.R.; Wright, M.H.; Tung, C.-W.; Maron, L.G.; McNally, K.L.; Fitzgerald, M.; Singh, N.; DeClerck, G.; Agosto-Perez, F.; Korniliev, P.; et al. Open access resources for genome-wide association mapping in rice. *Nat. Commun.* **2016**, *7*, 10532. [CrossRef] [PubMed]
34. Singh, N.; Jayaswal, P.K.; Panda, K.; Mandal, P.; Kumar, V.; Singh, B.; Mishra, S.; Singh, Y.; Singh, R.; Rai, V.; et al. Single-copy gene based 50 K SNP chip for genetic studies and molecular breeding in rice. *Sci. Rep.* **2015**, *5*, 11600. [CrossRef] [PubMed]
35. Gimhani, D.R.; Gregorio, G.B.; Kottearachchi, N.S.; Samarasinghe, W.L.G. SNP-based discovery of salinity-tolerant QTLs in a bi-parental population of rice (*Oryza sativa*). *Mol. Genet. Genom.* **2016**, *291*, 2081–2099. [CrossRef] [PubMed]
36. Chen, X.; Chern, M.; Canlas, P.E.; Jiang, C.; Ruan, D.; Cao, P.; Ronald, P.C. A conserved threonine residue in the juxtamembrane domain of the XA21 pattern recognition receptor is critical for kinase autophosphorylation and XA21-mediated immunity. *J. Biol. Chem.* **2010**, *285*, 10454–10463. [CrossRef]
37. Shobbar, Z.S.; Oane, R.; Gamuyao, R.; De Palma, J.; Malboobi, M.A.; Karimzadeh, G.; Javaran, M.J.; Bennett, J. Abscisic acid regulates gene expression in cortical fiber cells and silica cells of rice shoots. *New Phytol.* **2008**, *178*, 68–79. [CrossRef]
38. Xiang, Y.; Tang, N.; Du, H.; Ye, H.; Xiong, L. Characterization of OsbZIP23 as a Key Player of the Basic Leucine Zipper Transcription Factor Family for Conferring Abscisic Acid Sensitivity and Salinity and Drought Tolerance in Rice. *Plant Physiol.* **2008**, *148*, 1938–1952. [CrossRef]
39. Wang, C.; Zhang, Q.; Shou, H. Identification and expression analysis of OsHsfs in rice. *J. Zhejiang Univ. Sci. B* **2009**, *10*, 291–300. [CrossRef]
40. Wang, D.; Guo, Y.; Wu, C.; Yang, G.; Li, Y.; Zheng, C. Genome-wide analysis of CCCH zinc finger family in Arabidopsis and rice. *BMC Genom.* **2008**, *9*, 44. [CrossRef]
41. Panahi, B.; Ahmadi, F.S.; Mehrjerdi, M.Z.; Moshtaghi, N. Molecular cloning and the expression of the Na+/H+antiporter in the monocot halophyte *Leptochloa fusca* (L.) Kunth. *NJAS Wagening. J. Life Sci.* **2013**, *64–65*, 87–93. [CrossRef]
42. Reguera, M.; Bassil, E.; Blumwald, E. Intracellular NHX-Type Cation/H+ Antiporters in Plants. *Mol. Plant* **2014**, *7*, 261–263. [CrossRef]
43. He, F.; Liu, Q.; Zheng, L.; Cui, Y.; Shen, Z.; Zheng, L. RNA-Seq Analysis of Rice Roots Reveals the Involvement of Post-Transcriptional Regulation in Response to Cadmium Stress. *Front. Plant Sci.* **2015**, *6*, 1136. [CrossRef] [PubMed]
44. Sakai, H.; Lee, S.S.; Tanaka, T.; Numa, H.; Kim, J.; Kawahara, Y.; Wakimoto, H.; Yang, C.; Iwamoto, M.; Abe, T.; et al. Rice Annotation Project Database (RAP-DB): An Integrative and Interactive Database for Rice Genomics Special Focus Issue–Databases. *Plant Cell Physiol* **2013**, *54*, e6. [CrossRef]
45. Du, H.; Yu, Y.; Ma, Y.; Gao, Q.; Cao, Y.; Chen, Z.; Ma, B.; Qi, M.; Li, Y.; Zhao, X.; et al. Sequencing and de novo assembly of a near complete indica rice genome. *Nat. Commun.* **2017**, *8*, 15324. [CrossRef] [PubMed]
46. Singhabahu, S.; Wijesinghe, C.; Gunawardana, D.; Senarath Yapa, M.D.; Kannangara, M.; Edirisinghe, R.; Dissanayake, V.H.W. Whole Genome Sequencing and Analysis of Godawee, a Salt Tolerant Indica Rice Variety. *Rice Res. Open Access* **2017**, *5*, 1–9. [CrossRef]
47. Rathinasabapathi, P.; Purushothaman, N.; Vl, R.; Parani, M. Whole genome sequencing and analysis of Swarna, a widely cultivated indica rice variety with low glycemic index. *Sci. Rep.* **2015**, *5*, 1–10. [CrossRef] [PubMed]
48. Subbaiyan, G.K.; Waters, D.L.E.; Katiyar, S.K.; Sadananda, A.R.; Vaddadi, S.; Henry, R.J. Genome-wide DNA polymorphisms in elite indica rice inbreds discovered by whole-genome sequencing. *Plant Biotechnol. J.* **2012**, *10*, 623–634. [CrossRef]
49. Šmarda, P.; Bureš, P.; Horová, L.; Leitch, I.J.; Mucina, L.; Pacini, E.; Tichý, L.; Grulich, V.; Rotreklová, O. Ecological and evolutionary significance of genomic GC content diversity in monocots. *Proc. Natl. Acad. Sci. USA* **2014**, *111*, E4096–E4102. [CrossRef]
50. Li, S.; Wang, S.; Deng, Q.; Zheng, A.; Zhu, J.; Liu, H.; Wang, L.; Gao, F.; Zou, T.; Huang, B.; et al. Identification of genome-wide variations among three elite restorer lines for hybrid-rice. *PLoS ONE* **2012**, *7*, e30952. [CrossRef]
51. Tenaillon, M.I.; Sawkins, M.C.; Long, A.D.; Gaut, R.L.; Doebley, J.F.; Gaut, B.S. Patterns of DNA sequence polymorphism along chromosome 1 of maize (*Zea mays* ssp. mays L.). *Proc. Natl. Acad. Sci. USA* **2001**, *98*, 9161–9166. [CrossRef]

52. Batley, J.; Barker, G.; O'Sullivan, H.; Edwards, K.J.; Edwards, D. Mining for Single Nucleotide Polymorphisms and Insertions/Deletions in Maize Expressed Sequence Tag Data 1. *Plant Physiol.* **2003**, *132*, 84–91. [CrossRef]
53. Jiang, S.; Sun, S.; Bai, L.; Ding, G.; Wang, T.; Xia, T.; Jiang, H.; Zhang, X.; Zhang, F. Resequencing and variation identification of whole genome of the japonica rice variety "Longdao24" with high yield. *PLoS ONE* **2017**, *12*, e0181037. [CrossRef]
54. Zhang, Z.; Miteva, M.A.; Wang, L.; Alexov, E.; Diderot, P.; Cit, S.P. Analyzing Effects of Naturally Occurring Missense Mutations. *Comput. Math. Methods Med.* **2012**, *2012*, 805827. [CrossRef] [PubMed]
55. Needham, C.J.; Bradford, J.R.; Bulpitt, A.J.; Care, M.A.; Westhead, D.R. Predicting the effect of missense mutations on protein function: Analysis with Bayesian networks. *BMC Bioinform.* **2006**, *7*, 405. [CrossRef]
56. Wang, S.; Chu, B.; Lue, W.; Eimert, K.; Chen, J. adg2-1 represents a missense mutation in the ADPG pyrophosphorylase large subunit gene of Arabidopsis thaliana. *Plant J.* **1997**, *11*, 1121–1126. [CrossRef] [PubMed]
57. Wang, D.; Qin, Y.; Fang, J.; Yuan, S.; Peng, L.; Zhao, J.; Li, X. A Missense Mutation in the Zinc Finger Domain of OsCESA7 Deleteriously Affects Cellulose Biosynthesis and Plant Growth in rice. *PLoS ONE* **2016**, *11*, e0153993. [CrossRef] [PubMed]
58. Tang, X.; Wang, Y.; Zhang, Y.; Huang, S.; Liu, Z.; Fei, D.; Feng, H. A missense mutation of plastid RPS4 is associated with chlorophyll deficiency in Chinese cabbage (*Brassica campestris*). *BMC Plant Biol.* **2018**, *18*, 130. [CrossRef]
59. Ramonell, K.M.; Goff, K.E. The Role and Regulation of Receptor-Like Kinases in Plant Defense. *Gene Regul. Syst. Bio.* **2007**, *1*, 167–175.
60. Macdonald-Obermann, J.L.; Pike, L.J. The intracellular juxtamembrane domain of the epidermal growth factor (EGF) receptor is responsible for the allosteric regulation of EGF binding. *J. Biol. Chem.* **2009**, *284*, 13570–13576. [CrossRef]
61. Islam, M.S.; Ontoy, J.; Subudhi, P.K. Meta-analysis of quantitative trait loci associated with seedling-stage salt tolerance in rice (*Oryza Sativa* L.). *Plants* **2019**, *8*, 33. [CrossRef]
62. Xue, T.; Wang, D.; Zhang, S.; Ehlting, J.; Ni, F.; Jakab, S.; Zheng, C.; Zhong, Y. Genome-wide and expression analysis of protein phosphatase 2C in rice and Arabidopsis. *BMC Genom.* **2008**, *9*, 550. [CrossRef]
63. Yoshida, T.; Fujita, Y.; Sayama, H.; Kidokoro, S.; Maruyama, K.; Mizoi, J.; Shinozaki, K.; Yamaguchi-Shinozaki, K. AREB1, AREB2, and ABF3 are master transcription factors that cooperatively regulate ABRE-dependent ABA signaling involved in drought stress tolerance and require ABA for full activation. *Plant J.* **2010**, *61*, 672–685. [CrossRef] [PubMed]
64. Li, C.; Shen, H.; Wang, T.; Wang, X. ABA Regulates Subcellular Redistribution of OsABI-LIKE2, a Negative Regulator in ABA Signaling, to Control Root Architecture and Drought Resistance in *Oryza sativa*. *Plant Cell Physiol.* **2015**, *56*, 2396–2408. [CrossRef] [PubMed]
65. Kotchoni, S.O.; Jimenez-Lopez, J.C.; Gao, D.; Edwards, V.; Gachomo, E.W.; Margam, V.M.; Seufferheld, M.J. Modeling-dependent protein characterization of the rice aldehyde dehydrogenase (ALDH) superfamily reveals distinct functional and structural features. *PLoS ONE* **2010**, *5*, e11516. [CrossRef] [PubMed]
66. Hou, Q.; Bartels, D. Comparative study of the aldehyde dehydrogenase (ALDH) gene superfamily in the glycophyte Arabidopsis thaliana and Eutrema halophytes. *Ann. Bot.* **2015**, *115*, 465–479. [CrossRef] [PubMed]
67. Zong, W.; Tang, N.; Yang, J.; Peng, L.; Ma, S.; Xu, Y.; Li, G.; Xiong, L. Feedback regulation of ABA signaling and biosynthesis by a bZIP transcription factor targets drought resistance related genes. *Plant Physiol.* **2016**, *171*, 00469. [CrossRef]
68. Yi, S.Y.; Lee, H.Y.; Kim, H.; Lim, C.J.; Kim, W.B.; Jang, H.; Jeon, J.-S.; Kwon, S.-Y. Microarray Analysis of bacterial blight resistance 1 mutant rice infected with Xanthomonas oryzae pv. oryzae. *Plant Breed. Biotechnol.* **2013**, *1*, 354–365. [CrossRef]
69. Waters, E.R.; Aevermann, B.D.; Sanders-Reed, Z. Comparative analysis of the small heat shock proteins in three angiosperm genomes identifies new subfamilies and reveals diverse evolutionary patterns. *Cell Stress Chaperones* **2008**, *13*, 127–142. [CrossRef]
70. Neeraja, C.N.; Kulkarni, K.S.; Babu, P.M.; Rao, D.S.; Surekha, K.; Babu, V.R. Transporter genes identified in landraces associated with high zinc in polished rice through panicle transcriptome for biofortification. *PLoS ONE* **2018**, *13*, e0192362.

71. Tian, N.; Wang, J.; Xu, Z.Q. Overexpression of Na^+/H^+ antiporter gene AtNHX1 from Arabidopsis thaliana improves the salt tolerance of kiwifruit (*Actinidia deliciosa*). *S. Afr. J. Bot.* **2011**, *77*, 160–169. [CrossRef]
72. Yang, J.; Zhang, F.; Li, J.; Chen, J.P.; Zhang, H.M. Integrative analysis of the microRNAome and transcriptome illuminates the response of susceptible rice plants to rice stripe virus. *PLoS ONE* **2016**, *11*, e0146946. [CrossRef]
73. Wang, Y.; Wang, Q.; Zhao, Y.; Han, G.; Zhu, S. Systematic analysis of maize class III peroxidase gene family reveals a conserved subfamily involved in abiotic stress response. *Gene* **2015**, *566*, 95–108. [CrossRef] [PubMed]
74. Garg, R.; Jhanwar, S.; Tyagi, A.K.; Jain, M. Genome-wide survey and expression analysis suggest diverse roles of glutaredoxin gene family members during development and response to various stimuli in rice. *DNA Res.* **2010**, *17*, 353–367. [CrossRef] [PubMed]
75. Meyer, Y.; Belin, C.; Delorme-Hinoux, V.; Reichheld, J.-P.; Riondet, C. Thioredoxin and Glutaredoxin Systems in Plants: Molecular Mechanisms, Crosstalks, and Functional Significance. *Antioxid. Redox Signal.* **2012**, *17*, 1124–1160. [CrossRef] [PubMed]
76. Pittman, J.K.; Hirschi, K.D. Phylogenetic analysis and protein structure modelling identifies distinct Ca^{2+}/Cation antiporters and conservation of gene family structure within Arabidopsis and rice species. *Rice* **2016**, *9*, 3. [CrossRef]
77. Singh, A.K.; Kumar, R.; Tripathi, A.K.; Gupta, B.K.; Pareek, A.; Singla-Pareek, S.L. Genome-wide investigation and expression analysis of Sodium/Calcium exchanger gene family in rice and Arabidopsis. *Rice* **2015**, *8*, 21. [CrossRef]
78. Yun, K.Y.; Park, M.R.; Mohanty, B.; Herath, V.; Xu, F.; Mauleon, R.; Wijaya, E.; Bajic, V.B.; Bruskiewich, R.; de los Reyes, B.G. Transcriptional regulatory network triggered by oxidative signals configures the early response mechanisms of japonica rice to chilling stress. *BMC Plant Biol.* **2010**, *10*, 16. [CrossRef]
79. Kim, S.Y.; Chung, H.J.; Thomas, T. Isolation of a novel class of bZIP transcription factor that interact with ABA-responsive and embryo-specification elements in Dc3 promoter using a modified yeast one-hybrid system. *Plant J.* **1997**, *11*, 1237–1251. [CrossRef]
80. Fujiwara, T.; Beachy, R.N. Tissue-specific and temporal regulation of a beta-conglycinin gene: Roles of the RY repeat and other cis-acting elements. *Plant Mol. Biol.* **1994**, *24*, 261–272. [CrossRef]
81. Laloum, T.; Mita, D.; Gamas, P.; Niebel, A. CCAAT-box binding transcription factors in plants: Y so many? *Trends Plant Sci.* **2013**, *18*, 157–166. [CrossRef]
82. Menkens, A.; Schinder, U.; Cashmore, A. The G-box: A ubiquitous regulatory DNA element in plants bound by the GBF family of bZIP proteins. *Trends Biochem. Sci.* **1995**, *20*, 506–516. [CrossRef]
83. Santi, L.; Wang, Y.; Stile, M.R.; Berendzen, K.; Wanke, D.; Roig, C.; Pozzi, C.; Mu, K.; Mu, J.; Rohde, W.; et al. The GA octodinucleotide repeat binding factor BBR participates in the transcriptional regulation of the homeobox gene Bkn3. *Plant J.* **2003**, *34*, 813–826. [CrossRef] [PubMed]
84. Vandepoele, K.; Quimbaya, M.; Casneuf, T.; De Veylder, L.; Van de Peer, Y. Unraveling Transcriptional Control in Arabidopsis Using cis-Regulatory Elements and Coexpression Networks. *Bioinformatics* **2009**, *150*, 535–546.
85. Cooper, B.; Clarke, J.D.; Budworth, P.; Kreps, J.; Hutchison, D.; Park, S.; Guimil, S.; Dunn, M.; Luginbuhl, P.; Ellero, C.; et al. A network of rice genes associated with stress response and seed development. *Proc. Natl. Acad. Sci. USA* **2003**, *100*, 4945–4950. [CrossRef] [PubMed]
86. Golldack, D.; Li, C.; Mohan, H.; Probst, N. Tolerance to drought and salt stress in plants: Unraveling the signaling networks. *Front. Plant Sci.* **2014**, *5*, 151. [CrossRef]
87. Nigam, D.; Kumar, S.; Mishra, D.C.; Rai, A.; Smita, S.; Saha, A. Synergistic regulatory networks mediated by microRNAs and transcription factors under drought, heat and salt stresses in *Oryza sativa* spp. *Gene* **2015**, *555*, 127–139. [CrossRef] [PubMed]
88. Szabados, L.; Savouré, A. Proline: A multifunctional amino acid. *Trends Plant Sci.* **2010**, *15*, 89–97. [CrossRef]
89. Murray, M.G.; Thompson, W.F. Rapid isolation of high molecular weight plant DNA. *Nucleic Acids Res.* **1980**, *8*, 4321–4326. [CrossRef]
90. Li, H.; Durbin, R. Fast and accurate short read alignment with Burrows—Wheeler transform. *Bioinformatics* **2009**, *25*, 1754–1760. [CrossRef]
91. Gramene: A Comparative Resource for Plants. Available online: http://www.gramene.org/ (accessed on 12 October 2017).

92. National Center for Biotechnology Information (NCBI). Available online: https://www.ncbi.nlm.nih.gov/ (accessed on 15 October 2017).
93. NCBI Short Genetic Variations dbSNP. Available online: https://www.ncbi.nlm.nih.gov/SNP/snp_ref.cgi?locusId=4324197 (accessed on 12 November 2017).
94. NCBI ORFfinder. Available online: https://www.ncbi.nlm.nih.gov/orffinder/ (accessed on 15 December 2018).
95. Mansueto, L.; Fuentes, R.R.; Borja, F.N.; Detras, J.; Abriol-santos, M.; Chebotarov, D.; Sanciangco, M.; Palis, K.; Copetti, D.; Poliakov, A.; et al. Rice SNP-seek database update: New SNPs, indels, and queries. *Nucleic Acids Res.* **2017**, *45*, 1075–1081. [CrossRef]
96. Higo, K.; Ugawa, Y.; Iwamoto, M.; Korenaga, T. Plant cis-acting regulatory DNA elements (PLACE) database: 1999. *Nucleic Acids Res.* **1999**, *27*, 297–300. [CrossRef]
97. Xu, F.; Park, M.-R.; Kitazumi, A.; Herath, V.; Mohanty, B.; Yun, S.J.; de los Reyes, B.G. Cis-regulatory signatures of orthologous stress-associated bZIP transcription factors from rice, sorghum and Arabidopsis based on phylogenetic footprints. *BMC Genom.* **2012**, *13*, 497. [CrossRef] [PubMed]
98. Conesa, A.; Stefan, G. Blast2GO: A Comprehensive Suite for Functional Analysis in Plant Genomics. *Int. J. Plant Genom.* **2008**, *2008*. [CrossRef] [PubMed]

plants
MDPI

Article

Identification of Anther Length QTL and Construction of Chromosome Segment Substitution Lines of *Oryza longistaminata*

Takayuki Ogami, Hideshi Yasui, Atsushi Yoshimura and Yoshiyuki Yamagata *

Plant Breeding Laboratory, Faculty of Agriculture, Kyushu University. 744, Motooka, Nishi-ku, Fukuoka 819-0395, Japan; ohgammy9.30@gmail.com (T.O.); hyasui@agr.kyushu-u.ac.jp (H.Y.); ayoshi@agr.kyushu-u.ac.jp (A.Y.)
* Correspondence: yoshiyuk@agr.kyushu-u.ac.jp; Tel.: +81-92-802-4553

Received: 30 August 2019; Accepted: 29 September 2019; Published: 29 September 2019

Abstract: Life histories and breeding systems strongly affect the genetic diversity of seed plants, but the genetic architectures that promote outcrossing in *Oryza longistaminata*, a perennial wild species in Africa, are not understood. We conducted a genetic analysis of the anther length of *O. longistaminata* accession W1508 using advanced backcross quantitative trait locus (QTL) analysis and chromosomal segment substitution lines (CSSLs) in the genetic background of *O. sativa* Taichung 65 (T65), with simple sequence repeat markers. QTL analysis of the BC_3F_1 population (n = 100) revealed that four main QTL regions on chromosomes 3, 5, and 6 were associated to anther length. We selected a minimum set of BC_3F_2 plants for the development of CSSLs to cover as much of the W1508 genome as possible. The additional minor QTLs were suggested in the regional QTL analysis, using 21 to 24 plants in each of the selected BC_3F_2 population. The main QTLs found on chromosomes 3, 5, and 6 were validated and designated *qATL3*, *qATL5*, *qATL6.1*, and *qATL6.2*, as novel QTLs identified in *O. longistaminata* in the mapping populations of 94, 88, 70, and 95 BC_3F_4 plants. *qATL3*, *qATL5*, and *qATL6.1* likely contributed to anther length by cell elongation, whereas *qATL6.2* likely contributed by cell multiplication. The QTLs were confirmed again in an evaluation of the W1508ILs. In several chromosome segment substitution lines without the four validated QTLs, the anthers were also longer than those of T65, suggesting that other QTLs also increase anther length in W1508. The cloning and diversity analyses of genes conferring anther length QTLs promotes utilization of the genetic resources of wild species, and the understanding of haplotype evolution on the differentiation of annuality and perenniality in the genus *Oryza*.

Keywords: anther length; cell elongation; genetic architecture; outcrossing; perennial species; rice

---

1. Introduction

Life histories and breeding systems strongly affect the genetic diversity of seed plants. Annuals tend to allocate their resources to sexual reproduction to produce as many flowers as possible for the one-time dispersal of seeds. By contrast, perennial species tend to primarily allocate resources to vegetative growth because of the need to occupy physical space, and to use local water and nutrient resources over an extended life span depending on the ecological circumstances [1].

Perennial species tend to show higher heterozygosity when compared with the annual species or domesticated species [2]. Heterozygosity has been found to correlate with fitness-related traits, such as survival probability, reproductive success, and disease resistance [3–7]. Populations of perennials maintain high genetic diversity by producing few but large floral organs, and by promoting relatively high outcrossing rates with mechanisms to prevent self-pollination—such as self-incompatibility or monoecious flowers [8,9]. An understanding of the genetic architecture of a favorable trait that promotes outcrossing will help to answer a question: What set of polymorphisms or genes contribute to outcrossing characteristics? However, the genetic basis of breeding system-associated traits is not fully understood.

There are eight species of the genus *Oryza* with the AA genome: Two Asian wild species (*O. rufipogon* Griff. and *O. nivara* Sharma et Shastry), two wild species in Africa (*O. longistaminata* A. Chev. & Roehr. and *O. barthii* A. Chev.), one wild species in South America (*O. glumaepatula* Steud.), one wild species in Australia (*O. meridionalis* Ng.), and two cultivated species [10]. Of the two cultivated rice species, *O. sativa* L. (Asian rice) is domesticated from the Asian perennial wild species *O. rufipogon*, and *O. glaberrima* Steud. (African rice) is domesticated from the African annual wild species *O. barthii*. The Asian cultivated and wild species form a species complex in which the production of hybrid progeny is possible. The differentiation of annual (*O. nivara*) and perennial (*O. rufipogon*) species reflects adaptation to their ecological niches [11,12]. In Africa, the annual species *O. barthii* and the perennial species *O. longistaminata* occupy different ecological habitats. *Oryza longistaminata* is characterized by a rhizome [13], and has a particularly large anther [14], as indicated by the species name '*longistaminata*' (long stamen). Thus, the anther is one of the key traits of this species. Additionally, it has high heterozygosity [15] that is likely due to its self-incompatibility and high outcrossing rate, achieved by the large reproductive organs [16]. Cultivated rice is a self-pollinated species with an outcrossing rate of less than 4% [17]. The outcrossing rate of perennial wild species tends to be higher than that of cultivated species. That of *O. longistaminata* and *O. rufipogon* reaches 100% in certain combinations of hybridization, although the rate for wild species typically ranges from 3.2% to 50% [18,19]. To fully understand perenniality and their outcrossing characteristics, we need to investigate the genetic architectures of their reproductive systems.

Chromosomal segment substitution lines (CSSLs) facilitate the genetic analysis and characterization of quantitative traits of donor varieties, or species in the genetic background of a recurrent parent. CSSLs are powerful genetic tools for the identification of quantitative trait loci (QTLs) [20–23] in trials across different years and environments [24–26]. Studies of anther QTLs have been conducted with F_2 populations and recombinant inbred line (RIL) populations derived from the interspecific crosses between *O. rufipogon* and *O. sativa* [27–29], or intraspecific crosses between the *indica* and *japonica* subspecies of *O. sativa* [30]. However, QTLs that confer anther length in *O. longistaminata* have not been examined and validated using near-isogenic lines (NIL) in the genetic background of *O. sativa*.

In this study, we elucidated the genetic basis of anther length in *O. longistaminata*, by using advanced backcross QTL analysis and CSSLs in the genetic background of *O. sativa* L. cv. Taichung 65 (T65). QTLs conferring anther length identified in BC_3F_1 and BC_3F_2 populations were validated in the BC_3F_4 population, in which single QTL regions were segregating. We validated the QTLs again by constructing CSSLs of *O. longistaminata*. NILs were evaluated to characterize the histological cause of increased anther length. CSSLs chromosome segments derived from *O. longistaminata* acc. W1508 were named as W1508ILs, using the term 'introgression lines' (ILs) to refer to CSSLs based on interspecific hybridization.

2. Results

2.1. Genetic Variation in Anther Length in the Backcrossed Population

We measured the anther length of *O. barthii*, *O. glumaepatula*, and *O. longistaminata* (Figure 1) to evaluate the genetic variation in anther length. The anther length of *O. longistaminata* were longer than those of all *O. barthii* and *O. glumaepatula* accessions, except for *O. glumaepatula* W1183. To identify the QTLs derived from *O. longistaminata* in a uniform genetic background, we developed CSSLs carrying W1508 (*O. longistaminata*) chromosomal segments in the genetic background of *O. sativa* T65.

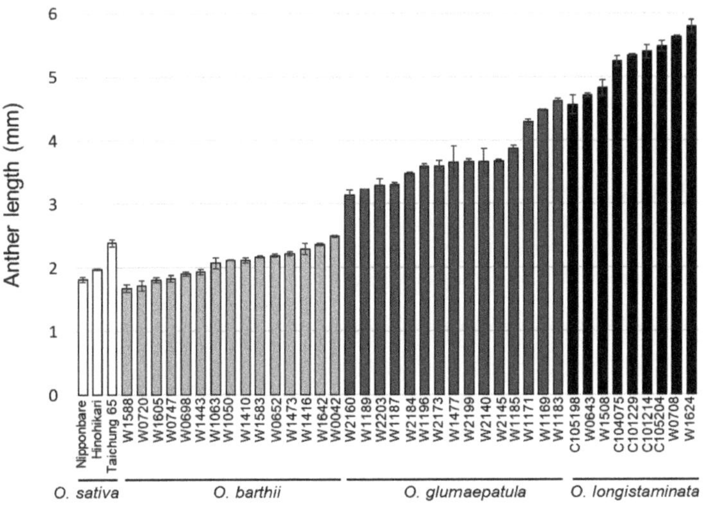

Figure 1. Distribution of anther length (mm) in cultivars of *O. sativa* and accessions of *O. barthii*, *O. glumaepatula*, and *O. longistaminata*. Mean ± SE, n = 3.

We developed 372 BC_3F_1 plants by recurrent backcrossing to F_1, BC_1, and BC_2 plants to develop the CSSLs of W1508 (Figure 2). We genotyped 100 BC_3F_1 plants, and selected 26 BC_3F_1 plants that covered as much of the W1508 genome as possible, to minimize the number of candidate CSSLs (Figure S1). Simple interval mapping (n = 100) suggested that *O. longistamina* alleles at one QTL on chromosome (Chr.) 3, one QTL on Chr. 5, and two QTLs on Chr. 6 increased the anther length (Table S1). The other peaks were below the experiment-wise threshold levels of the logarithm of odds (LOD) at $LOD_{\alpha = 0.05} = 2.65$ and $LOD_{\alpha = 0.01} = 3.40$. Multiple QTL analysis using forward/backward model selection also suggested that these four QTLs additively increase anther length without epistasis, which explains more than half (55.5%) of the phenotypic variation in the BC_3F_1 population (Figure S2, Table S1).

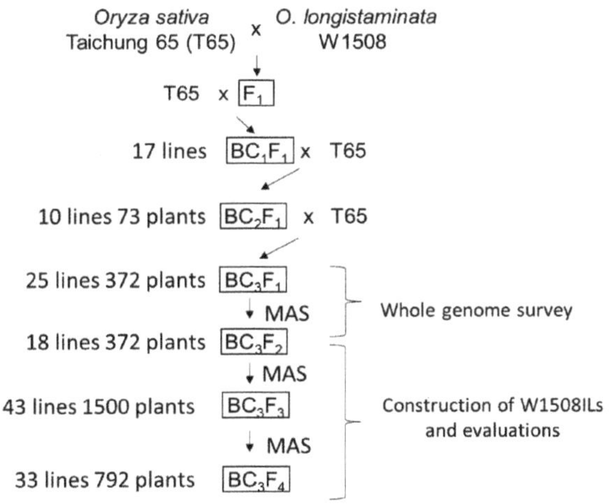

Figure 2. Breeding scheme of genetic materials used in this study. MAS represents marker-assisted selection.

To construct CSSLs, and to perform regional QTL analysis using 21 to 24 plants of the BC_3F_2 populations derived from the selected 26 BC_3F_1 plants covering the whole genome, we defined a heterozygous "target region" as a genomic region in BC_3F_1 plants for the fixation of the W1508-derived segments on the homozygous condition in the progeny (Figure S1). The anther length of T65 was 2.33 mm and that of W1508 was 4.33 mm. The mean anther length of the 26 BC_3F_2 lines ranged from 2.23 to 2.85 mm. Nine BC_3F_2 lines had a significantly higher mean anther length than T65, but none had a significantly lower mean anther length (Figure 3). Our working hypothesis was that the W1508 segments in the 9 BC_3F_2 lines with a higher mean anther length (boxed in Figure S1) most likely had QTLs associated with longer anther length. Simple interval mapping in the 26 BC_3F_2 populations at the target regions detected 13 minor QTLs on Chrs. 1, 2, 3, 4, 5, 6, 9, 10, and 11 (Table S2).

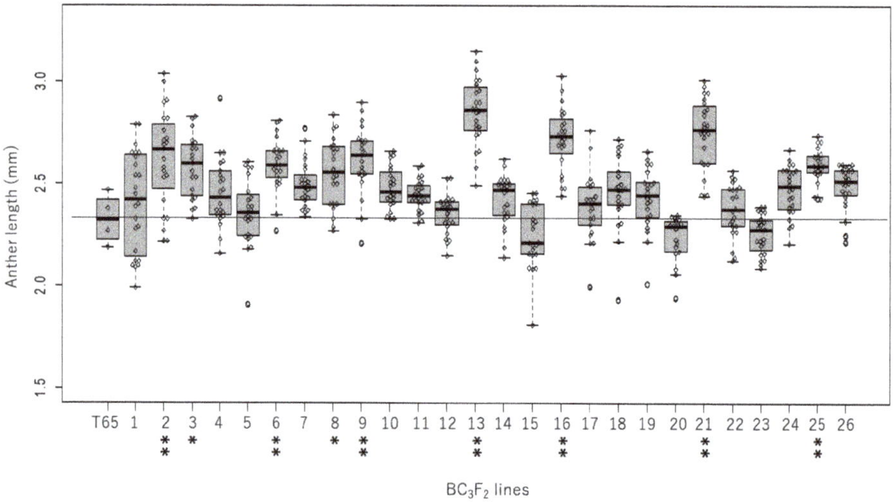

Figure 3. Box plots of anther length among the 26 BC_3F_2 populations. A thin horizontal line shows the average anther length in the Taichung 65 (T65) genetic background. Thick horizontal bars represent average values. Ranges between upper and lower quantiles are indicated by grey boxes. Maximum and minimum values excluding outlier are indicated by upper and lower whiskers, respectively. * and ** represent significant differences at P = 0.05 and P = 0.01, respectively.

2.2. Validation of QTL

For the identification of major QTLs conferring anther length, we conducted QTL analysis on BC_3F_2 13, 16, and 21 families that had the highest anther length means of 2.88 mm, 2.82 mm, and 2.68 mm, respectively (Figure 4). In the BC_3F_2 21 population (n = 44), there was segregation on chromosomes 3, 6, and 11 (Figure 4a). The QTL positively regulating anther length located between *RM3436* and *RM5959* on chromosome 3 was named *qATL3.3*. It was validated using the more advanced BC_3F_4 6 (n = 96) generation at 151.5 cM on chromosome 3, with the LOD value 18.27. *qATL3.3* explained 58.4% of the phenotypic variance in the BC_3F_4 population, with 0.17 mm of additive effect and 0.03 mm of dominant effect (Table S2).

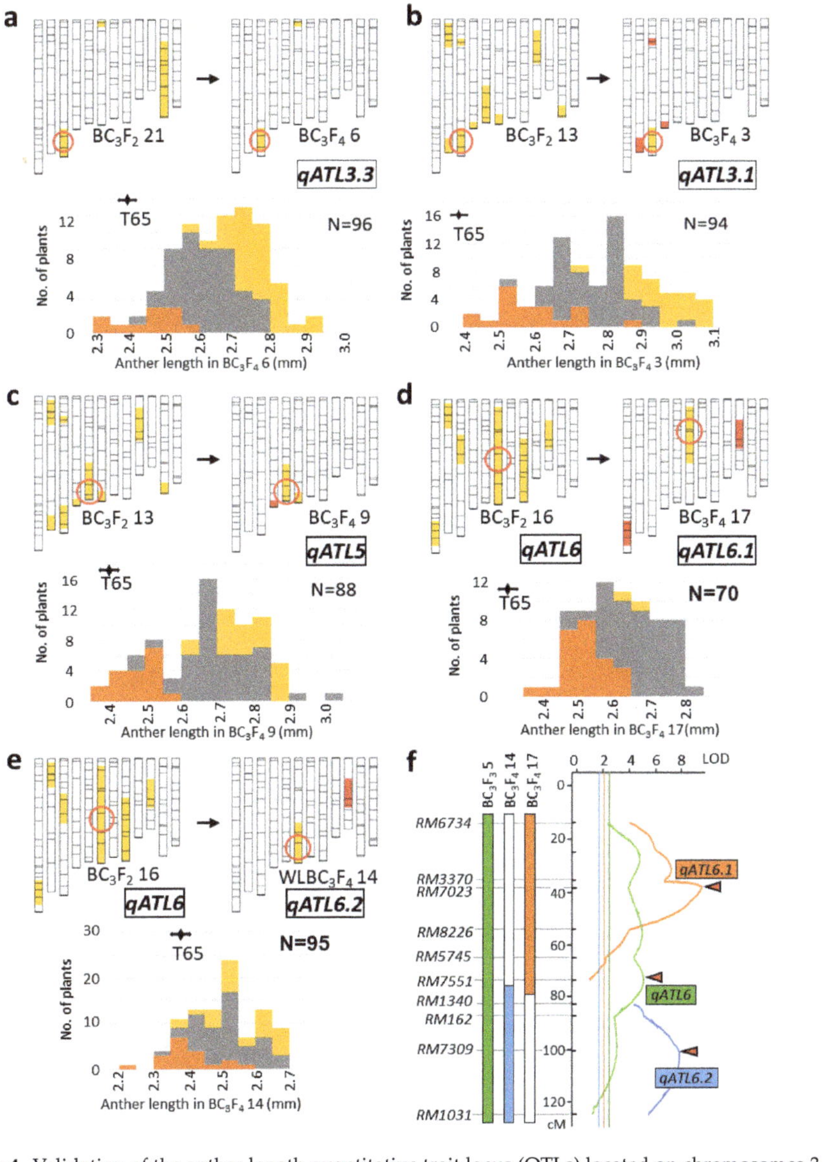

Figure 4. Validation of the anther length quantitative trait locus (QTLs) located on chromosomes 3, 5, and 6 in BC_3F_4 populations segregating at only one QTL region. (**a**)–(**e**) Graphical genotypes of the parents of BC_3F_3 and BC_3F_4 populations, and frequency distributions of anther length in BC_3F_4 populations for (**a**) *qATL3.3*, (**b**) *qATL3.1*, (**c**) *qATL5*, (**d**) *qATL6.1*, and (**e**) *qATL6.2*. Graphical genotypes: Yellow, heterozygous; red, homozygous for the W1508 allele. Frequency distributions shown in colors by genotypes at *RM5959* (**a**), *RM3525* (**b**), *RM1054* (**c**), *RM7023* (**d**), and *RM7309* (**e**): Orange, homozygous for T65; yellow, homozygous for W1508; gray, heterozygous; (**f**) logarithm of odds (LOD) curves in simple interval mapping of anther length for detection of *qATL6* (green) in BC_3F_3 5, *qATL6.1* (orange) in BC_3F_4 17, and *qATL6.2* (blue) in BC_3F_4 14.

Among the 71 plants in the BC_3F_2 13 population, the range of anther length was 2.26–3.32 mm, with a population mean of 2.82 mm. Simple interval mapping revealed two QTL: *qATL3.1* at 137.6 cM on Chr. 3, and *qATL5* at 122 cM on Chr. 5. Segregation distortion at *qATL5* was detected at the closest marker, *RM1054*, with a ratio of 22 T65 homozygotes: 43 T65 heterozygotes: 6 W1508 homozygotes. *qATL3.1* and *qATL5* were each independently validated in the more advanced BC_3F_4 3 and nine populations, respectively, derived from the BC_3F_2 13 population (Figure 4b,c). *qATL3.1* was detected again at 142.5 cM on Chr. 3, with an LOD score of 15.70, an additive effect of 0.19 mm, and a dominance effect of −0.002 mm, explaining 47.6% of the phenotypic variance (Table S2). *qATL3.3* and *qATL3.1* seemed to be identical QTLs, because substitution of the *O. longistaminata* chromosome segment ranged between the SSR markers *RM3525* and *RM5959*. We unified the QTL names *qATL3.3* and *qATL3.1* as the single QTL *qATL3.1*. In BC_3F_4 9, *qATL5* was detected again at 121.4 cM on the long arm of Chr. 5, with an LOD score of 17.28, an additive effect of 0.16 mm, and a dominance effect of 0.08 mm, explaining 59.6% of the phenotypic variation.

The BC_3F_2 16 population (n = 86) segregated at genomic regions on Chrs. 1, 2, 3, 6, 8, and 10, with anther lengths ranging from 2.21 to 3.10 mm and a population mean of 2.68 mm (Figure 4d,e). In the QTL analysis, the QTLs derived from *O. longistaminata* on Chr. 6 (*qATL6*) increased anther length. Because the genomic region of *qATL6* ranged widely from 13.8 to 114.8 cM, with LOD scores greater than the 5% significance level, we assumed that multiple QTLs on Chr. 6 were linked (Figure 4f). We examined two progeny lines of the BC_3F_2 16 population, BC_3F_4 17 and 14, that had the separated *O. longistaminata* chromosome segments on the short and long arms, respectively. In BC_3F_4 17, QTL *qATL6.1* was detected at 38.3 cM on Chr. 6S (Figure 4d,f), which, with an additive effect of 0.06 mm and a dominance effect of 0.08 mm, explained 48.8% of the phenotypic variation in anther length. *RM7023* segregation distortion of the nearest-neighbor marker, *RM7023*, gaved at a ratio of 25 T65 homozygotes: 47 T65 heterozygotes: 1 W1508 homozygotes. In BC_3F_4 14, a QTL designated as *qATL6.2* was detected on the long arm of Chr. 6 (Figure 4e,f) at *RM7309*, with an LOD score of 7.91, an additive effect of 0.08 mm, and a dominance effect of 0.04 mm.

2.3. Characterization of Anther Length QTL Using Near-Isogenic Lines

To understand the cytological characteristics of the anther length in near-isogenic lines (NILs) for *qATL3.1*, *qATL5*, *qATL6.1*, and *qATL6.2*, we investigated the longitudinal lengths of epidermal cells (cell length, CL) in the mid-anther region. The anther length of the NILs for *qATL3.1*, *qATL5*, *qATL6.1*, and *qATL6.2* were significantly longer than those of T65, according to Dunnett's multiple comparison (Figure 5a). The CL of the *qATL3.1* NIL was significantly longer than that of T65, but that of the *qATL5* and *qATL6.1* NILs was not different (Figure 5b). The average CL in the *qATL6.2* NIL was significantly smaller than that in T65. These results suggest that the positive effect of *qATL3.1* on anther length was due primarily to cell elongation in the longitudinal direction, whereas that of the other QTLs, particularly *qATL6.2*, might have been due to cell number. Next, we investigated the anther length and pollen grain numbers (PGN) of T65, W1508, and the QTL NILs. PGNs increased in the order of 1910.0 in T65, 2091.1 in the *qATL6B* NIL, 2598.1 in the *qATL6.1* NIL, 2690.1 in the *qATL5* NIL, 3058.8 in the *qATL3.1* NIL, and 6862.2 in W1508.

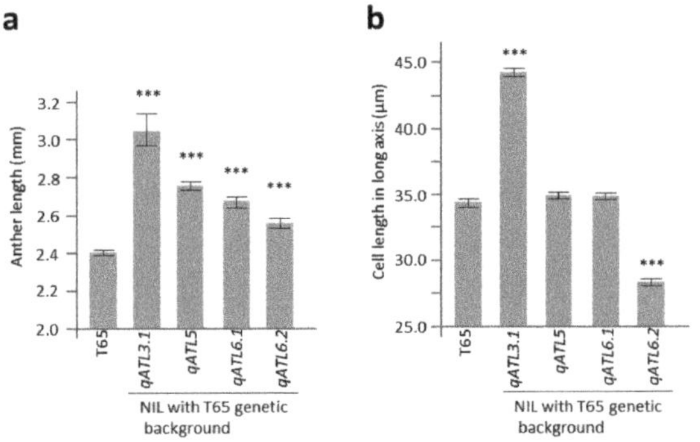

Figure 5. Characterization of QTLs using nearly isogenic lines (NIL) carrying W1508 chromosomal segments around QTL regions detected in this study. (**a**) Anther length (mm) in the NILs. Mean ± SE, n = 3; (**b**) cell length (μm) in the long axis of anthers in the NILs. * and ** represent significant differences in Dunnett's multiple comparisons to T65 control at 5% and 1% levels, respectively. The numbers of observed cells were 1460, 997, 953, 879, and 1033 in *qATL3.1* NIL, *qATL5* NIL, *qATL6.1* NIL, *qATL6.2* NIL, and T65 respectively.

2.4. Evaluation on W1508Ils

The target chromosome segments of W1508 selected in BC_3F_1 were fixed to homozygotes for the W1508 genes in BC_3F_2, BC_3F_3, BC_3F_4, and BC_3F_5, to construct the W1508ILs (Figure 6). We selected 32 lines covering the whole genomic region of W1508 using 125 SSR markers. Introgression of W1508 chromosomes did not occur at *RM3148* on Chr. 1; *RM5635* on Chr. 4; *RM2381* on Chr. 7; *RM8219* on Chr. 9; *RM7557*, *RM26090*, and *RM5918* on Chr. 11; or *RM3455*, *RM7195*, *RM1261*, and *RM309* on Chr. 12. The mean anther length in four or five plants increased significantly in W1508ILs 1, 2, 3, 7, 8, 9, 15, and 29, and decreased significantly in W1508IL 20 (Table S3). W1508ILs 8 and 9 carried *qATL3*, and W1508IL 15 carried *qATL5*. However, the mean anther length in W1508IL 17 (carrying *qATL6.1*) or 18 (carrying *qATL6.2*) did not increase significantly in five plant samples, suggesting that four or five replicates were not sufficient to detect the QTLs, despite being major QTL, for increased anther length. The minor QTLs detected in the BC_3F_2 populations were likely associated with the mean anther length in W1508ILs 1 and 2 at *RM3604*, and in W1508IL 7 at *RM5755*.

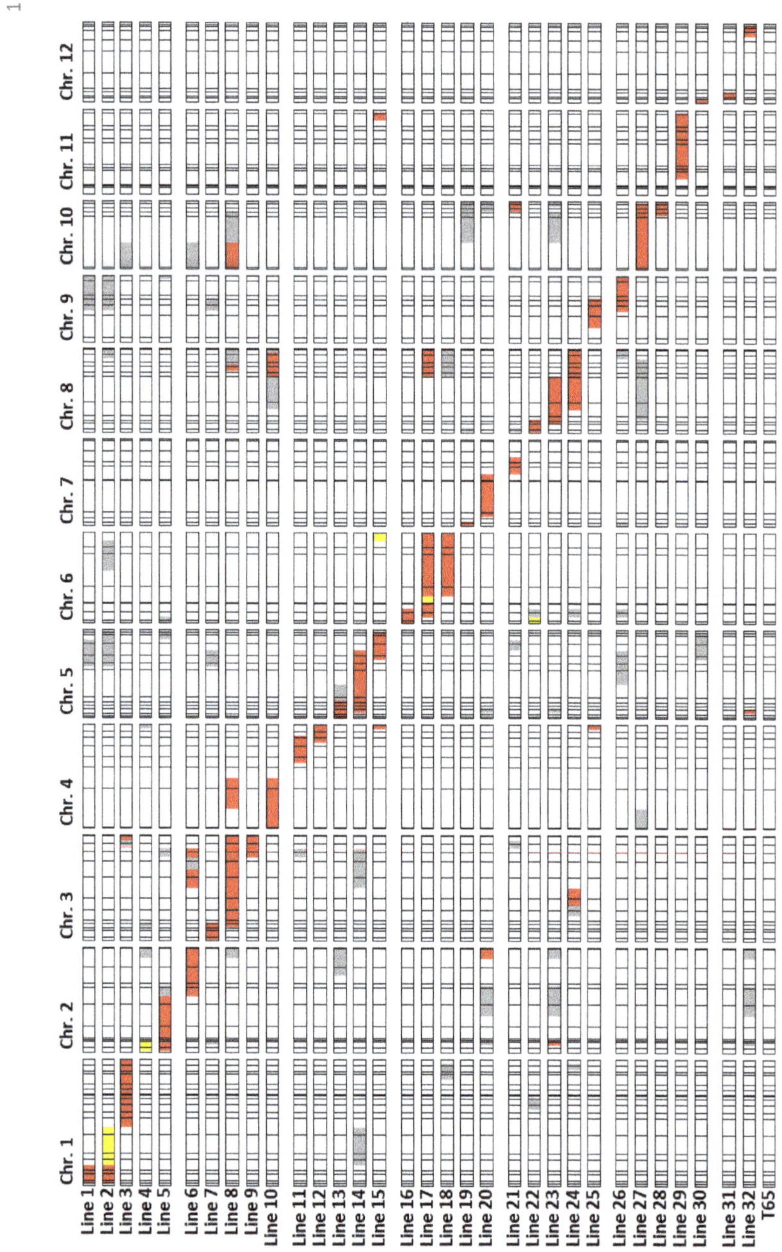

Figure 6. Construction of chromosome segment substitution lines of *O. longistaminata* accession W1508 (W1508ILs) in the genetic background of *O. sativa* L. cv. Taichung 65. Red and white boxes represent homozygous genotypes for W1508 and Taichung 65 alleles, respectively. Yellow boxes represent heterozygous genotypes. Missing genotypes at markers showing heterozygous genotypes at BC_3F_1 generation are indicated by grey.

3. Discussion

Polygenes or QTLs control most agronomic traits. For genetic analysis of quantitative traits, segregating populations such as F_2, RIL, and advanced backcross populations allow complicated genetic architectures to be disassembled to a few genetic elements, thereby increasing the power of detection. Because our objective was to construct a platform from which to analyze the genetic difference between *O. sativa* and *O. longistaminata*, we evaluated the phenotypic character anther length, which is a key character in *O. longistaminata*, and genotyped BC_3 populations and W1508ILs. We used genome-wide QTL exploration and model fitting procedures, and determined that phenotypic variation in anther length was explained by four QTLs on Chrs. 3, 5, and 6. Our subsequent single-factor segregation analyses of QTLs in BC_3F_4 populations validated *qATL3.1* (with an additive effect of 0.17 mm), *qATL3.3* (0.19 mm), *qATL5* (0.16 mm), *qATL6.1* (0.06 mm), and *qATL6.2* (0.08 mm). *qATL3.1* and *qATL3.3* were detected near one another on Chr. 3. Although we considered these two QTLs to be identical, their LOD peaks indicated different genomic positions, most likely owing to the slight segregation distortion around *qATL3.3* (df = 2, P = 0.006). QTLs regulating anther length were also reported around the same regions in *O. rufipogon* [27,28], and close to the *qATL5* region in F_2 populations derived from *O. sativa* × *O. rufipogon* [21], and in RILs derived from *O. sativa* subsp. *indica* × *O. rufipogon* W1944 [28,29]. QTLs that increase anther length have also been detected on Chr. 6 [27,28]. However, these previously identified QTLs have not been validated in *O. sativa* backcross populations in the absence of effects of other QTLs. Therefore, they may not be identical to the QTLs found here.

QTL analysis in NILs showed that the increase in anther length at flowering was likely due to cell elongation under the influence of *qATL3*, *qATL5*, and *qATL6.1*, and to cell division under the influence of *qATL6B*. The development of stamens is divided into two phases [31]. The early phase features the morphogenesis of floral organs, including stamen tissues, from the floral meristem. The late phase comprises pollen maturation, filament elongation, anther dehiscence, and pollen release, which are modulated by auxin (AUX), gibberellin (GA), and jasmonate (JA) [32–35]. AUX and GA can also regulate the early phase [34,36]. The expression of a MYB transcription factor, *HvGAMYB*, in barley, is upregulated by GA in early anther development in the nuclei of the epidermis, endothecium, middle layer, and tapetum, leading to sterility and shorter anther length [37]. Because AUX, GA, and JA affect stamen morphogenesis and male gametogenesis, the QTLs might be involved in epidermal cells via plant hormone signaling.

Because the mean anther length in T65 was 2.27 mm, we expected a plant homozygous for the *O. longistaminata* alleles at the four QTLs without epistatic effect among them to have an anther length of 3.21 mm (2.27 + 2 (0.17 + 0.16 + 0.06 + 0.08) mm), due to the additive effects of *qATL3.1*, *qATL5*, *qATL6.1*, and *qATL6.2*, or, of 3.25 mm (2.27 + 2 (0.19 + 0.16 + 0.06 + 0.08) mm) due to *qATL3.3*, *qATL5*, *qATL6.1*, and *qATL6.2*. As the anther length of W1508 was longer at 4.33 mm, additional, unidentified QTLs and interactions with minor effects are likely hidden in the *O. longistaminata* genome. QTL exploration in BC_3F_2 populations by simple interval mapping using only 21 to 24 plants in each population suggested 10 chromosomal regions with effects that increased anther length, and three that decreased anther length. We expected that the target regions independently segregated with *qATL3.1*, *qATL5*, *qATL6.1*, and *qATL6.2* in the QTL analysis of BC_3F_2 generations, and that the QTL analysis nested in each of the BC_3F_2 populations eliminated the genetic "noise" caused by *qATL3.1*, *qATL5*, *qATL6.1*, and *qATL6.2* in some of the BC_3F_2 populations, by fixation of the homozygous condition for the T65 allele (Table S2). The BC_3F_2 8, 10, and 11 populations, in which the BC_3F_1 parents did not retain the major QTL regions, might segregate QTLs additively with anther length increases of 0.15, 0.10, and 0.10 mm, respectively, at the SSR markers *RM5755*, *RM3317*, and *RM5635*. The BC_3F_2 25 population did not have the *O. longistaminata* segment at the major QTL regions, but the mean anther length of the population was significantly higher than that of T65 (Figure 3). Two populations—BC_3F_2 6 and 17—also did not have the major QTL region, and so most likely had another, unidentified QTL which could not be detected in the BC_3F_1 population. We also inferred chromosomal regions associated with increasing anther length at the major QTLs on Chrs. 3, 5, and 6, but the LOD peak positions were shifted, most likely owing to the few recombinations around QTL regions in populations with only 21 to 24 individuals. This result demonstrates that analysis in BC_3F_2 populations covering the whole genome of *O. longistaminata* increased the detection power, and that it was a practical way to perform the QTL scan where whole-genome genotyping is not available, despite reduced accuracy of the inferred map positions. This practical approach will be useful in the design of future experiments. The simple sum of the additive effects of the W1508 alleles at all QTLs reached 2.14 mm when homozygous (Table S2), which almost explained the difference between W1508 and T65 (4.33 mm − 2.27 mm = 2.06 mm). This suggested that multiple QTLs functioned additively, or by interacting with each other to form anthers as *O. longistaminata*.

The genetic analysis of anther length, which likely reflected on the lifestyle of the perennial wild rice species *O. longistaminata*, revealed that at least four major QTLs, *qATL3.1*, *qATL5*, *qATL6.1*, and *qATL6.2* increased anther length. Regional QTL analysis and CSSL analysis suggested that additional minor QTLs were also associated with regulation of anther length.

4. Materials and Methods

4.1. Plant Materials

We used *O. longistaminata* accession W1508, which was originally collected in Madagascar. *Oryza sativa* L. ssp. *japonica* Taichung 65 (T65) and W1508 were crossed, to develop F_1 plants with T65 cytoplasm. The National Institute of Genetics, Mishima, Japan, kindly provided *O. longistaminata* ratton and the seeds of *O. barthii* and *O. glumaepatula* accessions, and F_1 plants derived from a cross between T65 and W1508. T65 was used as the male parent in recurrent backcrosses to develop BC_3F_1 plants (Figure 2). The BC_3F_2 population was grown in 2015 and 2016.

4.2. Measurement of Anther Length

Panicles were collected at the heading stage and fixed and preserved in 70% ethanol. All anthers from 1 spikelet, just before anthesis, were removed and placed on a glass slide and then photographed under a microscope (Axioplan, Carl Zeiss, Jena, Germany) with a digital camera (DMX-1200, Nikon, Tokyo, Japan). The lengths of 4 to 6 anthers were measured in ImageJ software version 1.8 [38]. The average anther length was used as the phenotypic value of an individual plant.

4.3. Genotyping

Adult leaves were freeze-dried (FDU-1200, Eyela, Tokyo, Japan) and ground with a Multi-bead shocker (Yasui Kikai, Osaka, Japan). Genomic DNA was extracted by the potassium acetate method [39] from ground samples. The plants were genotyped for 124 simple sequence repeat (SSR) markers evenly distributed across the rice genomes (Table S1). For each marker, a 15 μL reaction mixture consisted of 50 mM KCl, 10 mM Tris (pH 9.0), 1.5 mM $MgCl_2$, 200 μM dNTPs, 0.2 μM primers, 0.75 U of GoTaq DNA polymerase (Promega, Fitchburg, WI, USA), and ~10 ng of genomic DNA template. PCR was performed on a GeneAmp PCR System 9700 (Applied Biosystems, Foster City, CA, USA). The thermal profile was an initial denaturation at 95 °C for 5 min: 35 cycles of 95 °C for 30 s, 55 °C for 30 s, and 72 °C for 30 s; and then a final elongation step at 72 °C for 7 min. The amplified products were electrophoresed in 4% agarose gel in 0.5× TBE buffer.

4.4. QTL Analysis

Simple interval mapping by an R/qtl library was used to detect QTLs conferring anther length [40]. LOD score thresholds with a significance level of 5% were empirically estimated by 1000 permutation tests. LOD peaks that exceeded the thresholds defined the QTLs. For multiple QTL mapping in the BC_3F_1 population, a forward/backward stepwise search was used for model selection, including epistasis of two loci by using the stepwiseqtl function in the R/qtl library, with a penalized LOD score criterion to balance model fitting and model complexity.

4.5. Measurement of Numbers of Pollen Grains

Six ethanol-fixed anthers from one spikelet per plant were crushed in a 1.5-mL tube containing 10 μL of 1% I_2–KI solution, and then suspended in 390 μL of distilled water. One μL of the suspension was applied to a glass slide, and pollen grains in 10 fields were photographed under a light microscope (Axioplan, Carl Zeiss, Jena, Germany). The pollen grains were counted in each field, and the numbers were multiplied by the dilution ratio to estimate the pollen grain number per anther.

4.6. Observation of Epidermal Cells of Anthers

Anthers were put in a drop of distilled water on a glass slide, and cut at both ends, and pollen was removed by pipetting back and forth. The epidermal cells of three anthers from one spikelet per plant were photographed at the center in the long axis of the anthers, through an optical microscope (Axioplan, Carl Zeiss, Jena, Germany) with a digital camera (DMX-1200, Nikon, Tokyo, Japan), and were counted in ImageJ. Three plants per line were evaluated as a replicate each, and the average value of epidermal cell length of nine anthers was calculated.

5. Conclusions

Genetic analysis of anther length, which is likely to relate to an outcrossing lifestyle in the perennial wild rice species *O. longistaminata*, revealed that at least four major QTLs, *qATL3*, *qATL5*, *qATL6.1*, and *qATL6.2* increase anther length. The regional QTL analysis and constructed CSSL series (designated 'W1508IL') suggested additional minor QTLs associated with the regulation of anther length. The cloning and diversity analysis of genes conferring anther length QTLs promotes utilization of the genetic resources of wild species, and the understanding of haplotype evolution on the differentiation of annuality and perenniality in the genus *Oryza*.

Supplementary Materials: The following are available online at http://www.mdpi.com/2223-7747/8/10/388/s1, Figure S1: Graphical genotypes of BC_3F_1 for candidates for introgression lines of W1508 in T65 genetic background; Figure S2: Model selections of QTLs including main and interaction effects for anther length; Table S1: QTLs detected by simple interval mapping and multiple QTL mapping in BC_3F_1 population; Table S2: Summary of simple interval mapping for anther length in BC_3F_1, BC_3F_2, and BC_3F_4 populations derived from a cross between T65 and W1508; Table S3: Anther length of W1508ILs; Table S4: SSR markers for genotyping in this study.

Author Contributions: Funding acquisition, A.Y. and H.Y.; investigation, T.O. and Y.Y.; resources, A.Y. and H.Y.; supervision, H.Y.; validation, T.O.; writing, original draft, T.O.; writing, review and editing, Y.Y.

Funding: This work was supported by a Grant-in-Aid from the Japan Agency for Medical Research and Development (National Bioresource Project (Rice)) JP19km0210105j0003 and JSPS KAKENHI Grant Number 18K05576. This work was partially supported by the Science and Technology Research Partnership for Sustainable Development (SATREPS) Grant number JPMJSA1706.

Acknowledgments: Wild rice species accessions used in this study were distributed from the National Institute of Genetics supported by the National Bioresource Project (NBRP), AMED, Japan, and were kindly shared from Maekawa, Institute of Plant Science and Resources, Okayama University, Japan.

Conflicts of Interest: The authors declare no conflicts of interest.

References

1. Friedman, J.; Rubin, M.J. All in good time: Understanding annual and perennial strategies in plants. *Am. J. Bot.* **2015**, *102*, 497–499. [CrossRef] [PubMed]
2. Kuroda, Y.; Urajrong, H.; Sato, Y.I. Population genetic structure of wild rice (*Oryza rufipogon*) in mainland southeast Asia as revealed by microsatellite polymorphisms. *Topics* **2002**, *12*, 159–170. [CrossRef]
3. Annavi, G.; Newman, C.; Buesching, C.D.; Macdonald, D.W.; Burke, T.; Dugdale, H.L. Heterozygosity-fitness correlations in a wild mammal population: Accounting for parental and environmental effects. *Ecol. Evol.* **2014**, *4*, 2594–2609. [CrossRef] [PubMed]
4. Charpentier, M.J.E.; Williams, C.V.; Drea, C.M. Inbreeding depression in ring-tailed lemurs (*Lemur catta*): Genetic diversity predicts parasitism, immunocompetence, and survivorship. *Conserv. Genet.* **2008**, *9*, 1605–1615. [CrossRef]

5. Slate, J.; Kruuk, L.E.B.; Marshall, T.C.; Pemberton, J.M.; Clutton-Brock, T.H. Inbreeding depression influences lifetime breeding success in a wild population of red deer (*Cervus elaphus*). *Proc. R. Soc. Lond. B* **2000**, *267*, 1657–1662. [CrossRef] [PubMed]
6. Acevedo-Whitehouse, K.; Vicente, J.; Gortazar, C.; Hofle, U.; Fernandez-de-Mera, I.G.; Amos, W. Genetic resistance to bovine tuberculosis in the Iberian wild boar. *Mol. Ecol.* **2005**, *14*, 3209–3217. [CrossRef]
7. David, P. Heterozygosity-fitness correlations: New perspectives on old problems. *Heredity* **1998**, *80*, 531–537. [CrossRef] [PubMed]
8. Charlesworth, D. Evolution of plant breeding systems. *Curr. Biol.* **2006**, *16*, 726–735. [CrossRef]
9. Hamrick, J.L.; Godt, M.J.W. Effects of life history traits on genetic diversity in plant species. *Philos. Trans. R Soc. Biol. Sci.* **1977**, *351*, 1291–1298. [CrossRef]
10. Vaughan, D.A.; Lu, B.R.; Tomooka, N. The evolving story of rice evolution. *Plant Sci.* **2008**, *174*, 394–408. [CrossRef]
11. Banaticla-Hilario, M.C.N.; Berg, R.G.; Hamilton, N.R.S.; McNally, K.L. Local differentiation amidst extensive allele sharing in *Oryza nivara* and *O. rufipogon*. *Ecol. Evol.* **2013**, *3*, 3047–3062. [CrossRef]
12. Liu, R.; Zheng, X.M.; Zhou, L.; Zhou, H.F.; Ge, S. Population genetic structure of *Oryza rufipogon* and *Oryza nivara*: Implications for the origin of *O. nivara*. *Mol. Ecol.* **2015**, *24*, 5211–5228. [CrossRef] [PubMed]
13. Reuscher, S.; Furuta, T.; Bessho-Uehara, K.; Cosi, M.; Jena, K.K.; Toyoda, A.; Fujiyama, A.; Kurata, N.; Ashikari, M. Assembling the genome of the African wild rice *Oryza longistaminata* by exploiting synteny in closely related *Oryza* species. *Commun. Biol.* **2018**, *1*, 162. [CrossRef] [PubMed]
14. Hiroi, K.; Mamun, A.A.; Wada, T.; Takeoka, Y. A Study on the interspecific variation of spikelet structure in the genus *Oryza*. *Jpn. J. Crop Sci.* **1991**, *60*, 153–160. [CrossRef]
15. Zhang, Y.; Zhang, S.; Liu, H.; Fu, B.; Li, L.; Xie, M.; Song, Y.; Li, X.; Cai, J.; Wan, W.; et al. Genome and comparative transcriptomics of African wild rice *Oryza longistaminata* provide insights into molecular mechanism of rhizomatousness and self-incompatibility. *Mol. Plant* **2015**, *8*, 1683–1686. [CrossRef]
16. Chu, Y.E.; Morishima, H.; Oka, H.I. Partial self-incompatibility found in *Oryza perennis* subsp. *barthii*. *Gene Genet. Sys.* **1969**, *44*, 225–229. [CrossRef]
17. Sahadevan, P.C.; Namboodiri, K.M.N. Natural crossing in rice. *Proc. Indian Acad. Sci. Sect. B* **1963**, *58*, 176–185.
18. Oka, H.I.; Morishima, H. Variations in the breeding systems of wild rice, *Oryza perennis*. *Evolution* **1967**, *21*, 249–258. [CrossRef]
19. Sakai, K.I.; Narise, T. Studies on the breeding behavior of wild rice. *Annu. Rep. Natl. Inst. Genet. Jpn.* **1959**, *9*, 64–65.
20. Eshed, Y.; Zamir, D. A genomic library of *Lycopersicon pennellii* in *L. esculentum*: A tool for fine mapping genes. *Euphytica* **1994**, *79*, 175–179. [CrossRef]
21. Eshed, Y.; Zamir, D. An introgression line population of *Lycopersicon pennellii* in the cultivated tomato enables the identification and fine mapping of yield associated QTL. *Genetics* **1995**, *141*, 1147–1162. [PubMed]
22. Eshed, Y.; Zamir, D. Less-than-additive epistatic interactions of quantitative trait loci in tomato. *Genetics* **1996**, *143*, 1807–1817. [PubMed]
23. Chetelat, R.T.; Meglic, V. Molecular mapping of chromosome segments introgressed from *Solanum lycopersicoides* into cultivated tomato (*Lycopersicon esculentum*). *Theor. Appl. Genet.* **2000**, *100*, 232–241. [CrossRef]
24. Grandillo, S.; Ku, H.M.; Tanksley, S.D. Characterization of *fs8.1*, a major QTL influencing fruit shape in tomato. *Mol. Breed.* **1996**, *2*, 251–260. [CrossRef]
25. Yamamoto, T.; Kuboki, Y.; Lin, S.Y.; Sasaki, T.; Yano, M. Fine mapping of quantitative trait loci *Hd-1*, *Hd-2* and *Hd-3*, controlling heading date of rice, as single Mendelian factor. *Theor. Appl. Genet.* **1998**, *97*, 37–44. [CrossRef]
26. Monforte, A.J.; Tanksley, S.D. Fine mapping of a quantitative trait locus (QTL) from *Lycopersicon hirsutum* chromosome 1 affecting fruit characteristics and agronomic traits: Breaking linkage among QTLs affecting different traits and dissection of heterosis for yield. *Theor. Appl. Genet.* **2000**, *100*, 471–479. [CrossRef]
27. Xiong, L.Z.; Liu, K.D.; Dai, X.K.; Xu, C.G.; Zhang, Q.F. Identification of genetic factors controlling domestication related traits of rice using an F_2 population of a cross between *Oryza sativa* and *O. rufipogon*. *Theor. Appl. Genet.* **1999**, *98*, 243–251. [CrossRef]

28. Cai, W.; Morishima, H. (2002) QTL clusters reflect character associations in wild and cultivated rice. *Theor. Appl. Genet.* **2002**, *104*, 1217–1228. [CrossRef]
29. Uga, Y.; Fukuta, Y.; Ohsawa, R.; Fujimura, T. Variations of floral traits in Asian cultivated rice (*Oryza sativa* L.) and its wild relatives (*O rufipogon* Griff.). *Breed. Sci.* **2003**, *53*, 345–352. [CrossRef]
30. Uga, Y.; Siangliw, M.; Nagamine, T.; Ohsawa, R.; Fujimura, T.; Fukuta, Y. Comparative mapping of QTLs determining glume, pistil and stamen sizes in cultivated rice (*Oryza sativa* L.). *Plant Breed.* **2010**, *129*, 657–669. [CrossRef]
31. Goldberg, R.B.; Beals, T.P.; Sanders, P.M. Anther development: Basic principles and practical applications. *Plant Cell* **1993**, *5*, 1217–1229. [CrossRef] [PubMed]
32. McConn, M.; Browse, J. The critical requirement for linolenic acid is pollen development, not photosynthesis, in an Arabidopsis Mutant. *Plant Cell* **1996**, *8*, 403–416. [CrossRef] [PubMed]
33. Cecchetti, V.; Altamura, M.M.; Falasca, G.; Costantino, P.; Cardarelli, M. Auxin regulates *Arabidopsis* anther dehiscence, pollen maturation, and filament elongation. *Plant Cell* **2008**, *20*, 1760–1774. [CrossRef] [PubMed]
34. Cheng, Y.; Dai, X.; Zhao, Y. Auxin biosynthesis by the YUCCA flavin monooxygenases controls the formation of floral organs and vascular tissues in *Arabidopsis*. *Genes Dev.* **2006**, *20*, 1790–1799. [CrossRef] [PubMed]
35. Jewell, J.B.; Browse, J. Epidermal jasmonate perception is sufficient for all aspects of jasmonate-mediated male fertility in *Arabidopsis*. *Plant J.* **2016**, *85*, 634–647. [CrossRef] [PubMed]
36. Plackett, A.R.; Thomas, S.G.; Wilson, Z.A.; Hedden, P. Gibberellin control of stamen development: A fertile field. *Trends Plant Sci.* **2011**, *16*, 568–578. [CrossRef] [PubMed]
37. Murray, F.; Kalla, R.; Jacobsen, J.; Gubler, F. A role for HvGAMYB in anther development. *Plant J.* **2003**, *33*, 481–491. [CrossRef]
38. Schneider, C.A.; Rasband, W.S.; Eliceiri, K.W. NIH Image to ImageJ: 25 years of image analysis. *Nat. Methods* **2012**, *9*, 671–675. [CrossRef]
39. Dellaporta, S.L.; Wood, J.; Hicks, J.B. A plant DNA minipreparation: Version II. *Plant Mol. Biol. Rep.* **1983**, *1*, 19–21. [CrossRef]
40. Broman, K.W.; Wu, H.; Sen, S.; Churchill, G.A. R/qtl: QTL mapping in experimental crosses. *Bioinformatics* **2003**, *19*, 889–890. [CrossRef]

plants

MDPI

Article

HWA1- and *HWA2*-Mediated Hybrid Weakness in Rice Involves Cell Death, Reactive Oxygen Species Accumulation, and Disease Resistance-Related Gene Upregulation

Kumpei Shiragaki [1], Takahiro Iizuka [1], Katsuyuki Ichitani [2], Tsutomu Kuboyama [3], Toshinobu Morikawa [1], Masayuki Oda [1] and Takahiro Tezuka [1,4,*]

1 Graduate School of Life and Environmental Sciences, Osaka Prefecture University, 1-1 Gakuen-cho, Nakaku, Sakai, Osaka 599-8531, Japan; ma201027@edu.osakafu-u.ac.jp (K.S.); ddlfte@gmail.com (T.I.); d-morikawa@hannan-u.ac.jp (T.M.); masa-oda@d.email.ne.jp (M.O.)

2 Faculty of Agriculture, Kagoshima University, 1-21-24 Korimoto, Kagoshima, Kagoshima 890-0065, Japan; ichitani@agri.kagoshima-u.ac.jp

3 College of Agriculture, Ibaraki University, 3-21-1 Chuo, Ami, Ibaraki 300-0393, Japan; tsutomu.kuboyama.a@vc.ibaraki.ac.jp

4 Education and Research Field, College of Life, Environment, and Advanced Sciences, Osaka Prefecture University, 1-1 Gakuen-cho, Nakaku, Sakai, Osaka 599-8531, Japan

* Correspondence: tezuka@plant.osakafu-u.ac.jp; Tel.: +81-72-254-8457

Received: 28 August 2019; Accepted: 24 October 2019; Published: 25 October 2019

Abstract: Hybrid weakness is a type of reproductive isolation in which F_1 hybrids of normal parents exhibit weaker growth characteristics than their parents. F_1 hybrid of the *Oryza sativa* Indian cultivars 'P.T.B.7' and 'A.D.T.14' exhibits hybrid weakness that is associated with the *HWA1* and *HWA2* loci. Accordingly, the aim of the present study was to analyze the hybrid weakness phenotype of the 'P.T.B.7' × 'A.D.T.14' hybrids. The height and tiller number of the F_1 hybrid were lower than those of either parent, and F_1 hybrid also exhibited leaf yellowing that was not observed in either parent. In addition, the present study demonstrates that SPAD values, an index correlated with chlorophyll content, are effective for evaluating the progression of hybrid weakness that is associated with the *HWA1* and *HWA2* loci because it accurately reflects degree of leaf yellowing. Both cell death and H_2O_2, a reactive oxygen species, were detected in the yellowing leaves of the F_1 hybrid. Furthermore, disease resistance-related genes were upregulated in the yellowing leaves of the F_1 hybrids, whereas photosynthesis-related genes tended to be downregulated. These results suggest that the hybrid weakness associated with the *HWA1* and *HWA2* loci involves hypersensitive response-like mechanisms.

Keywords: *Oryza sativa*; hybrid weakness; cell death; reactive oxygen species; leaf yellowing; SPAD; hypersensitive response

1. Introduction

The traits of existing crop cultivars can be improved by crossing cultivars or lines to introduce beneficial traits, such as resistance or tolerance to disease or stress, to susceptible cultivars. However, because reproductive isolation mechanisms can hinder the production of hybrids, methods must be developed to overcome the underlying mechanisms of such reproductive isolation.

One type of post-zygotic reproductive isolation, namely hybrid weakness (i.e., hybrid lethality or hybrid necrosis). F_1 hybrids that exhibit this phenomenon are characterized by weaker growth than their parents, and the phenomenon has been reported to occur in the offspring of crosses involved a number of species, including *Oryza sativa* [1–4], *Nicotiana* spp. [5], *Capsicum* spp. [6], *Arabidopsis thaliana* [7], *Triticum* spp. [8,9], *Gossypium* spp. [10], and *Phaseolus vulgaris* [11]. The genetic mechanisms of hybrid weakness are explained by the Bateson–Dobzhanzky–Muller model [12–14], which posits that the reduced hybrid vigor is driven by deleterious interactions between genes at different loci. In many cases, one of the causal genes is related to disease resistance (*R*), and interactions between the *R* gene and other causal gene cause autoimmune responses in the hybrid offspring [2,7]. The autoimmune responses include the accumulation of reactive oxygen species such as H_2O_2, cell death, upregulation of disease resistance-related genes, and downregulation of photosynthesis-related genes [2,7,15,16].

In rice, hybrid weakness has been reported to result from interactions between the *HWI1* locus, which encodes the LRR-RLK gene (*R* gene), and the *HWI2* locus, which encodes a subtilisin-like protease, and hybrids have been reported to exhibit localized programmed cell death (PCD), the high accumulation of salicylic and jasmonic acids, and amplified heat-related weakness symptoms [2]. These results demonstrate that the interaction of causal genes can activate downstream immune responses, such as hypersensitive response-like mechanisms [2,7].

The hybrid weakness that results from the interaction of *Hwa1-1*, a dominant allele of the *HWA1* locus, and *Hwa2-1*, a dominant allele of the *HWA2* locus, was firstly reported in rice by Oka [4]. In that study, F_1 hybrid seedlings that exhibited hybrid weakness were reported to exhibit normal germination and seedling growth until developing three to four leaves, after which plant growth halted and the leaves yellowed. Then, unless the environment was particularly favorable, the plants died before reaching anthesis. The distributions of the *Hwa1-1* and *Hwa2-1* alleles were limited to Indian cultivars [4], and both the *HWA1* and *HWA2* loci were located in a 1637-kb region of the long arm of chromosome 11 [17]. However, the causal genes have not been identified, and the molecular mechanism underlying the hybrid weakness associated with the *HWA1* and *HWA2* loci remain unclear.

Accordingly, the aim of the present study was to characterize the phenotypes of the hybrid weakness that is associated with the *HWA1* and *HWA2* loci, in order to understand the system's underlying mechanisms. The effectiveness of SPAD values, an index correlated with chlorophyll content [18], for determining the progression of the hybrid weakness was also evaluated. The occurrence of cell death and H_2O_2 accumulation was also evaluated, and the expression of disease resistance and photosynthesis-related genes in the leaves of F_1 hybrids exhibiting hybrid weakness were analyzed.

2. Results

2.1. Hybrid Weakness Phenotypes

The Oryza sativa Indian cultivars 'A.D.T.14' and 'P.T.B.7' carry homozygous Hwa1-1 and Hwa2-1 alleles, respectively. All the F_1 hybrids of a cross between 'A.D.T.14' and 'P.T.B.7' exhibited dwarfism, reduced tiller number, and leaf yellowing (Figure 1). Increases in the height of the F_1 hybrids nearly halted at 50 days after sowing (DAS), whereas that of the parents continued increasing (Figure 2A). The progression of plant age in leaf number in the F_1 hybrids was the same as that in both parents (Figure 2B). The tiller number of both parents continued increasing and reached >35 tillers at 70 DAS, whereas that of the F_1 hybrids increased little and only reached five tillers by 70 DAS (Figure 2C). In addition, both parents headed by 80 DAS, whereas none of the F_1 hybrids had started heading after 140 DAS (Figure 1).

Leaf yellowing was first observed in the F_1 hybrids that had developed seventh or eighth leaves at 30 DAS. Afterward, the leaves turned yellow sequentially, from the lower to the upper leaves. At 60 DAS, the fourth, fifth, and sixth leaves of the F_1 hybrids turned yellow, starting from the leaf tip, and progressing toward the leaf base, whereas those of both parents remained green (Figure 3A–C). Meanwhile, the SPAD values of the fourth, fifth, and sixth leaves of the F_1 hybrids were lower than those of the parents (Figure 3D–F). Furthermore, in the leaves of the F_1 hybrids, the SPAD values of the lower leaves were lower than those of the upper leaves (fourth vs. fifth and sixth leaves and fifth vs. sixth leaves), and within each leaf, the SPAD values of the leaf tips were lower than those of the leaf bases (Figure 3D–F).

Figure 1. Parental and F_1 hybrid phenotypes at 80 days after sowing. (**A**) *Oryza sativa* 'A.D.T.14'; (**B**) F_1 hybrid; and (**C**) *O. sativa* 'P.T.B.7'. Arrows indicate emerging panicles. Scale bars indicate 50 cm.

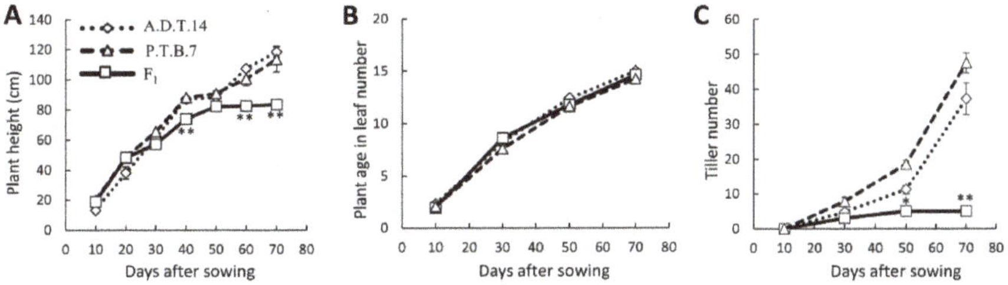

Figure 2. Phenotypic traits of parental and F_1 hybrid rice. (**A**) plant height; (**B**) plant age in leaf number; and (**C**) tiller number. Values and error bars indicate mean ± SE values (n = 5), although some error bars are hidden by the symbols. Mid-parental and hybrids values were compared using two-tailed Student's t-test. Significance: ** $P < 0.01$, * $P < 0.05$.

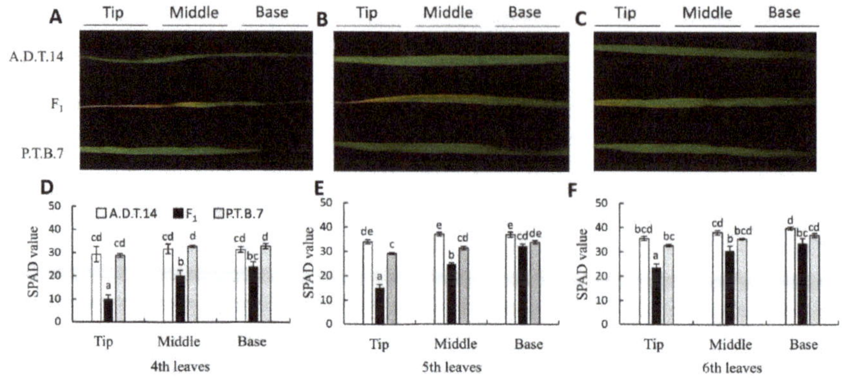

Figure 3. Phenotypes and SPAD values of leaves (fourth to sixth) from parental and F_1 hybrid rice. The phenotypes (**A–C**) and SPAD values (**D–F**) of fourth (**A, D**), fifth (**B, E**), and sixth (**C, F**) leaves were assessed at 60 days after sowing. SPAD value was measured at the tip, middle, and base of each leaf. Values and error bars indicate mean ± SE values (n = 3). Different lowercase letters in each plot (**D–F**) indicate significant differences (Tukey HSD test, $P < 0.05$).

2.2. *Cell Death and H_2O_2 Accumulation*

The physiological changes that accompanied leaf yellowing were surveyed by analyzing F_1 hybrid leaves that had been classified into four stages based on degree of yellowing (Figure 4A). Chlorophyll content was assessed by SPAD analysis and spectrophotometry. The SPAD values of Stage-1, -2, and -3 leaf tips, Stage-3 leaf middles, and Stage-3 leaf bases were lower than those of Stage-0 leaves (Figure 4B). The SPAD values of Stage-3 leaf tips, Stage-3 leaf middles, and Stage-3 leaf bases were lower than those of Stage-1 leaves (Figure 4B). The SPAD values of Stage-3 leaf middles and Stage-3 leaf bases were lower than those of Stage-2 leaves (Figure 4B). Total chlorophyll content also decreased in all leaf parts (tip, middle, and base) as yellowing progressed (Figure 4C). Because of the usefulness of SPAD value as discussed later, the progression of yellowing of leaves used in the subsequent experiments were evaluated based on SPAD value.

To determine whether cell death occurred in F_1 hybrid leaves, cellular ion leakage, owing to ion permeability by cell death, was measured. Ion leakage increased slightly and significantly in the tips of Stage-2 and Stage-3 leaves, respectively (Figure 4D). Cell death in F_1 hybrid leaves was also evaluated using trypan blue staining, which is used to identify the highly permeable membranes of dead cells. Only Stage-3 leaves contained dead cells (Figure 5A), and the analysis of transverse sections of Stage 3 revealed that the dead cells were located around vascular and epidermal cells (Figure 5B).

Meanwhile, 3,3-diaminobenzidine (DAB) staining revealed the presence of H_2O_2, which, as a reactive oxygen species, is an important regulator of cell death. Plant tissue is stained brown when DAB is oxidized by H_2O_2 into an insoluble polymer. Hydrogen peroxide (H_2O_2) was detected in the leaves of all stages, except Stage 0 (Figure 6).

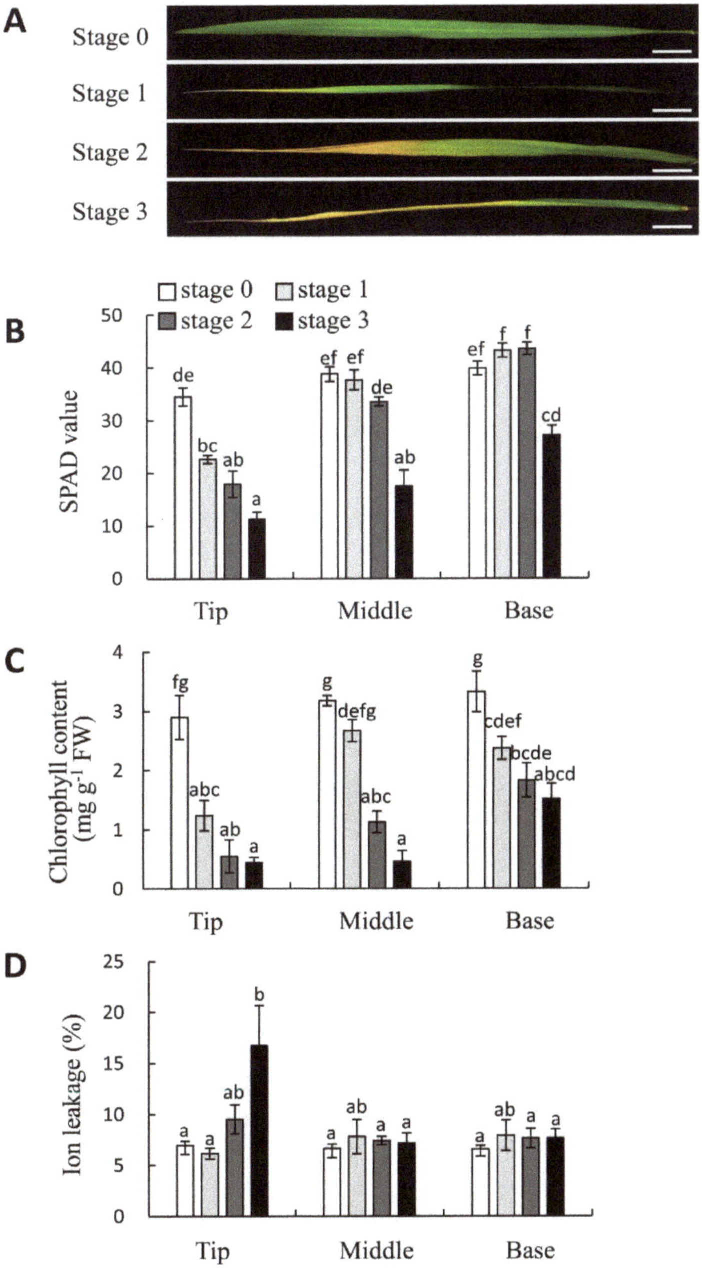

Figure 4. Physiological changes of yellowing hybrid leaves. (**A**) Stages of yellowing: Stage 0, no yellowing; Stage 1, 1/4 of leaf yellow; Stage 2, 1/2 of leaf yellow; Stage 3, 3/4 of leaf yellow. Scale bars indicate 2 cm. (**B**) Changes in the SPAD values at the tip, middle, and base of leaves during the progression of yellowing. (**C**) Changes in chlorophyll content during the progression of yellowing. (**D**) Changes in ion leakage during the progression of yellowing. Values and error bars (**B–D**) indicate mean ± SE values (n = 3), and different lowercase letters in each plot (**B–D**) indicate significant differences (Tukey HSD test, $P < 0.05$).

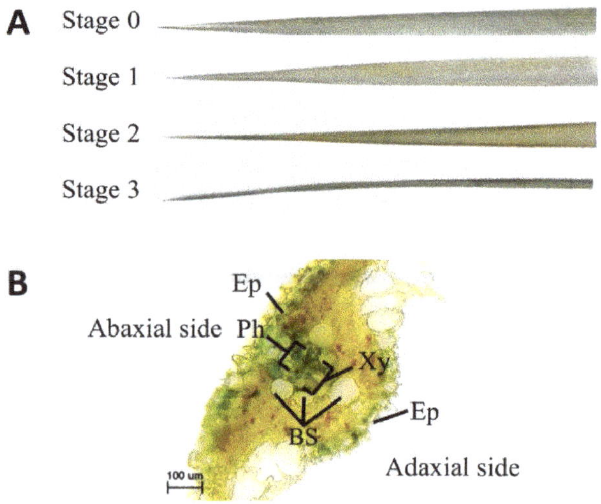

Figure 5. Trypan blue staining of dead cells in F_1 leaves. (**A**) Stained leaves from each yellowing stage. (**B**) Transverse section of a stained Stage-3 leaf. Scale bar indicates 100 μm. Xy: Xylem; Ph: Phloem; Ep: Epidermal cell; BS: Bundle sheath cell.

Figure 6. Presence of reactive oxygen species in hybrid leaves. 3,3-diaminobenzidine (DAB) staining was used to detect H_2O_2. Scale bars indicate 1 cm.

2.3. Hybrid Weakness-Related Gene Expression

At 70 DAS, the 11th (Stage 3) and 13th (Stage 0) leaves of the parents and F_1 offspring were collected for gene expression analysis (Figure 7). The 13th leaves of both the parents and hybrids were entirely green, as indicated by high SPAD values, even though the SPAD values of the F_1 hybrids were somewhat lower than those of either parent (Figure 7A,C). Meanwhile, the 11th leaves of the F_1 hybrids exhibited significant yellowing, as indicated by low SPAD values, whereas those of both parents were entirely green, as indicated by high SPAD values (Figure 7B,D).

The expression of 11 disease resistance-related genes and four photosynthesis-related genes were surveyed (Table 1 and Table S1). The PR1 genes (PR1A and PR1B), the expression of which is induced by salicylic acid [19,20], were upregulated in the 11th leaves of the F_1 hybrids (Figure 8), as were several PR2 genes (Gns5, Gns2, and OsEGL2), which encode glucanase-related proteins that degrade fungal cell walls [21] (Figure 8). Meanwhile, of several genes that encode chitinase-related proteins (PR4, CHT9, CHT11, and RIXI), which also degrade fungal cell walls [22], PR4, CHT9, and RIXI were all upregulated in the 11th leaves of the F_1 hybrids; however, only PR4 was upregulated significantly (Figure 8). The expression of ACO2, which encodes an enzyme related to ethylene production [23], was similar in the 11th and 13th leaves of the F_1 hybrids (Figure 8). Finally, PDC1, the expression of which is induced by jasmonic acid [24], was upregulated in the 11th leaves of the F_1 hybrids, although not significantly (Figure 8).

Of the four photosynthesis-related genes, PSAF, LHCB, and OsRbcL were somewhat downregulated in the 11th leaves of the F_1 hybrids; however, none of these trends were significant (Figure 8).

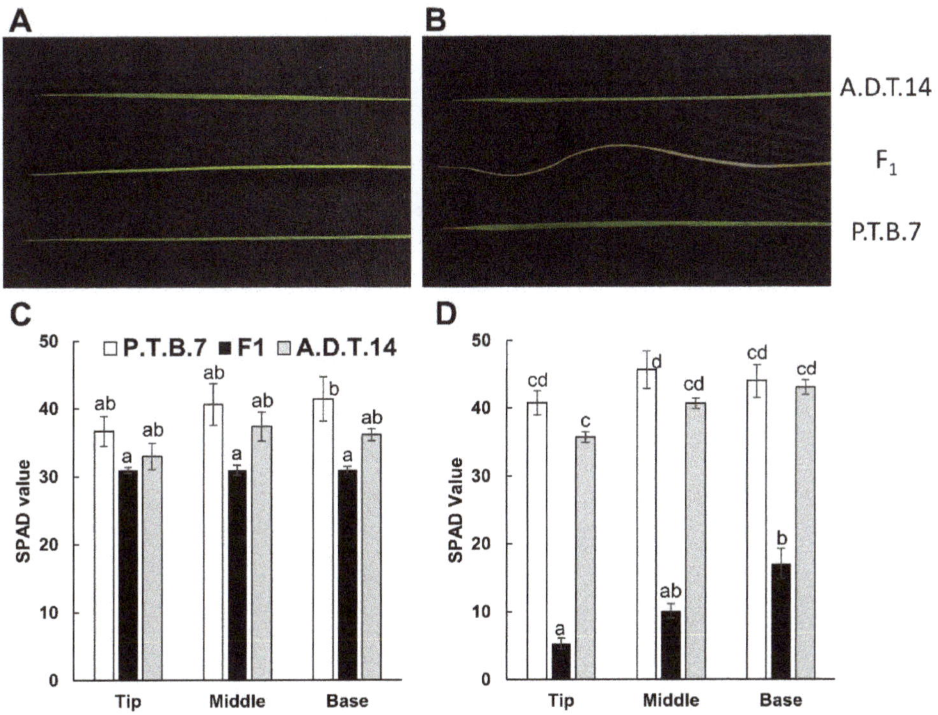

Figure 7. Phenotypes and SPAD values of the parental and F_1 leaves (11th and 13th) used for gene expression analysis. The phenotypes (**A**, **B**) and SPAD values (**C**, **D**) of 11th (**A**, **C**) and 13th (**B**, **D**) leaves were assessed at 70 d after sowing. SPAD value was measured at the tip, middle, and base of each leaf. Values and error bars indicate mean ± SE values (n = 3), and different lowercase letters in each plot (**C**, **D**) indicate significant differences (Tukey HSD test, $P < 0.05$).

Table 1. Genes analyzed for RT-PCR.

Gene Symbol	RAP-DB locus ID [a]	CGSNL Gene Name [b]	Description
PR1A	Os07g0129200	*PATHOGENESIS-RELATED GENE 1A*	Similar to Pathogenesis-related protein PR1a
PR1B	Os01g0382000	*PATHOGENESIS-RELATED GENE 1B*	Similar to Pathogenesis-related protein PRB1-2 precursor
Gns5	Os01g0940700		Similar to Glucan endo-1,3-beta-glucosidase GII precursor
Gns2	Os01g0944900		Similar to Glucan endo-1,3-beta-D-glucosidase
OsEGL2	Os01g0942300		Similar to Beta glucanase precursor
PR4	Os11g0592200	*PATHOGENESIS-RELATED GENE 4*	Similar to Chitin-binding allergen Bra r 2
CHT9	Os05g0399400	*CHITINASE 9*	Chitinase 9
CHT11	Os03g0132900	*CHITINASE 11*	Similar to Chitinase 11
RIXI	Os11g0701800		Chitinase III C10701-rice (Class III chitinase homologue)
ACO2	Os09g0451000	*AMINOCYCLOPROPANE-1-CARBOXYLIC ACID OXIDASE 2*	Similar to 1-aminocyclopropane-1-carboxylase 1
PDC1	Os05g0469600	*PYRUVATE DECARBOXYLASE 1*	Similar to Pyruvate decarboxylase
PSAF	Os03g0778100	*PHOTOSYSTEM I SUBUNIT*	Similar to Photosystem-1 F subunit
LHCB	Os03g0592500		Similar to Photosystem II type II chlorophyll a/b binding protein
OsRbcL	Os06g0598500		Similar to Ribulose bisphosphate carboxylase large chain precursor
Fd1	Os08g0104600		Ferredoxin I, chloroplast precursor
Actin	Os05g0438800		Similar to Actin1

[a] Identity of each gene was referenced using the Rice Annotation Project database (https://rapdb.dna.affrc.go.jp/); [b] CGSNL (Committee on Gene Symbolization, Nomenclature and Linkage, Rice Genetics Cooperative) gene names were referenced using Oryzabase (https://shigen.nig.ac.jp/rice/oryzabase/) [25].

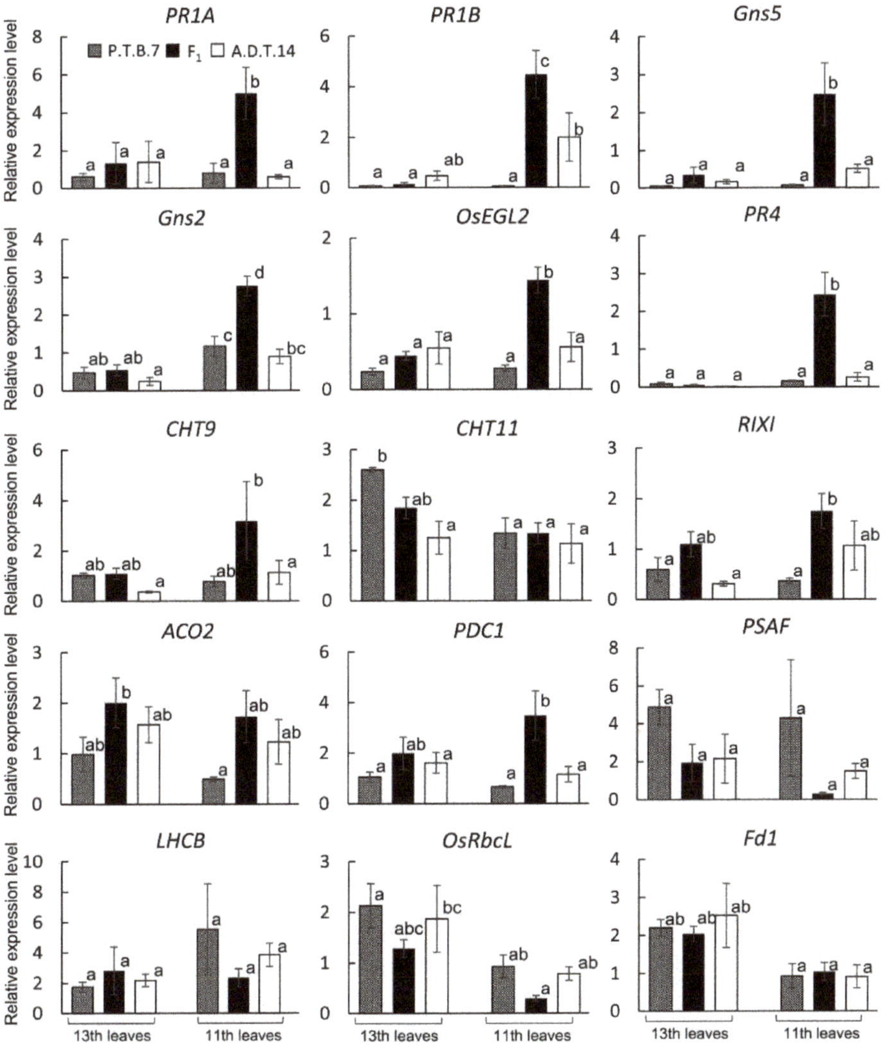

Figure 8. Relative gene expression levels of parental and F_1 hybrid leaves between 'P.T.B.7' and 'A.D.T.14'. Values and error bars indicate mean ± SE values ($n = 3$), and different lowercase letters in each plot indicate significant differences (Tukey HSD test, $P < 0.05$).

3. Discussion

Oka [4] reported that F_1 hybrids that exhibit hybrid weakness associated with the *HWA1* and *HWA2* loci exhibit growth termination and leaf yellowing after the seedlings developed three or four leaves. However, the plant growth phenotypes were not described in detail. In contrast, the present study determined that F_1 hybrids from the cross of 'A.D.T.14' and 'P.T.B.7' rice exhibited limited growth and tiller number, as well as and leaf yellowing (Figures 1–3). Even though leaf yellowing was also reported by Oka [4], the timing of the yellowing process was different [4]. In the present study, leaf yellowing was observed in F_1 hybrids that had developed seven or eight leaves at 30 DAS and, furthermore, was associated with the downregulation of photosynthesis-related genes (Figure 8).

In *O. sativa*, three other gene sets have been reported to cause hybrid weakness, and the phenotypes associated with each system are different. More specifically, the hybrid weakness associated with the *HWC1* and *HWC2* loci is characterized by short stature, short roots, and rolled leaves [26], whereas that associated with the *HWI1* and *HWI2* loci is characterized by short stature and impaired root formation [2], and that associated with the *HW3* and *HW4* loci is characterized by short culms, fewer panicles, pale green leaves, and chlorotic leaf spots [3]. Remarkably, leaf yellowing has only been reported for the hybrid weakness associated with the *HWA1* and *HWA2* loci. Together, these reports suggest that either the causal genes of each system have different functions or the processes downstream of the causal gene interactions are different.

In the present study, the usefulness of SPAD value for determining the progression of leaf yellowing during hybrid weakness associated with the *HWA1* and *HWA2* loci were evaluated. SPAD values generally corresponded with leaf yellowing (Figure 4A,B) but failed to identify significant differences between the chlorophyll content of Stage-0 leaves and that of either the bases of Stage-1 leaves or the middles or bases of Stage-2 leaves. These results indicate that spectrophotometry is more sensitive than SPAD values to changes in chlorophyll content (Figure 4B,C). However, it is important to note that, because SPAD value accurately reflected degree of leaf yellowing and because spectrophotometry requires leaf destruction (Figure 4), SPAD value measurement is an effective and nondestructive method that can be used to quickly and easily evaluate hybrid weakness associated with the *HWA1* and *HWA2* loci.

The hybrid weakness phenotype that was studied by the present study also exhibited cell death in the yellow leaves (Figures 4D and 5). Similarly, the hybrid weakness associated with the *HWI1* and *HWI2* loci involved cell death at the basal nodes [2], and the hybrid weakness associated with the *HW3* and *HW4* loci involved cell death in leaves [3]. Cell death has been also detected in the leaves of intraspecific *Arabidopsis* hybrids that exhibit hybrid necrosis [7,27] and in the leaves, stems, and roots of interspecific *Nicotiana* hybrids that exhibit hybrid lethality [28,29]. Therefore, despite differences in localization, cell death appears to be a common feature of hybrid weakness in plants.

In the present study, cell death was only detected in Stage-2 and Stage-3 leaves, which indicates that the timing of cell death does not coincide with that of either leaf yellowing or reductions in SPAD value or chlorophyll content (Figure 4). On the other hand, H_2O_2 was detected in Stage-1, -2, and -3 leaves that exhibited yellow leaf tips and low SPAD values (Figure 6). H_2O_2 commonly triggers plant cell death during hypersensitive reactions, senescence, abiotic stress responses, and development [30–32]. We detected that reactive oxygen species would lead to cell death on hybrid weakness by *HWA1* and *HWA2*.

In many cases of hybrid weakness, one of the causal genes encodes an *R* gene, and the interaction of the *R* gene with another causal gene triggers an autoimmune response [7,27,33,34]. In the hybrid weakness associated with the *HWI1* and *HWI2* loci, the causal genes include the LRR-RLK gene (*R* gene) and a subtilisin-like protease gene, respectively, and the interaction of the causal genes results in an autoimmune response [2]. Meanwhile, in the hybrid weakness associated with the *HW3* and *HW4* loci, *HW3* encodes a calmodulin-binding protein, and even though the gene is a defense-response gene, not an *R* gene, the interaction of *HW3* with *HW4* results in an autoimmune response [3]. Furthermore, in the hybrid weakness associated with the *HWA1* and *HWA2* loci, a candidate region, which harbored both loci, also contained 12 *R* genes, along with many other genes [17]. During hypersensitivity reactions, reactive oxygen species are produced, thereby mediating cell death, chloroplast disruption, and the upregulation of defense-related genes [35,36]. In the present study, cell death was detected in the F_1 leaves after H_2O_2 generation and leaf yellowing (Figure 3, Figure 5, and Figure 6), the yellow leaves of the F_1 hybrids exhibited upregulated defense-related genes (Figure 8). These results suggest that the hybrid weakness associated with the *HWA1* and *HWA2* loci involves hypersensitive reaction-like responses. However, because leaf senescence is also associated with H_2O_2 production, cell death, the upregulation of certain defense genes, and leaf yellowing [37], it is possible that the hybrid weakness associated with the *HWA1* and *HWA2* loci involves premature senescence. Additional molecular studies will reveal the exact mechanism underlying the hybrid weakness associated with the *HWA1* and *HWA2* loci.

4. Materials and Methods

4.1. Plant Materials and Growth Conditions

The *Oryza sativa* Indian cultivars 'A.D.T.14' (*indica* [17]), which is homozygous for the *Hwa1-1* allele, and 'P.T.B.7' (*aus* [17]), which is homozygous for the *Hwa2-1* allele, were crossed to generate F_1 hybrid offspring. The genotypes of the two cultivars were previously reported by Oka [4]. F_1 seeds were obtained by crossing 'P.T.B.7' (♀) and 'A.D.T.14' (♂) parents. Seeds of 'A.D.T.14', 'P.T.B.7', and the F_1 offspring were sown on 17 July 2011. After the seeds were germinated on moistened filter paper in Petri dishes, the seedlings were transplanted to soil (Sukoyaka-Jinko-Baido; Yanmar Co., Ltd., Osaka, Japan) in Wagner pots of 1/5000 a. The seedlings were grown under natural light conditions in a greenhouse at Osaka Prefecture University, Sakai, Japan. The temperature and humidity of the greenhouse were recorded using a data logger (Ondotori; T&D Co., Ltd., Matsumoto, Japan), and the plants were fertilized weekly using Otsuka-A prescription (OAT Agrio Co., Ltd., Tokyo, Japan), which contained 18.6 mM N, 5.1 mM P, 8.6 mM K, 8.2 mM Ca, and 0.4 mM Mg. The plants were cultivated for 140 DAS to survey plant height, plant age in leaf number, tiller number, days to heading, and SPAD value. The plant height was measured from the surface of the soil to the tip of the tallest leaves. To evaluate the relationship between leaf yellowing and physiology, the leaves were classified according to degree of leaf yellowing. Leaves in which 0%, 25%, 50%, or > 75% of the blade had turned yellow were assigned to Stages 0, 1, 2, and 3, respectively (Figure 4A). These leaves classified according to degree of leaf yellowing were used to measured SPAD value and chlorophyll content, as well as to detect dead cells and H_2O_2. Parts of seedlings were cultivated in Wagner pots of 1/10,000 a in an incubator (14 h natural light and 10 h dark, 28 °C, light intensity: 512 μmol m^{-2} s^{-1}), and at 70 DAS, these plants were used as material for gene expression analysis.

4.2. SPAD and Chlorophyll Measurement

A SPAD meter (SPAD-502; Konica Minolta, Inc., Tokyo, Japan) was used to measure the SPAD values of the leaves without causing damage. SPAD values were obtained from the tip, middle, and base of each leaf. Meanwhile, total chlorophyll content was measured using a previously described spectrophotometric method [38]. Briefly, the leaves were cut into small pieces, weighed, treated with 20 mL 80% acetone, and ground using a pestle until bleached. The resulting solutions were transferred to 1.5 mL tubes and centrifuged at 10,000 g for 5 min. Each supernatant was transferred to a cuvette, and the absorbance of each supernatant was measured at 663.6 and 646.6 nm, after the spectrophotometer (V-530; JASCO Corp., Hachioji, Japan) was zeroed at 750 nm. Total chlorophyll concentration (mg g^{-1} FW) was calculated using the following equation: $[(17.76 \times OD_{646.6} + 7.34 \times OD_{663.6}) \times$ extraction volume in a cuvette]/fresh weight (g).

4.3. Ion leakage Measurement

Ion leakage was measured, as described previously [39]. Leaf disks (3 cm^2) were taken from the tips, middles, and bases of the leaves, floated for 5 min in water that contained 0.2% (v/v) Tween 20 for removing ion generating on making leaf disks, transferred to Petri dishes that contained fresh water with Tween 20 (0.2%), and incubated for 3 h for leaking out ions by cell death. Their conductivity (value A) of the solutions was measured using a conductivity meter (Twin Cond B-173; Horiba, Ltd., Kyoto, Japan). The leaf disks were then incubated at 95 °C for 25 min, for leaking out ions of whole leaf disks by destroying whole organization, and cooled to room temperature, and their conductivity of the solutions was also measured (value B). Finally, ion leakage (%) was calculated using the following equation: (value A/value B) × 100%.

4.4. *Trypan Blue Staining*

Trypan blue staining was performed as described previously [40]. Detached leaves were stained by boiling for 8 min in a 1:1 (v:v) mixture of ethanol and lactophenol (i.e., alcoholic lactophenol) that contained 0.1 mg ml^{-1} trypan blue, cleared in 70% chloral hydrate solution overnight, and then preserved in 70% glycerol. Trypan blue stains dead cells. Transverse slices were prepared using a hand-section method and visualized using a light microscope (Olympus BX50; Olympus, Co. Ltd., Tokyo, Japan).

4.5. *Detection of Hydrogen Peroxide Accumulation*

Hydrogen peroxide was detected visually, using previously described methods [41]. Briefly, leaves were soaked in a 3,3-diaminobenzidine (DAB) solution for 24 h, transferred to boiling 96% ethanol until bleaching, and then visualized. The presence of H_2O_2 was indicated by brown staining.

4.6. *Real-Time qRT-PCR*

Total RNA was isolated from leaves using an RNAiso PLUS kit (Takara Bio, Inc., Shiga, Japan), according to the manufacturer's protocol and then treated with RNase-free DNase (Promega Co., Madison, USA), and first-strand cDNA was synthesized from total RNA (2 μg) using oligo $(dT)_{18}$ primers and ReverTra Ace (Toyobo Co., Ltd., Osaka, Japan). Real-time RT-PCR was carried out to analyze the expression of 11 defense-related genes and four photosynthesis-related genes (Table 1), using *Actin* as an internal control. The primers used to amplify *PR1A, PR1B, Gns5, PR4*, and *Actin* had been reported previously [42], and the other primers were designed based on RAP-DB locus ID using the Primer-BLAST design tool [43] (Table S1). Real-time RT-PCR was performed in 20 μL reaction mixtures that contained 10 μL KAPA SYBR FAST qPCR Master Mix (2×) ABI PRISM (Takara Bio), 10 μM of each forward and reverse primer (0.4 μL each), and 1 μL cDNA template, and the real-time PCR amplification was performed under the following conditions: Initial denaturation at 94 °C for 10 min, followed by 40 cycles of 15 s at 94 °C and 1 min at 60 °C, with a final 30 s extension at 72 °C using an Applied Biosystems 7300 Real-Time PCR System (Applied Biosystems, Foster, CA, USA). The results were analyzed using ABI Prism software (Applied Biosystems). Each gene expression level was divided by the expression level of *Actin* to calculate relative expression level.

4.7. *Statistical Analysis*

Data were analyzed using SPSS (version 22; IBM, Co., Armonk, USA). Tukey HSD tests were used to compare SPAD, chlorophyll content, and ion leakage values, and two-tailed Student's t-tests were used to compare mid-parental (mean of 'A.D.T.14' and 'P.T.B.7') and hybrid values of plant height, foliar age, and tiller number.

Supplementary Materials: The following are available online at http://www.mdpi.com/2223-7747/8/11/450/s1, Table S1: Real-time RT-PCR primers.

Author Contributions: Conceptualization, K.I., T.K., and T.T.; formal analysis, K.S.; funding acquisition, K.I., T.K., and T.T.; investigation, K.S. and T.I.; resources, K.I.; supervision, K.I., T.K., T.M., M.O., and T.T.; writing—original draft, K.S.; writing—review and editing, T.T.

Funding: This research was funded by JSPS KAKENHI (grant no. JP24580009) from the Japan Society for the Promotion of Science.

Acknowledgments: We are grateful to the Genebank of the National Institute of Agrobiological Sciences (Tsukuba, Japan) for providing seeds of the parent lines ('A.D.T.14' and 'P.T.B.7'). We would like to thank Editage (www.editage.com) for English language editing.

Conflicts of Interest: The authors declare no conflict of interest.

References

1. Amemiya, A.; Akemine, H. Biochemical genetic studies on the root growth inhibiting complementary lethal in rice plant (Studies on the embryo culture in rice plant. 3). *Bull. Natl. Inst. Agric. Sci.* **1963**, *10*, 139–226.
2. Chen, C.; Chen, H.; Lin, Y.S.; Shen, J.B.; Shan, J.X.; Qi, P.; Shi, M.; Zhu, M.Z.; Huang, X.H.; Feng, Q.; et al. A two-locus interaction causes interspecific hybrid weakness in rice. *Nat. Commun.* **2014**, *5*, 3357. [CrossRef] [PubMed]
3. Fu, C.Y.; Wang, F.; Sun, B.R.; Liu, W.G.; Li, J.H.; Deng, R.F.; Liu, D.L.; Liu, Z.R.; Zhu, M.S.; Liao, Y.L.; et al. Genetic and cytological analysis of a novel type of low temperature-dependent intrasubspecific hybrid weakness in rice. *PLoS ONE* **2013**, *8*, e73886. [CrossRef] [PubMed]
4. Oka, H.I. Phylogenetic differentiation of cultivated rice. XV Complementary lethal genes in rice. *Jpn. J. Genet.* **1957**, *32*, 83–87. [CrossRef]
5. Tezuka, T.; Marubashi, W. Genomic factors lead to programmed cell death during hybrid lethality in interspecific hybrids between *Nicotiana tabacum* and *N. debneyi*. *SABRAO J. Breed Genet.* **2006**, *38*, 69–81.
6. Inai, S.; Ishikawa, K.; Nunomura, O.; Ikehashi, H. Genetic analysis of stunted growth by nuclear-cytoplasmic interaction in interspecific hybrids of *Capsicum* by using RAPD markers. *Theor. Appl. Genet.* **1993**, *87*, 416–422. [CrossRef]
7. Bomblies, K.; Lempe, J.; Epple, P.; Warthmann, N.; Lanz, C.; Dangl, J.L.; Weigel, D. Autoimmune response as a mechanism for a Dobzhansky-Muller-type incompatibility syndrome in plants. *PLoS Biol.* **2007**, *5*, e236. [CrossRef]
8. Tsunewaki, K. Monosomic and conventional analyses in common wheat. III Lethality. *Jpn. J. Genet.* **1960**, *35*, 71–75. [CrossRef]
9. Tsunewaki, K. Aneuploid analysis of hybrid necrosis and hybrid chlorosis in tetraploid wheats using the D genome chromosome substitution lines of durum wheat. *Genome* **1992**, *35*, 594–601. [CrossRef]
10. Silow, R.A. The comparative genetics of *Gossypium anomalum* and the cultivated Asiatic cottons. *J. Genet.* **1941**, 259–358. [CrossRef]
11. Reiber, J.M.; Neuman, D.S. Hybrid weakness in *Phaseolus vulgaris* L. II. Disruption of root-shoot integration. *J. Plant Growth Regul.* **1999**, *18*, 107–112. [CrossRef] [PubMed]
12. Bateson, W. Heredity and variation in modern science. In *Darwin and Modern Science*; Seward, A.C., Ed.; Cambridge University Press: Cambridge, UK, 1909; pp. 85–101.
13. Dobzhansky, T. *Genetics and the Origin of Species*; Columbia University Press: New York, NY, USA, 1937.
14. Muller, H.J. Isolating mechanisms, evolution, and temperature. *Biol. Symp.* **1942**, *6*, 71–124.
15. Mizuno, N.; Hosogi, N.; Park, P.; Takumi, S. Hypersensitive response-like reaction is associated with hybrid necrosis in interspecific crosses between tetraploid wheat and *Aegilops tauschii* Coss. *PLoS ONE* **2010**, *5*, e11326. [CrossRef] [PubMed]
16. Okada, M.; Yoshida, K.; Takumi, S. Hybrid incompatibilities in interspecific crosses between tetraploid wheat and its wild diploid relative *Aegilops umbellulata*. *Plant Mol. Biol.* **2017**, *95*, 625–645. [CrossRef]
17. Ichitani, K.; Taura, S.; Tezuka, T.; Okiyama, Y.; Kuboyama, T. Chromosomal location of *HWA1* and *HWA2*, complementary hybrid weakness genes in rice. *Rice* **2011**, *4*, 29–38. [CrossRef]
18. Markwell, J.; Osterman, J.C.; Mitchell, J.L. Calibration of the Minolta SPAD-502 leaf chlorophyll meter. *Photosyn. Res.* **1995**, *46*, 467–472. [CrossRef]
19. Agrawal, G.K.; Rakwal, R.; Jwa, N.S.; Agrawal, V.P. Signalling molecules and blast pathogen attack activates rice *OsPR1a* and *OsPR1b* genes: A model illustrating components participating during defence/stress response. *Plant Physiol. Biochem.* **2001**, *39*, 1095–1103. [CrossRef]
20. Van Loon, L.C.; Rep, M.; Pieterse, C.M.J. Significance of inducible defense-related proteins in infected plants. *Annu. Rev. Phytopathol.* **2006**, *44*, 135–162. [CrossRef]
21. Romero, G.O.; Simmons, C.; Yaneshita, M.; Doan, M.; Thomas, B.R.; Rodriguez, R.L. Characterization of rice endo-β-glucanase genes (Gns2–Gns14) defines a new subgroup within the gene family. *Gene* **1998**, *223*, 311–320. [CrossRef]
22. Klarzynski, O.; Plesse, B.; Joubert, J.-M.; Yvin, J.-C.; Kopp, M.; Kloareg, B.; Fritig, B. Linear β-1,3 glucans are elicitors of defense responses in tobacco. *Plant Physiol.* **2000**, *124*, 1027–1038. [CrossRef]
23. Kende, H. Ethylene biosynthesis. *Annu. Rev. Plant. Biol.* **1993**, *44*, 283–307. [CrossRef]

24. Sugimori, M.; Kiribuchi, K.; Akimoto, C.; Yamaguchi, T.; Minami, E.; Shibuya, N.; Sobajima, H.; Cho, E.M.; Kobashi, N.; Nojiri, H.; et al. Cloning and characterization of cDNAs for the jasmonic acid-responsive genes *RRJ1* and *RRJ2* in suspension-cultured rice cells. *Biosci. Biotechnol. Biochem.* **2002**, *66*, 1140–1142. [CrossRef]
25. McCouch, S.R. Gene nomenclature system for rice. *Rice* **2008**, *1*, 72–84. [CrossRef]
26. Saito, T.; Ichitani, K.; Suzuki, T.; Marubashi, W.; Kuboyama, T. Developmental observation and high temperature rescue from hybrid weakness in a cross between Japanese rice cultivars and Peruvian rice cultivar 'Jamaica'. *Breed. Sci.* **2007**, *57*, 281–288. [CrossRef]
27. Alcázar, R.; García, A.V.; Kronholm, I.; de Meaux, J.; Koornneef, M.; Parker, J.E.; Reymond, M. Natural variation at Strubbelig Receptor Kinase 3 drives immune-triggered incompatibilities between *Arabidopsis thaliana* accessions. *Nat. Genet.* **2010**, *42*, 1135–1139. [CrossRef]
28. Tezuka, T.; Marubashi, W. Apoptotic cell death observed during the expression of hybrid lethality in interspecific hybrids between *Nicotiana tabacum* and *N. suaveolens*. *Breed. Sci.* **2004**, *54*, 59–66. [CrossRef]
29. Yamada, T.; Marubashi, W.; Niwa, M. Apoptotic cell death induces temperature-sensitive lethality in hybrid seedlings and calli derived from the cross of *Nicotiana suaveolens* × *N. tabacum*. *Planta* **2000**, *211*, 614–622. [CrossRef]
30. Huysmans, M.; Lema, A.S.; Coll, N.S.; Nowack, M.K. Dying two deaths—Programmed cell death regulation in development and disease. *Curr. Opin. Plant Biol.* **2017**, *35*, 37–44. [CrossRef]
31. Petrov, V.; Hille, J.; Mueller-Roeber, B.; Gechev, T.S. ROS-mediated abiotic stress-induced programmed cell death in plants. *Front Plant Sci.* **2015**, *6*, 69. [CrossRef]
32. Woo, H.R.; Kim, H.J.; Nam, H.G.; Lim, P.O. Plant leaf senescence and death—Regulation by multiple layers of control and implications for aging in general. *J. Cell Sci.* **2013**, *126*, 4823. [CrossRef]
33. Huang, X.; Li, J.; Bao, F.; Zhang, X.; Yang, S. A gain-of-function mutation in the *Arabidopsis* disease resistance gene *RPP4* confers sensitivity to low temperature. *Plant Physiol.* **2010**, *154*, 796–809. [CrossRef] [PubMed]
34. Yang, S.; Hua, J. A haplotype-specific resistance gene regulated by *BONZAI1* mediates temperature-dependent growth control in *Arabidopsis*. *Plant Cell* **2004**, *16*, 1060–1071. [CrossRef] [PubMed]
35. Coll, N.S.; Epple, P.; Dangl, J.L. Programmed cell death in the plant immune system. *Cell Death Differ.* **2011**, *18*, 1247. [CrossRef] [PubMed]
36. Overmyer, K.; Brosché, M.; Kangasjärvi, J. Reactive oxygen species and hormonal control of cell death. *Trends Plant Sci.* **2003**, *8*, 335–342. [CrossRef]
37. Jajic, I.; Sarna, T.; Strzalka, K. Senescence, stress, and reactive oxygen species. *Plants* **2015**, *4*, 393–411. [CrossRef]
38. Porra, R.J.; Thompson, W.A.; Kriedemann, P.E. Determination of accurate extinction coefficients and simultaneous equations for assaying chlorophylls a and b extracted with four different solvents: Verification of the concentration of chlorophyll standards by atomic absorption spectroscopy. *Biochim. Biophys. Acta* **1989**, *975*, 384–394. [CrossRef]
39. Rizhsky, L.; Shulaev, V.; Mittler, R. Measuring programmed cell death in plants. In *Apoptosis Methods and Protocols*; Brady, H.J.M., Ed.; Humana Press: Totowa, NJ, USA, 2004; pp. 179–189.
40. Zhang, H.K.; Zhang, X.; Mao, B.Z.; Li, Q.; He, Z.H. Alpha-picolinic acid, a fungal toxin and mammal apoptosis-inducing agent, elicits hypersensitive-like response and enhances disease resistance in rice. *Cell Res.* **2004**, *14*, 27–33. [CrossRef]
41. Chandru, H.K.; Kim, E.; Kuk, Y.; Cho, K.; Han, O. Kinetics of wound-induced activation of antioxidative enzymes in *Oryza sativa*: Differential activation at different growth stages. *Plant Sci.* **2003**, *164*, 935–941. [CrossRef]
42. Yamamoto, E.; Takashi, T.; Morinaka, Y.; Lin, S.; Wu, J.; Matsumoto, T.; Kitano, H.; Matsuoka, M.; Ashikari, M. Gain of deleterious function causes an autoimmune response and Bateson–Dobzhansky–Muller incompatibility in rice. *Mol. Genet. Genomics* **2010**, *283*, 305–315. [CrossRef]
43. Ye, J.; Coulouris, G.; Zaretskaya, I.; Cutcutache, I.; Rozen, S.; Madden, T.L. Primer-BLAST: A tool to design target-specific primers for polymerase chain reaction. *BMC Bioinform.* **2012**, *13*, 134. [CrossRef]

plants

MDPI

Article

Molecular and Morphological Divergence of Australian Wild Rice

Dinh Thi Lam [1,2], Katsuyuki Ichitani [3], Robert J. Henry [4] and Ryuji Ishikawa [5,*]

1 United Graduate School of Agricultural Sciences, Iwate University, Morioka, Iwate 020-8550, Japan; lamiasvn@gmail.com
2 Institute of Agricultural Science for Southern Vietnam, District 1, Ho Chi Minh City 121, Vietnam
3 Faculty of Agriculture, Kagoshima University, 1-21-24 Korimoto, Kagoshima, Kagoshima 890-0065, Japan; ichitani@agri.kagoshima-u.ac.jp
4 Queensland Alliance for Agriculture and Food Innovation, University of Queensland, Brisbane QLD 4072, Australia; robert.henry@uq.edu.au
5 Faculty of Agriculture and Life Science, Hirosaki University, 3 Bunkyo-cho, Hirosaki, Aomori 036-8561, Japan
* Correspondence: ishikawa@hirosaki-u.ac.jp; Tel.: +81-172-39-3778

Received: 10 December 2019; Accepted: 4 February 2020; Published: 10 February 2020

Abstract: Two types of perennial wild rice, Australian *Oryza rufipogon* and a new taxon Jpn2 have been observed in Australia in addition to the annual species *Oryza meridionalis*. Jpn2 is distinct owing to its larger spikelet size but shares *O. meridionalis*-like morphological features including a high density of bristle cells on the awn surface. All the morphological traits resemble *O. meridionalis* except for the larger spikelet size. Because Jpn2 has distinct cytoplasmic genomes, including the chloroplast (cp), cp insertion/deletion/simple sequence repeats were designed to establish marker systems to distinguish wild rice in Australia in different natural populations. It was shown that the new taxon is distinct from Asian *O. rufipogon* but instead resembles *O. meridionalis*. In addition, higher diversity was detected in north-eastern Australia. Reproductive barriers among species and Jpn2 tested by cross-hybridization suggested a unique biological relationship of Jpn2 with other species. Insertions of retrotransposable elements in the Jpn2 genome were extracted from raw reads generated using next-generation sequencing. Jpn2 tended to share insertions with other *O. meridionalis* accessions and with Australian *O. rufipogon* accessions in particular cases, but not Asian *O. rufipogon* except for two insertions. One insertion was restricted to Jpn2 in Australia and shared with some *O. rufipogon* in Thailand.

Keywords: *Oryza*; speciation; divergence; life history; phylogenetic relation; Australian continent

1. Introduction

The *Oryza* genus is comprised of 23 species with varying genome compositions and ploidy levels [1]. The two cultivated species, *Oryza sativa* and *Oryza glaberrima*, belong to AA genome species, and their progenitors were wild *Oryza rufipogon* and *Oryza barthii*, respectively. The AA genome species are dispersed across the major continents and were once classified as a single species, *Oryza perennis*, comprising Asian, American, African, and Oceanian forms [2]. The Asian species, *O. rufipogon* represents different life histories and varies from annual to perennial. Their life history is a continuum with annual, intermediate, and perennial forms [3,4]. The American species, *Oryza glumaepatula* also varies from annual to perennial. African species, *O. barthii* and *Oryza longistaminata*, however, are exclusively annual and perennial types, respectively.

Oceanian species had been known as *O. perennis* (later changed to the current species nomenclature) including annual and perennial types as a continuum within a single species [3]. After rearrangement of the species classification, an annual type was defined as an Oceanian endemic species, *Oryza meridionalis* and the perennial form as *O. rufipogon* [4]. Their distributions in Australia are well studied [5,6]. Speciation of these species has been confirmed using retrotransposon insertions [7,8] and crossing ability [9–11].

In general, annual and perennial species have different adaptive strategies to allocate their energy resources [3,12]. Annual species tend to have higher seed productivity than perennial species. *O. meridionalis*, the Australian annual species, produces plenty of seeds and disperses these seeds. *O. meridionalis* inhabits ponds or the periphery of ponds, ditches, or lakes during the rainy season. Water levels in wild rice habitats recede and water in the peripheral areas of annual species disappears during the dry season [13]. Annual species produce large amounts of seed for the next generation. In contrast, the life history of Australian perennial species is similar to Asian perennial species except for a unique taxon known as Jpn2 or taxon B [6,14]. In addition, Jpn2 type wild rice exhibits different morphological and genetic characteristics [14]. Including the new wild rice type, Australian perennial and annual rice chloroplast (cp) genomes have been completely sequenced in order to understand the uniqueness in evolutionary relationships among other wild rice [15,16]. This showed that the cp genome of Australian *O. rufipogon*, Jpn1 (taxon A) has a closer relationship to *O. meridionalis* than to Asian *O. rufipogon*, although its nuclear type tended to show higher similarity to Asian *O. rufipogon*. Another perennial species, Jpn2 (taxon B), also shared similarity not only with the cp genome to *O. meridionalis* but also the nuclear type [14,17,18]. This analysis showed that all Australian wild rice shared some cp genetic similarity with *O. meridionalis*. Nuclear genomes in Australia showed huge variation never seen in Asian wild rice. These findings with ecological observations confirmed that there were two types of perennial rice. Their distribution in northern Queensland and their unique morphological traits were also reported [6,14].

In this paper, we further characterized these two taxa at morphological and reproductive levels, which enabled us to determine how they have diverged at the species level. Cytoplasmic markers to distinguish them were developed and variation among natural populations was evaluated. These findings will help to distinguish these taxa in field research for further analysis and also give clues to their evolutionary origins. Retro-transposable elements were also used to screen the species examined in this study. Some of these provide clear evidence of phylogenetic relationships because of the unique mechanism of transposition insertion.

2. Results

2.1. Morphological Features

Two types of Australian perennial wild rice were collected (Table 1). Based on our previous report [14], identifying two types of perennials: Jpn1 (taxon A) and Jpn2 (taxon B), morphological traits were able to be discriminated between the Australian perennials. Bristle cells have a thorn-like architecture along the awns (Figure 1). SEM enabled us to compare the density of these cells. They varied from 2.33 to 5.33 per 200 μm square among Asian wild rice (Table 2). In *O. meridionalis*, W1299 and W1300 had 12.67 and 14.67 per 200 μm^2, respectively. The density in Jpn1 was similar to that in Asian wild rice. That in Jpn2 was similar to *O. meridionalis*. There were significant differences between the two groups, W1299/W1300/Jpn2 and W106/W0120/W0137/Jpn1. Other traits, such as anther length, suggested that Jpn2 shared short anthers with other annual accessions such as W0106 in *O. rufipogon*, and W1299 and W1300 in *O. meridionalis*.

Table 1. Samples collected in Australia and control core collections developed in NBRP.

		No. of	GPS Data		
Sites	**Populations**	**Plants**	**S**	**E**	**Year**
P26	P26a	8	S16.5332	E145.2138	2011
	P26b	8	S16.5529	E145.2136	2011
	P26c	8	S16.5541	E145.2130	2011
	P26d	8	S16.5539	E145.2128	2011
	P26e	8	S16.5536	E145.2125	2011
	P26f	8	S16.5536	E145.2124	2011
	P26g	8	S16.5535	E145.2123	2011
	P26h	8	S16.5534	E145.2122	2011
	P26i	8	S16.5533	E145.2122	2011
Jpn1	Jpn1	8	S16.3809	E145.1936	2011
Jpn2	Jpn2	8	S15.2622	E144.1239	2011
Jpn3	Jpn3	8	S15.0431	E143.4321	2011
P6	P6	8	S15 41519	E145 02473	2011
P7	P7E	8	S15 42003	E145 04219	2011
	P7L	8	S15 42003	E145 04219	2011
P8	P8W	4	S15 41302	E145 07237	2011
	P8R	4	S15 41302	E145 07237	2011
P10	P10H	8	S15.41416	E145.09040	2011
	P10L	8	S15.41416	E145.09040	2011
P12	P12	8	S15.31486	E144.22564	2011
P17	P17	8	S15.09256	E143.49481	2011
P21	P21	8	S15.23047	E144.07228	2011
P22	P22	8	S15.78099	E144.14333	2011
P23	P23	8	S15.31198	E144.22079	2011
P26j	P26j	8	S16.55118	E145.2136	2011
P26PL	P26PL	8	S16.15579	E145.2060	2011
P27	P27				
P5	P5O	8	S15 45329	E144 59419	2010
	P5N	8	S15 45329	E144 59419	2010
	P5W	8	S15 45329	E144 59419	2010
Core collection of NBRP					
	Oryza meridinoalis	19*			
	Oryza rufipogon	32			

*W1299 noted as no rank in Oryza database, was added to the core collection in this study.

Table 2. Comparison of morphological traits, anther length, pancle length, paniclewidth, and density of bristle cells.

				Anther Length (mm)				Spikelet Length (mm)				Spikelet Width (mm)				Density of Bristle Cells (per 200 μm²)			
Species	**Accession**	**Life History**	**n=**	**mean**	**±**	**SD**		**mean**	**±**	**SD**		**mean**	**±**	**SD**		**mean**	**±**	**SD**	
O.rufipogon																			
	W0106	Annual	5	1.50	±	0.07	**	7.01	±	0.63	**	2.42	±	0.20		4.67	±	1.92	**
	W0120	Perennial	5	1.88	±	0.10	**	7.06	±	0.16	**	2.34	±	0.11		2.67	±	1.92	**
	W0137	Perennial	5	2.31	±	0.05	**	6.95	±	0.25	**	2.60	±	0.11		3.67	±	1.92	**
O.meridionalis																			
	W1299	Annual	5	1.42	±	0.05	**	6.47	±	0.27	**	2.32	±	0.12		12.67	±	3.42	
	W1300	Annual	5	ND				6.91	±	0.19	**	2.13	±	0.05	**	14.67	±	1.60	
	Australian wild rice																		
Greenhouse																			
	Jpn1	Perennial	5	3.48	±	0.20		7.13	±	0.15	**	2.02	±	0.16	**	2.67	±	3.42	**
	Jpn2	Perennial	5	1.64	±	0.07	**	8.28	±	0.14	*	2.62	±	0.11		12.00	±	2.34	
Field																			
	Jpn1	Perennial	5	3.79	±	0.20		7.62	±	0.17	**	2.00	±	0.07	**	ND			
	Jpn2	Perennial	5	1.74	±	0.19	**	8.60	±	0.23		2.40	±	0.23		ND			

*,**: Significant differences compared with the longest anther of Jpn1, the largest panicle of Jpn2, the widest width of W0137, and density of bristle cells of W1300 at 5 and 1% levels, respectively.

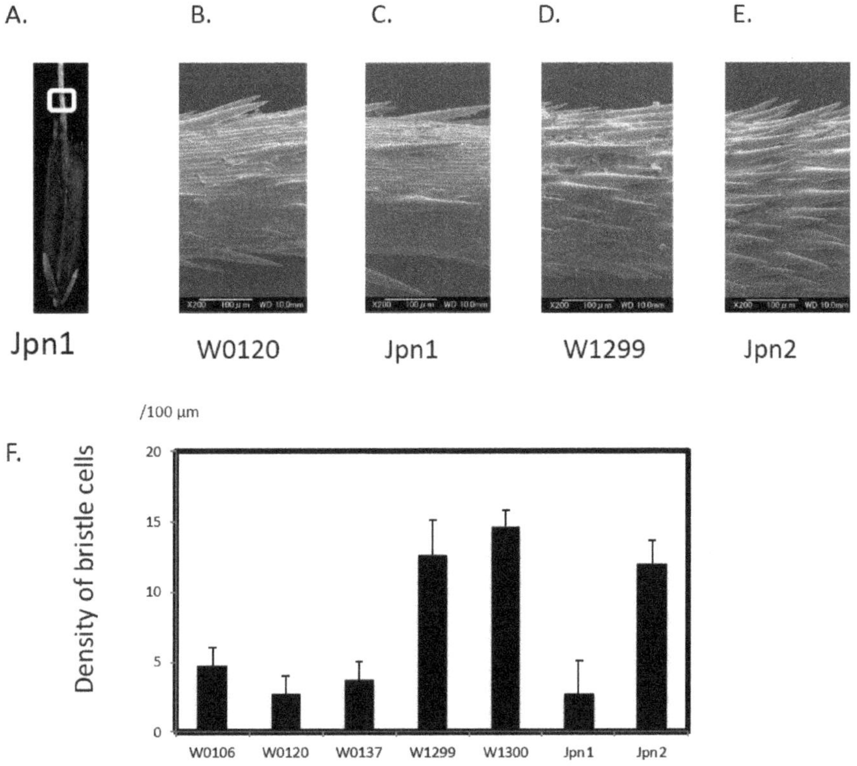

Figure 1. Variation in the density of bristle cells in awns. Panel **A**: spikelet of Jpn1, Panels B to E: enlarged SEM photos of W0120 (Panel **B**), Jpn1 (Panel **C**), W1299 (Panel **D**), Jpn2 (Panel **E**). Panel **F**: density of bristle cells per 200 μm^2. Bars indicating standard error (n = 3).

2.2. Maternal Lineages

In order to trace maternal lineages, next-generation sequencing data obtained from Jpn1 and Jpn2 were used for re-sequencing and comparison with the Nipponbare complete cp genome sequence. More than 53 million reads were obtained from the two accessions. Two genome sequences of *O. meridionalis*, and *O. rufipogon* were added for comparison. In all cases, 100% coverage was achieved with 733 to 2002 mean depth. When the nuclear genome was used as a reference genome, 66%–88% coverage with 7.6 to 11.4 mean depth was obtained.

Simple sequence repeats were found at 20 loci in the cp genomes. Simple insertions or deletions (INDELs) were also found at 21 loci (Table 3). Two loci were not amplified, and six loci were not confirmed because of difficulty of primer design for these fragments. One region ranging from nucleotide 17,336 to 17,392 of the Nipponbare cp genome was amplified as a single amplicon because of its short size. In total, 29 insertions/deletions (INDELs)/simple sequence repeats (SSRs) in the cp genome were polymorphic. Australian rice accessions including *O. meridionalis*, Jpn1, and Jpn2 shared the same genotype at 26 out of the 29 loci developed by plastid INDELs and SSRs.

Table 3. INDEL and SSR markers in chloroplast genomes and developed markers.

Marker Type	Genotype (Based on Relative Migration Distance)								INDEL	INDEL Start Position in Nipponbare cp Genome	Forward	Reverse
	Nipponbare	W0106	W0120	W0137	W1299	W1300	Jpn1	Jpn2				
INDEL1	2	2	2	2	2	2	1	2	-C	1605	CTATTCCGAAGAGGAAGTCTAC	TCTCCGTATCAATGATCTGGTG
SSR1	1	1	1	1	2	2	3	2	+AA	3535	CTTTTGACTTTGGGATACAGTC	GATTAGTGCCTGATGTAGGG
INDEL2	2	2	2	2	1	1	1	1	-CAATC	5852	GGAATTTCCATCCTCAACAGA	GTTTTGTTACGGAAAAATGGTATG
SSR2	1	1	1	1	2	2	2	2	+A	6098	TTCTCGTATTTCTTCGACTCG	GATAAGAACTGCTCGTTAGATAG
INDEL3	1	1	1	1	2	2	2	2	+AGAAA	8192	GCCGCTTTAGTCCACTCAGCCATC	TCAATGCCTTTTTTCAATGGTCTC
SSR3	1	1	1	1	2	2	2	2	+A	11441	CTGGCTCGGTTATTCTATC	GAAAACCGGTATAGTTCTAGG
INDEL4	2	1	2	1	1	1	1	1	-AGGG	12669	GCAACAGGGTTCCCTAAACCG	GCCAAATTGAGCAGGTTGCG
INDEL5	2	2	2	2	1	1	1	1	-T	13566	GCTTCGCGACTCTGTACTCA	TACTTAAGGCGTCCTTAAGG
INDEL6	2	1	2	1	1	1	1	1	-AC	14011	GAAATCTGGGCCATAGAGAA	CTAAGCAGAGACATTCAGAATC
INDEL7	NA								-TATTTCTAAGA	14527	-	-
SSR4	2	2	2	2	1	1	1	1	-A	17099	GAAAAAATCCATGGAGGGAGAG	CCCAACATATCGCACATTTTCC
SSR5	2	1	1	1	3	3	3	3	-TTTCTA	17336	GGTCGCTTCTAGTAGCGATTATG	TGCCGAACTTTATTCTTTCTCTC
SSR6	including SSR5 to INDEL10								-TTTCTA	17358		
INDEL8									-ATAGAA	17379		
INDEL9									+AGAATTAT	17385		
INDEL10									+GAATTATATAGAAC	17392		
INDEL11	1	1	1	1	2	2	1	2	+TGG	19001	GAATATCATAAACTGTAAGTGGCAG	CACATGAAATTCTCGGGAACTCC
SSR7	1	1	1	1	2	2	2	2	+T	41464	GAGGCAAGTGTTCGGATCTATTATG	CTATATTATGCTCAAGGAAAGTAGA
SSR8	1	1	2	1	1	1	1	1	-TATAT	46086	CTCTAATTCGCAAATCTATTTTTC	CAAGAAATTCGCATGTTCTC
SSR9	2	1	2	1	1	1	1	1	-T	46174	GAGAACATGCGAATTTCTTG	CATACTATAACGCTTGATATTC
INDEL12	1	2	1	2	2	2	2	2	+T	47211	GTCGTGAGGGTTCAAGT	CGAGTTAATAATCGACATTCCTTGCC
INDEL13	1	1	1	1	2	2	1	2	+AGGAC	50351	GCCTGTCCAGTCTATAAACAAG	GGGTCTTTGAAACAGTTCG
SSR10	1	1	1	1	2	2	2	2	+T	53999	CATAGAATGTACACAGGGTGTACCC	CTCACAACGACAGGGTCTAC
INDEL14	INDEL14 and SSR11								-CTTTTTTTTTAGAATA	57017	GGATAGAAAGGCCGCGAG including	GACTATTGTATTTTTGAGTTTGC
SSR11									-A	57061		
INDEL15	2	2	2	2	1	1	1	1	-CTTTTCAAT	64815	CCAGATGCTTTGTCATTCCC	TCATGACTCTAAGGTCCAACC
INDEL16	1	1	1	1	2	2	2	2	+TTCCTATTTAATA	65452	GTCGTTATTGTCGTAAGCATACGA	GATGAATACCCTCGATACATATG
SSR12	NA								+T	65615	-	-
INDEL17	1	1	1	1	2	2	2	2	+t	66896	CCAATGGCTTTTGCTACTATAACC	GAAAGAAAGGGCTCCGGTG
SSR13	1	1	1	1	2	2	2	2	+A	71377	GCACCTGTTATCTCTATCAAG	GTCTGGTTGCGAGGTCTGAATAG
SSR14	2	2	2	2	1	1	1	1	-TTTCTA	75980	GATATCCGTTTCAGGGTAAA	CTGATTCGTAGGCGTGGAC
SSR15	2	1	2	1	1	1	1	1	-A	76232	CAAATTTTACGAACAGAAGCTC	CCGAAGACTCGAAGGATACC

Table 3. *Cont.*

Marker Type	Genotype (Based on Relative Migration Distance) Nipponbare	W0106	W0120	W0137	W1299	W1300	Jpn1	Jpn2	INDEL	INDEL Start Position in Nipponbare cp Genome	Forward	Reverse
SSR16	NA								-T	76574	CATAACTAAACCCTCGAAAGTAA	CCCGCCTATAGCGGTAATC
INDEL18	2	3	2	3	1	1	1	1	-T	77728	GCTACATTTAAAAGGGTCTGAGG	CTGCCAGCAAAATGCCC
SSR17	NA								-T	78423	-	-
INDEL19	1	1	1	1	2	2	2	2	+T	80090	GGGTTGTACCAAGTCTGAA	GCTCGAGGACGTAGTTCTCCCATAA
SSR18	NA								-C	93004	-	-
SSR19	1	1	1	1	2	2	2	2	+C	93534	GTTCGTCCTCAATGGGAAAATG	GGGAAGTCCTATTGATTGCTG
INDEL20	1	2	2	2	2	2	2	2	+AACA	104530	GATCATTTTCTGGCGTCAGCG	GAATATTGTACCGAGGAATTCG
INDEL21	1	1	1	1	2	2	2	2	+G	121618	AAGGCTCGAATGGTACGATC	CTTCTCGAGAATCCATACATCCC
SSR20	NA								-G	122142	-	-

NA: not amplified.

Five chloroplast markers, INDEL1, INDEL11, INDEL13, INDEL18, and INDEL19, represented polymorphisms among natural populations (Table S2 (Supplementary Materials)). Plastid types were defined as distinct combinations of each genotype. In total, nine plastid types (Type 1 to 9, r1, and r2) with r1 and r2 types in the control *O. rufipogon*, were recognized. Asian *O. rufipogon* and *O. sativa* accessions, were obviously different from the Australian wild rices.

Three accessions in PNG *O. rufipogon*, W1235, W1238, and W1239, and W2109 in Australian *O. rufipogon* shared the Type 5 plastid type with *O. meridionalis*. W1230 in Papua New Guinea *O. rufipogon* shared the r2 plastid type with the Asian type. W1236 carried a unique plastid type. Jpn2 shared Type 1 with *O. meridionalis*. Other *O. meridionalis* in the core collection divided into three types, Types 1, 5, and 8. Only two types, Types 1 and 2, were detected in the Northern Territory and in Western Australia. Newly collected accessions from Queensland carried seven types. Five of them were newly detected.

2.3. Reproductive Isolation

Biological species can be detected by the pollen fertility of hybrids. Jpn1 and Jpn2 were crossed with Asian wild rice and *O. meridionalis*. Each F_1 plant was grown in a greenhouse, and leaf samples were used to check whether they were hybrids originating from the cross. Anthers were taken to check pollen fertility by staining with I_2–KI. Well-stained pollen grains were counted.

Seed fertility was also assessed but this may not reflect reproductive ability of the respective plants (Table 4). W0106, W0120, and W1299 showed more than 95% pollen fertility. However, except for W0120, they showed lower seed fertility of 19.5% in W0106 and 22.3% in W1299. The panicles were bagged to prevent out-crossing and this might explain the low seed fertility. In combinations with Jpn1 and Asian *O. rufipogon*, F_1 plants with W0106 and W0120 had more than 90% pollen fertility. However, seed fertility was relatively low, similar to self-pollination of W0106 and W1299. We relied on data from pollen fertility rather than seed fertility and concluded that by this criterion, Jpn1 is related to Asian *O. rufipogon*, and that Jpn2 is not close to either *O. rufipogon* or *O. meridionalis*.

Table 4. Pollen and seed fertility of self pollinated plants and F_1 plants among Asian *O. rufipogon*, *O. meridionalis* and alternative perennials in Australia.

		Pollen and Seed Fertility (%)							
Female	**Male**	**Self Pollinated**		**Crossed with Jpn1**		**Crossed with Jpn2**		**Crossed with W1297**	
		Pollen	**Seed**	**Pollen**	**Seed**	**Pollen**	**Seed**	**Pollen**	**Seed**
W0106		97.2	30.1	87.2	15.7	10.2	0.0	0.1	0.0
		98.8	8.9	95.7	18.3	0.8	0.0	0.1	0.0
	Mean*	98.0	19.5	91.4	17.0	5.5	0.0	0.1	0.0
W0120		98.1	96.8	84.0	24.5	10.2	0.0	21.3	0.0
		98.7	74.5	78.3	21.6	12.7	0.0	20.2	0.0
	Mean	98.4	85.6	81.2	23.1	11.5	0.0	20.8	0.0
W1299		95.7	20.0	0.0	0.0	29.4	0.0	99.3	42.9
		96.1	24.5	4.5	0.0	39.8	0.3	-	-
	Mean	95.9	22.3	2.2	0.0	34.6	0.1	-	-
W1297		95.7	54.4	5.3	0.0	53.2	0.0	-	-
		96.9	75.9	6.4	0.0	33.8	0.0	-	-
	Mean	96.3	65.2	2.2	0.0	34.6	0.1	-	-

*Mean: data obtained from multiple plants were averaged and noted the mean.

2.4. Unique Insertion of Retrotransposable Element in Jpn2

In total, six presumed insertions were confirmed only in the Jpn2 genome but not in Nipponbare (Table 5). Two *pSINE1* insertions were shared among Jpn2 and 19 *O. meridionalis* accessions. Another insertion amplified with Chr3-10559212-r (w/L) and pSINE1-L showed an insertion shared among Jpn1, Jpn2, and 19 *O. meridionalis* accessions (Figure 2). Chr1-4067055-f (w/L) and pSINE1-L amplified the same amplicons not only from Jpn2 and 19 *O. meridionalis* accessions but also with W0106, which originated in India, suggesting that some parts of the Jpn2 genome share the insertion with wild rice from India. No *O. rufipogon* accessions in the core collection except for W2266 and W2267 were tested because of lack of DNA, and 19 *O. meridionalis* showed these insertions. Results suggested that the insertion was probably shared among *O. meridionalis* and W0106. Chr3-10203820-f (w/L) can amplify with pSINE1-L only in Jpn2 and no other *O. meridionalis* showed any amplicons. In screening for the insertion among 30 *O. rufipogon* accessions in the core collection, W0180 and W1921, both of which originated from Thailand, showed amplicons. The insertion sequence in Jpn2 was screened from the raw reads and 53 bp were recovered. When aligned with *pSINE1*, 92.4% high similarity was retained. When *pSINE3* insertions were examined, three of the presumed insertions were amplified only among Jpn2 and 19 *O. meridionalis* accessions.

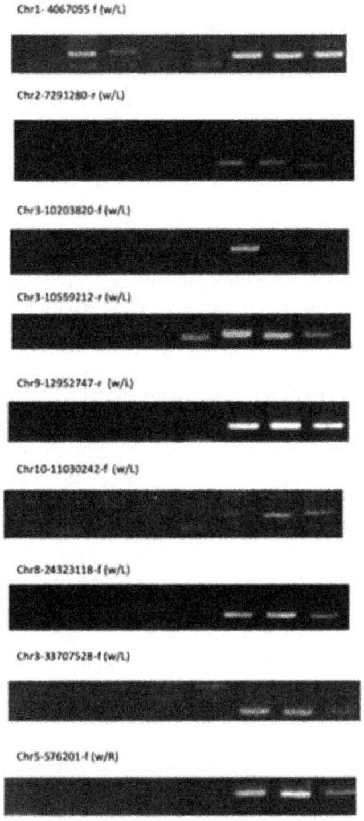

Figure 2. *pSINE1* and *pSINE3* insertions amplified with flanking primers and outward primers from *pSINE* consensus sequences. From lane 1 to 8, Nipponbare, W0106, W0120, W0137, Jpn1, Jpn2, W1299, and W1300 were used as each DNA template.

Table 5. Screening retrotransposable element insertions.

Flanking Primers to Confirm *pSINE1/3* Insertions		Genome	Insertion (+) / no-Insertion (-)								
with Inner Primers toward Outside	**Sequence**	**Position**	**Nipponbare**	**W0106**	**W0120**	**W0137**	**Jpn1**	**Jpn2**	**W1299**	**W1300**	**Remarks**
pSINE1											
Jpn2-meridionalis class											
Chr2-7291280-r (w/L)	TCTCTCTACAGATAATGCTC	7291535	-	-	-	-	-	+	+	+	Other *O. meridionalis*
Chr9-12952747-r (w/L)	CACACCCATCTACATCGATG	12953007	-	-	-	-	-	+	+	+	Other *O. meridionalis*
Chr10-11030242-f (w/L)	GATTGCCGGCTTCTTTACTAG	11029860									
Australia class											
Chr3-10559212-r (w/L)	ACCTATAACAACTGAGAGAC	10559538	-	-	-	-	+	+	+	+	Other *O. meridionalis*
Jpn2-meridionalis-W0106 class											
Chr1-4067055-f (w/ L)	GAAAGAGATCACAGGTAAAC	4066713	-	+	-	-	-	+	+	+	Other *O. meridionalis*
Jpn2-W0180,W1921 class											
Chr3-10203820-f (w/L)	TCCACCGACTTATAAATCAC	10203450	-	-	-	-	-	+	-	-	No other *O. meridionalis*, but two *O. rufipogon*,W0180, W1921
Inner primer toward outside											
pSINE1-L	GAAGACCCCTGGGCGTTTCT										
(paired primer to amplify *pSINE1* insertion)											
pSINE3											
Chr3-33707528-f (w/L)	GTGTAAATATGTATTGTACC	33702014	-	-	-	-	-	+	+	+	Other *O. meridionalis*
Chr8-24323118-f (w/L)	GCCTATTACTATCAATCACC	24322813	-	-	-	-	-	+	+	+	Other *O. meridionalis*
Chr5-576201-f (w/R)	GATAACTAGGGTAAATGAC	575868		-	-	-	-	+	+	+	Other *O. meridionalis*
Inner primer toward outside											
pSINE3-R	TCCTTCCTAGATTGGTCCC										
pSINE3-L	TGCTAGCCGGGAAGACC										
(paired primer to amplify *pSINE3* insertion)											

3. Discussion

3.1. Unique Morphological Traits in Australian Wild Rice

O. rufipogon is composed of a continuum of annual and perennial strains in Asia. They represent different life history traits related to the r-K strategy to maximize fitness [4,12]. Perennial and annual types are regarded as K- and r-strategists, respectively. Intermediates represented the r–K continuum. K selection works for individuals to increase their life span and r selection works to produce more offspring. Thus, perennials spend more energy on vegetative organs before the flowering stage. Annuals spend energy to produce more panicles and seeds. Because anther size is related to preference for outcrossing, perennials tend to carry longer anthers than annuals and produce more pollen to maximize the chance of outcrossing [3,4,12]. Such resource allocation was also confirmed in three Asian *O. rufipogon*, the Australian perennial Jpn1, and the Oceanian annual *O. meridionalis*. In order to adapt to the dry season, *O. meridionalis* plants produce many seeds and die after scattering their seeds. *O. meridionalis* has short anthers and slender panicles. The appearance of Jpn1 is similar to Asian *O. rufipogon*, with similar long anthers and open panicles. Our measurements also suggested a trend. In our previous paper, we reported that Jpn2 represents a perennial life history [14]. It generated shoots and roots from its stems to follow water in peripheral areas, growing to the inner side because of the shrinking water mass during the dry season. The morphological appearance of Jpn2 was quite different to Jpn1. Anthers length is a unique characteristic in the morphology of this type, being shorter in *O. meridionalis* [14]. *O. rufipogon* W0106, an annual type, also shared this short anther characteristic. In this study, we also demonstrated another morphological trait characteristic of Jpn2. Jpn2 has a high density of bristle cells along the awn. Genome sequencing also suggested that Jpn2 shared higher similarity to *O. meridionalis* than *O. rufipogon* [16]. These characteristics of Jpn2 infer that this species/taxon has diverged from *O. meridionalis*.

3.2. Maternal Variation

Cytoplasmic marker systems can be developed for the mitochondrial genome as suggested in this report. Other markers were also developed based on whole cp genome sequences. Whole cp genome sequences have been determined for several Australian accessions [15–19]. The maternal genome data clearly showed Jpn1 and Jpn2 shared high similarity to *O. meridionalis* with some variation. INDELs and SSRs were designed based on the cp genome sequences. Core collections and natural populations were examined to determine the distribution of cp variation. Higher variation was found among accessions in Queensland compared with others accessions collected from the Northern Territory and Western Australia. Variations in cp genomes were distinguished at high resolution using single nucleotide polymorphisms [17,18]. This study showed that easily scored INDELs and SSRs also detected higher diversity in the northern Queensland accessions. These marker systems with whole cp genome screening will provide more clues about the maternal relationships among these related species/taxa and how they diverged.

3.3. Reproductive Barriers Among Australian Wild Rice

Reproductive barriers among *Oryza species* including the Australian species have been confirmed and numerical characteristics supported speciation reproductive barriers [2]. *O. meridionalis* already has high genetic reproductive barriers and sterility detected in F_1 lines of crosses with Asian wild rice. Even among *O. meridionalis*, some sterile lines were reported [9]. In our study, F_1 between Jpn2 and *O. meridionalis* displayed reproductive sterility. Jpn2 in particular, developed a reproductive barrier with both Asian wild rice and *O. meridionalis*. Because the extent of the reproductive barrier corresponded to that of two different organisms, it is concluded that Jpn2 does not belong to *O. rufipogon* or *O. meridionalis*. We have not yet determined when they diverged from each other. Clade analysis of cp genomes suggested that a clade including *O. meridionalis* diverged at a date estimated as 0.86–11.99 million years ago [18]. Similar estimation has also been reported based on sequences among *Oryza* genus [19–22]. Such a long time since divergence has allowed the accumulation of quite diverse genomes in the north-eastern part of Australia and created Jpn2 and various wild rice found at the P5 site.

F_1 hybrids between Jpn2 and *O. meridionalis* showed relatively high pollen fertility with *O. meridionalis*, although the F_1 showed complete seed sterility. It was suggested that the divergence between these two plants is a relatively recent event compared with the divergence from other species.

3.4. Retrotransposable Elements

Retrotransposable elements are well known markers for examining evolutionary pathways. This is mainly due to the unique mechanisms of transpositions. Two retrotransposable elements, *pSINE1* and *pSINE3*, were recognized in species divergence among AA genome and between Asian wild rice and *O. meridionalis* [3,6,23]. These have offered researchers a powerful tool for phylogenetic analysis. On the other hand, when there are no genome sequences, new markers to distinguish particular genomes are not available. In fact, there was no genomic data on insertions in a novel taxon such as Jpn2. Thus, we established a screening methodology to extract retrotransposable elements from raw reads. Recent developments in sequencing technology offer huge numbers of reads to increase target sites. Even with our limited volume of data, we succeeded in picking up insertions in the Jpn2 genome. The uniqueness of Jpn2 was also found with an insertion of the *pSINE1* retrotransposon, which was detected in Jpn2 only and in none of the other accessions of *O. meridionalis*. Two accessions of *O. rufipogon* in Thailand may provide key information on how Australian wild rice originated. This tool will open the way to draw a more precise evolutionary pathway and to understand valuable genetic resources among wild rice.

4. Materials and Methods

4.1. Plant Materials

Wild rice was collected in Australia with permission from the Queensland government, EcoAccess. We developed these collections as de novo resources, which can be accessed repeatedly from the same site with accurate GPS data allowing us to reconfirm their life cycles. Successive observations were made from 2009 until 2011, and the life history traits at the collection sites were reconfirmed year by year. This field research was supported by overseas scientific research funds (JSPS) and collaborative research with the Queensland Herbarium and Queensland Alliance for Agriculture and Food Innovation (QAAFI), University of Queensland. Thirty populations were collected from their natural habitat. Observation of the ecological habitats and life cycle of each population in April 2008, August 2009, and September 2009 were used to determine their life history such as annual or perennial behavior especially for Jpn1 and Jpn2 populations. Jpn1 and Jpn2 were typical perennial sites and individuals survived as living plants in a swamp (Jpn1) or a pond (Jpn2). In order to compare these accessions with cultivated rice, *Oryza sativa*, and wild species, *O. rufipogon* (W0106, W0120, and W0137) and *O. meridionalis* (W1297, W1299 and W1300) were compared. All plant materials were grown in greenhouse conditions at Kagoshima University. Samples collected from nature were compared with Jpn1 and Jpn2 grown from seeds collected to compare environmental effects on anther length and lemma size. Jpn1 and Jpn2 were crossed with W0106, W0120, W1297, and W1299 to test these relationships. The density of bristle cells on the surface of awns per 100 μm^2 was counted using a scanning electric microscope (JSM-7000F, JEOL co., Japan). A core collection derived from the National Bio-Resource (NBR) Project in Japan was kindly provided, as shown in Table S1 (Supplementary Materials) [24].

4.2. Crossing and Fertility Test

Jpn1 and Jpn2 were crossed with Asian wild rice, W106 and W120, and to Australian *O. meridionalis*, W1299 and W1300. F_1 plants were grown in a greenhouse at Kagoshima University. Anthers were taken and stained with I_2–KI solution and well-stained grains were counted as fertile pollen. The remaining panicles were wrapped in paraffin bags to prevent outcrossing. Fully filled grains were counted to calculate seed fertility.

4.3. Data Mining from Whole Genome Sequences

Whole genome sequences of Jpn1 and Jpn2 were obtained using Illumina GAIIx to develop INDEL markers of the cpDNA and retrotransposon INDEL markers. The total numbers of pair-end reads were 52,087,744 for Jpn1 and 54,749,858 for Jpn2. Total nucleotides sequenced were 3.9 Gb for Jpn1 and 4.1 Gb for Jpn2. Mean depth was 2022 in Jpn1 and 2002.5 in Jpn2. With our draft data, we aligned these raw reads to the cpDNA of Nipponbare (GenBank: GU592207.1) using CLC-work bench genomics version 6.0. Several INDELs were grouped together to screen for using single PCR reactions with the designed markers listed in Table 5.

Retrotransposable elements *pSINE1* and *pSINE3* have been reported to be uniquely found in either species, *O. rufipogon* or *O. meridionalis* [7,8,23]. Consensus sequences of the 5′ and 3′ termini were used to design a consensus probe to screen the draft sequence data. Based on the alignment of *pSINE1* elements, consensus sequences were presumed [8]. Two probe sequences were applied for *pSINE1* to screen the data: CCA.CA.CTTGTGGAGCTAGCCGG, in which the periods indicate degenerate nucleotides, for the 5′ termini, and TAGGT.TTCCCTAATATTCGCG for the 3′ termini. These degenerate probes were applied to screen raw reads of Jpn1 and Jpn2. 5′ termini which were confirmed to carry AAGACCCCTGGGCATTTCTC as the complementary sequence. Then, internal sequences of the 5′ probe were obtained from the read to confirm whether it shares homology, ranging from 74% to 82%. In these cases, we adopted the outside of 5′ termini as flanking sequences of *pSINE1* insertions. 3′ termini carried TAG followed by a poly T stretch. We adopted the downstream sites as flanking sequences of *pSINE1* insertions.

Based on *pSINE3* family elements, *r3004*, *r3005*, *r3012*, and *r3024*, the 5′ terminal consensus probe GCCGGGAAGACCCCGGGCC was used to screen internal sequences. The internal sequences were used to design an internal probe, CTAGCTCAGCTTGTGCTA. In order to examine the flanking sequences of insertions, the consensus probe was applied. After confirming the 5′ end shared the 5′ terminus of *pSINE3*, the outside sequences from TTTCTC were regarded as pSINE3 insertions.

When multiple reads were obtained as single locations, we specified the genome position based on the Nipponbare genome and detected 14 and 16 insertions that did not overlap. Of these, 21 could be aligned to the Nipponbare genome without *pSINE1* insertions at the site. Flanking sequences in the Nipponbare genome were applied to design primers to amplify the presumed insertions of either *pSINE1* or *pSINE3*. Outward primers inside *pSINE1* or *pSINE3* were also designed as shown in Table 5. Preliminary screening was performed with Nipponbare, three *O. rufipogon*, W0106, W0120, and W0137, two *O. meridionalis*, W1299 and W1300, and Jpn1, and Jpn2.

4.4. Data Analysis

Dendrograms were constructed using the neighbor-joining method based on Nei's unbiased genetic distances by Populations1.2.30 beta2 program, which was downloaded from http://bioinformatics.org/~{}tryphon/populations/#ancre_bibliographie. All dendrograms were drawn by the TreeExplorer software used to show and edit population dendrograms as supplied with MEGA [25].

5. Conclusions

These data suggested that Jpn2 (taxon B) may be a distinct new species belonging to the *Oryza* genus and isolated from other species by reproductive barriers.

Supplementary Materials: The following are available online at http://www.mdpi.com/2223-7747/9/2/224/s1, Table S1: Lists of core collections in the NBRP (National Bio-resource Project). Table S2: plastid types among core collections and natural populations in Australia.

Author Contributions: Conceptualization, R.I., R.J.H., and K.I.; methodology, R.I.; software, R.I.; validation, D.T.L., R.I., R.J.H., and K.I.; formal analysis, D.T.L., R.J.H., K.I., and R.I.; investigation, D.T.L., R.J.H., K.I., and

R.I.; resources, R.I., R.J.H., K.I., and R.I.; writing—original draft preparation, R.I.; writing—review and editing, D.T.L., R.I., R.J.H., and K.I.; visualization, D.T.L. and R.I.; supervision, R.I.; project administration, R.I.; funding acquisition, R.I. All authors have read and agreed to the published version of the manuscript.

Funding: This research was funded by a Grant-in-aid B (Overseas project. No. 16H05777) and partly by a Grant-in-aid for Scientific Research on Innovative Areas (15H05968), partly by a Grant-in-aid for Scientific Research A (19H00542), and partly by a Grant-in-aid for Scientific Research A (19H00549).

Acknowledgments: We acknowledge Bryan Simon, who supported our mission in Australia.

Conflicts of Interest: The authors declare no conflicts of interest.

Abbreviations

Cp	chloroplast
INDEL	insertion/deletion
SSR	simple sequence repeat

References

1. Brar, D.S.; Khush, G.D. Wild relative of rice: A valuable genetic resources for genomics and breeding research. In *The Wild Oryza Genomes*; Mondal, T.K., Henry, R.J., Eds.; Springer International Publishing AG: Basel, Switzerland, 2018; pp. 1–25.
2. Morishima, H. Phenetic similarity and phylogenetic relationships among strains of *Oryza perennis*, estimated by methods of numerical taxonomy. *Evolution* **1969**, *23*, 429–443. [CrossRef]
3. Oka, H.-I.; Morishima, H. Variations in the breeding systems of a wild rice, *Oryza perennis*. *Evolution* **1967**, *21*, 249–258. [CrossRef]
4. Morishima, H.; Oka, H.I.; Chang, W.T. Directions of differentiation in populations of wild rice, *O. perennis* and O. sativa f. spontanea. *Evolution* **1961**, *15*, 326–339. [CrossRef]
5. Ng, N.Q.; Hawkes, J.G.; William, J.T.; Chang, T.T. The recognition of a new species of rice (Oryza) from Australia. *Bot. J. Linn Soc.* **1981**, *82*, 327–330. [CrossRef]
6. Henry, R.J.; Rice, N.; Waters, D.L.E.; Kasem, S.; Ishikawa, R.; Hao, Y.; Dillon, S.; Crayn, C.; Wing, R.; Vaughan, D. Australian Oryza: Utility and Conservation. *Rice* **2010**, *3*, 235–241. [CrossRef]
7. Cheng, C.; Tsuchimoto, S.; Ohtsubo, H.; Ohtsubo, E. Evolutionary relationships among rice species with AA genome based on SINE insertion analysis. *Genes Genet. Syt.* **2002**, *77*, 323–334. [CrossRef]
8. Xu, J.-H.; Osawa, I.; Tsuchimoto, S.; Ohtsubo, H.; Ohtsubo, E. Two new SINE elements, *p-SINE2* and *p-SINE3*, from rice. *Genes Genet. Syst.* **2005**, *80*, 161–171. [CrossRef]
9. Juliano, A.B.; Elizabeth, M.; Naredo, B.; Lu, B.-R.; Jackson, M.T. Genetic differentiation in *Oryza meridionalis* Ng based on Molecular and Crossability Analyses. *Genet. Resour. Crop Evol.* **2005**, *52*, 435–445. [CrossRef]
10. Elizabeth, M.; Naredo, B.; Juliano, A.B.; Lu, B.-R.; Jackson, M.T. Hybridization of AA genome rice species from Asia and Australia, I. Crosses and development of hybrids. *Genet. Resour. Crop Evol.* **1997**, *44*, 17–23.
11. Lu, B.-R.; Elizabeth, M.; Naredo, B.; Juliano, A.B.; Jackson, M.T. Hybridization of AA genome rice species from Asia and Australia II. Meiotic analysis of *Oryza meridionalis* and its hybrids. *Genet. Resour. Crop Evol.* **1997**, *44*, 25–31. [CrossRef]
12. Sano, Y.; Morishima, H. Variation in resources allocation and adaptive strategy of a wild rice, *Oryza perennis* Moench. *Bot. Gaz.* **1982**, *143*, 518–523. [CrossRef]
13. Vaughan, D.A. *The Wild Relatives of Rice-A Genetic Resources Handbook*; IRRI: Manila, Philippines, 1994; p. 64.
14. Sotowa, M.; Ootsuka, K.; Kobayashi, Y.; Hao, Y.; Tanaka, K.; Ichitani, K.; Flowers, J.M.; Purugganan, M.D.; Nakamura, I.; Sato, Y.-I.; et al. Molecular relationships between Australian annual wild rice, Oryza meridionalis, and two related perennial forms. *Rice* **2013**, *6*, 26. [CrossRef]
15. Waters, D.; Nock, C.J.; Ishikawa, R.; Rice, N.; Henry, R.J. Chloroplast genome sequence confirms distinctiness of Australian and Asian wild rice. *Ecol. Evol.* **2012**, *2*, 211–217. [CrossRef]
16. Brozynska, M.; Omar, E.S.; Furtado, A.; Crayn, D.; Simon, B.; Ishikawa, R.; Henry, R.J. Chloroplast Genome of Novel Rice Germplasm Identified in Northern Australia. *Trop. Plant Biol.* **2014**, *7*, 111–120. [CrossRef]
17. Brozynska, M.; Copetti, D.; Furtado, A.; Wing, R.; Crayn, D.; Fox, G.; Ishikawa, R.; Henry, R.J. Sequencing of Australian wild rice genomes reveals ancestral relationships with domesticated rice. *Plant Biot. J.* **2016**, *15*, 765–774.

18. Moner, A.M.; Furtado, A.; Chivers, I.; Fox, G.; Crayn, D.; Henry, R.J. Diversity and evolution of rice progenitors in Australia. *Ecol. Evol.* **2018**, *8*, 4360–4366. [CrossRef]
19. Wambugu, P.W.; Brozynska, M.; Furtado, A.; Waters, D.; Henry, R.J. Relationships of wild and domesticated rices (Oryza AA genome species) based upon whole chloroplast genome sequences. *Sci. Rep.* **2015**, *5*, 13957. [CrossRef]
20. Tang, L.; Zou, X.H.; Achoundong, G.; Potgieter, C.; Second, G.; Zhang, D.Y.; Ge, S. Phylogeny and biogeography of the rice tribe (*Oryzeae*): Evidence from combined analysis of 20 chloroplast fragments. *Mol. Phylogenetics Evol.* **2010**, *54*, 266–277. [CrossRef]
21. Kim, K.; Lee, S.C.; Lee, J.; Yu, Y.; Yang, K.; Choi, B.S.; Koh, H.-J.; Waminal, N.E.; Choi, H.-I.; Kim, N.-H.; et al. Complete chloroplast and ribosomal sequences for 30 accessions elucidate evolution of Oryza AA genome species. *Sci. Rep.* **2015**, *5*, 15655. [CrossRef]
22. Zhang, Q.-J.; Zhua, T.; Xiaa, E.-H.; Shia, C. Rapid diversification of five Oryza AA genomes associated with rice adaptation. *Proc. Natl. Acad. Sci. USA* **2014**, *46*, E4954–E4962. [CrossRef]
23. Xu, J.-H.; Kurta, N.; Akimoto, M.; Ohtsubo, H.; Ohtsubo, E. Identification and characterization of Australian wild strains of *Oryza meridionalis* and *Oryza rufipogon* by SINE insertion polymorphism. *Genes Genet. Syst.* **2005**, *80*, 129–134. [CrossRef]
24. Nonomura, K.I.; Morishima, H.; Miyabayashi, T.; Yamaki, S.; Eiguchi, M.; Kubo, T.; Kurata, N. The wild Oryza collection in National BioResource Project (NBRP) of Japan: History, biodiversity and utility. *Breed. Sci.* **2010**, *60*, 502–508. [CrossRef]
25. Kumar, S.; Dudley, J.; Nei, M.; Tamura, K. MEGA: A biologist-centric software for evolutionary analysis of DNA and protein sequences. *Brief Bioinform.* **2008**, *9*, 299–306. [CrossRef]

plants

MDPI

Article

Segregation Distortion Observed in the Progeny of Crosses Between *Oryza sativa* and *O. meridionalis* Caused by Abortion During Seed Development

Daiki Toyomoto [1], Masato Uemura [2], Satoru Taura [3], Tadashi Sato [4], Robert Henry [5], Ryuji Ishikawa [6] and Katsuyuki Ichitani [1,2,*]

1 United Graduate School of Agricultural Sciences, Kagoshima University, 1-21-24 Korimoto, Kagoshima, Kagoshima 890-0065, Japan
2 Faculty of Agriculture, Kagoshima University, 1-21-24 Korimoto, Kagoshima, Kagoshima 890-0065, Japan
3 Institute of Gene Research, Kagoshima University, 1-21-24 Korimoto, Kagoshima, Kagoshima 890-0065, Japan
4 Graduate School of Agricultural Science, Tohoku University, Sendai, Miyagi 980-8577, Japan
5 Queensland Alliance for Agriculture and Food Innovation, University of Queensland, Brisbane, Queensland 4072, Australia
6 Faculty of Agriculture and Life Science, Hirosaki University, Hirosaki, Aomori 036-8561, Japan
* Correspondence: ichitani@agri.kagoshima-u.ac.jp; Tel.: +8199-285-8547

Received: 31 August 2019; Accepted: 3 October 2019; Published: 8 October 2019

Abstract: Wild rice relatives having the same AA genome as domesticated rice (*Oryza sativa*) comprise the primary gene pool for rice genetic improvement. Among them, *O. meridionalis* and *O. rufipogon* are found in the northern part of Australia. Three Australian wild rice strains, Jpn1 (*O. rufipogon*), Jpn2, and W1297 (*O. meridionalis*), and one cultivated rice cultivar Taichung 65 (T65) were used in this study. A recurrent backcrossing strategy was adopted to produce chromosomal segment substitution lines (CSSLs) carrying chromosomal segments from wild relatives and used for trait evaluation and genetic analysis. The segregation of the DNA marker RM136 locus on chromosome 6 was found to be highly distorted, and a recessive lethal gene causing abortion at the seed developmental stage was shown to be located between two DNA markers, KGC6_10.09 and KGC6_22.19 on chromosome 6 of W1297. We name this gene as *SEED DEVELOPMENT 1* (gene symbol: *SDV1*). *O. sativa* is thought to share the functional dominant allele *Sdv1-s* (s for *sativa*), and *O. meridionalis* is thought to share the recessive abortive allele *sdv1-m* (m for *meridionalis*). Though carrying the *sdv1-m* allele, the *O. meridionalis* accessions can self-fertilize and bear seeds. We speculate that the *SDV1* gene may have been duplicated before the divergence between *O. meridionalis* and the other AA genome *Oryza* species, and that *O. meridionalis* has lost the function of the *SDV1* gene and has kept the function of another putative gene named *SDV2*.

Keywords: reproductive barrier; segregation distortion; abortion; wild rice; *O. meridionalis*; *O. sativa*; gene duplication

1. Introduction

Rice (*Oryza sativa*) is one of the most important staple crops in the world. It feeds about one-third of the world population. Wild rice relatives having the same AA genome as domesticated rice comprise the primary gene pool for rice genetic improvement and include the following species; *O. rufipogon*, *O. meridionalis*, *O. glumaepatula*, *O. barthii*, *O. longistaminata*. Another domesticated *Oryza* species *O. glaberrima* (African rice) also has an AA gemome, and contributes to rice improvement. Though there are several reproductive barriers among these species as described below, transfer of useful genes such as disease resistance gene from AA genome *Oryza* species to rice has been successful via hybridization.

O. meridionalis and *O. rufipogon* are found in the northern part of Australia [1]. *O. rufipogon* is inferred to be the direct progenitor of *O. sativa* [2], and widely distributed not only in Australia but also in South and South East Asia and New Guinea. On the other hand, the distribution of *O. meridionalis* is confined to the northern parts of Australia and Irian Jaya, Indonesia [1]. Molecular data provides support for the divergence of *O. meridionalis* from the other AA genome *Oryza* species [3–7]. This is reflected by low pollen fertility of the hybrids between *O. meridionalis* and the other AA genome species [8,9], with almost no progeny being produced from the selfing of the hybrids. To utilize the rice breeding potential of wild relatives of rice, a recurrent backcrossing strategy has been adopted to produce chromosomal segment substitution lines (CSSLs) carrying chromosomal segments from wild relatives of rice in the genetic background of cultivated rice [10–13]. Subsequent backcrossing with *O. sativa* as pollen parent was successful, because the F_1 plants between *O. sativa* and its wild relatives retained female fertility.

To elucidate the genetic potential for the improvement of cultivated rice using these wild species, we produced three kinds of CSSLs with different Australian wild rice strains in the same genetic background. As a model agronomic trait, we selected late-heading, because the wild rice strains in this study head later than the recurrent parent Taichung 65 by about 50 days, and heading-time is easily scored. We have succeeded in mapping the late-heading time genes from these wild rice strains (see below) and found a new genetic distortion phenomenon in the *Oryza*. In this study, we report the genetic mechanism of the new distortion phenomenon.

2. Results

2.1. Mapping of Photoperiod Sensitivity Gene

Three wild rice strains, Jpn1, Jpn2, and W1297, and one cultivated rice cultivar Taichung 65 (T65) were used in this study. We bred various CSSLs in a T65 genetic background incorporating the three Australian wild rice strains, W1297, Jpn1, and Jpn2, chromosomal segments by recurrent backcrossing (see Material and Methods). Hereafter, the backcrossing populations using Jpn1, Jpn2, and W1297 as donor parent are described as BCnFm (Jpn1), BCnFm (Jpn2), and BCnFm (W1297), respectively. "n" and "m" represent numbers of backcrossing and selfing, respectively. The frequency distributions of days to heading of the three BC_3F_2 populations are shown in Figure 1. All populations showed bimodal distributions. A total of 94 DNA markers covering the whole 12 chromosomes and showing polymorphism between T65 and the three wild rice strains were subjected to preliminary linkage analysis using bulked DNA from the three BC_3F_2 populations. Only one marker RM136, located 568 kbp away from a photoperiod sensitivity gene *HD1* [14], showed heterozygosity in all the bulk DNAs. Chi square values for the independence between genotypes of RM136 and days to heading (early and late heading divided by the dotted line in Figure 1) were 26.880, 81.073, and 86.693 for Jpn1, Jpn2, and W1297, respectively, all highly significant ($P < 0.0001$). These results suggest that the three strains from Australia carry photoperiod sensitive alleles of the *HD1* locus, because heterozygotes and homozygotes of these strains at the RM136 locus headed much later than the homozygotes of T65. This cultivar proved to carry a photoperiod insensitive allele at the *HD1*(= *Se1*) locus [15], which behaved as an early heading-time allele in a usual cropping season in Japan [16–18].

In the BC_3F_2 (W1297), the segregation of the RM136 locus was highly distorted: very few homozygotes of W1297 appeared. To check if this phenomenon was specific to the cross with W1297 as donor parent, and to evaluate this phenomenon more clearly under a more uniform genetic background, BC_4F_2 populations with the same cross combinations were subject to further study. As for W1297, the BC_3F_1 plants producing the BC_3F_2 population for the analysis was backcrossed again to produce BC_4F_1 plants. Among them, late flowering plants were selected to produce BC_4F_2 populations. As for Jpn1 and Jpn2, different BC_3F_1 plants from that producing the BC_3F_2 population for the above experiment were backcrossed to produce BC_4F_1 plants. Among them, late flowering plants were selected to produce BC_4F_2 populations.

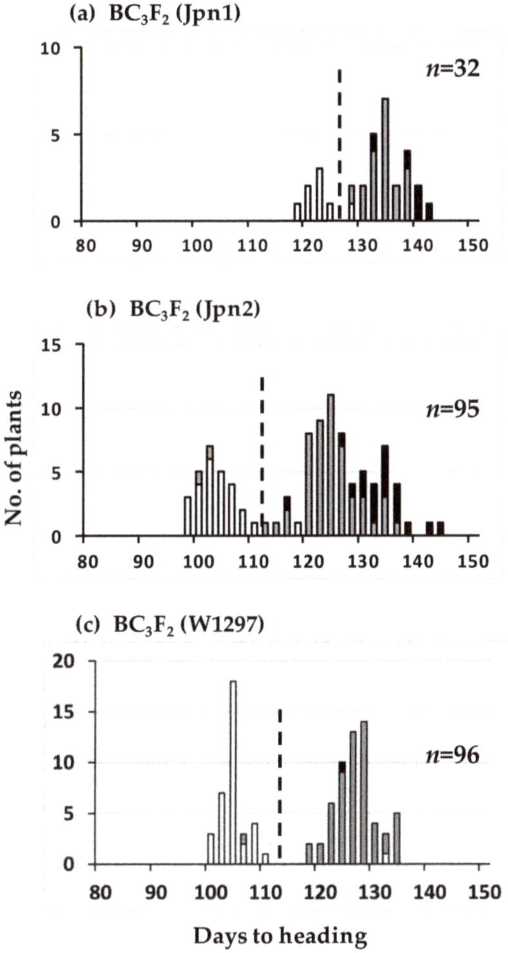

Figure 1. Frequency distributions of days to heading of the three BC_3F_2 populations using T65 as the recurrent parent. Jpn1, Jpn2, and W1297 were respectively used as donor parent in subfigure (**a**), (**b**), and (**c**). Three classified genotypes were assessed for RM136 as indicated: white, homozygous for T65, grey, heterozygous, black, homozygous for wild rice strains, Jpn1 (**a**), Jpn2 (**b**), and W1297 (**c**). Dotted lines dividing each population into early heading and late heading were drawn for chi-square analysis (see text).

2.2. Mapping of Segregation Distortion Gene

In the BC_4F_2 (W1297), the genotype of RM136 was distorted again (data not shown). In our preliminary experiment, among the published DNA markers around RM136, RM314 [19] located at 4,845kb, RM276 [19] at 6,231kb, RM7023 [20] at 6972kb, RM3628 [20] at 23,738kb and RM5314 [20] at 24,843kb on the IRGSP 1.0 pseudomolecule for chromosome 6 were fixed for the T65 allele. On the other hand, RM6818 and RM193 (Table 1) were segregating. These results suggest that the cause of segregation distortion is located between RM7023 and RM3628. Because other published DNA markers in our stocks failed in amplification of W1297 or did not distinguish T65 from W1297, we designed new DNA markers (Table 1), and performed linkage analysis. For the five consecutive markers from KGC6_12.02 to KGC6_19.48, only homozygotes of T65 and heterozygotes appeared (Figure 2), and no recombination occurred among the five markers (Table 2). The ratio of 64 homozygotes of T65: 119 heterozygotes fitted very well to 1:2 ($\chi^2(1:2) = 0.221$, $P = 0.638$).

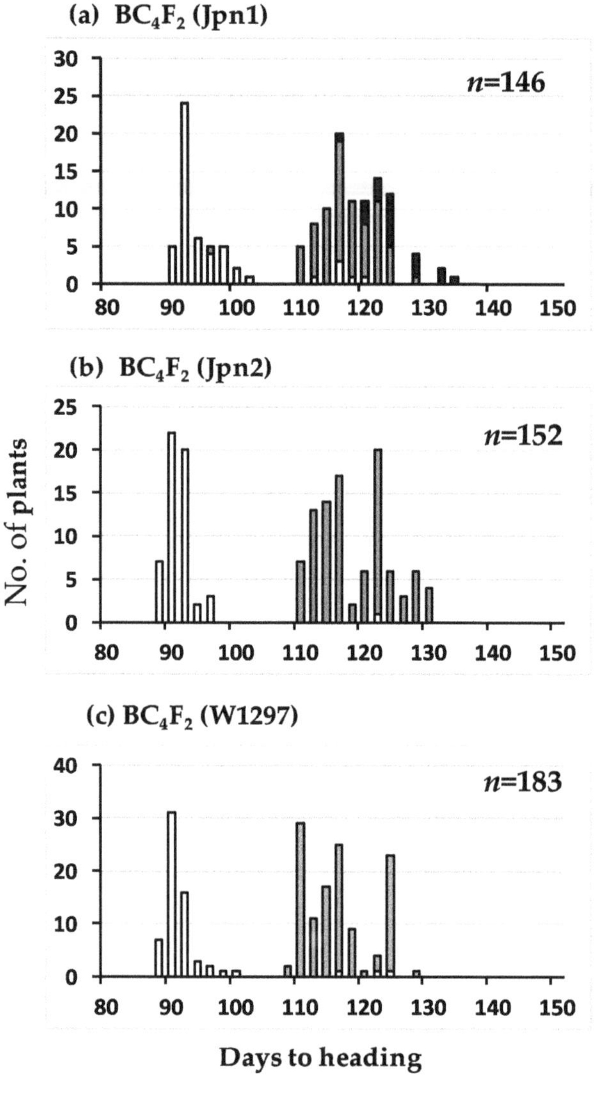

Figure 2. Frequency distributions of days to heading of the three BC_4F_2 populations using T65 as recurrent parent. Jpn1, Jpn2, and W1297 were respectively used as donor parent in subfigure (**a**), (**b**) and (**c**). Three classified genotypes were assessed for KGC6_12.02 as indicated: white, homozygous for T65, grey, heterozygous, black, homozygous for wild rice strains, Jpn1 (**a**), Jpn2 (**b**), and W1297 (**c**).

Table 1. Primer sequences of DNA markers designed or redesigned for linkage analysis of *SDV1* gene.

Marker Name	Kind of DNA Marker		Primer Sequence	Location on IRGSP 1.0 pseudomolecule chromosome 6		
				From	To	Source
KGC6_8.73	Indel	F	GAAGAGGAACATATGTGGTGTAAGC	8731826	8731914	This study
		R	AAAATTTATACTCTTGGTGACGTGA			
RM136	SSR	F	GAGAGCTCAGCTGCTGCCTCTAGC	8752461	8752562	Temnykh et al. [19]
		R	GAGGAGCGCCACGGTGTACGCC			
KGC6_8.82	Indel	F	TCTCTACCACACTCATCATCTGC	8820385	8820484	This study
		R	CCCTCGAGTAATAAACGATCCAG			
KGC6_10.09	Indel	F	TAGTCCTACGAAAACCCCTACTAGA	10090008	10090165	This study
		R	TTCCACGCACTAATACTACTACCTC			
KGC6_12.02	Indel	F	TTGATTTTGGGAAACATCAGGTAGC	12020476	12020625	This study
		R	AGCATGGTAATTTCATCGGATTCAA			
KGC6_13.00	Indel	F	CATTCGCATGGTAGCCTTTTCTTAT	13008017	13008169	This study
		R	CATAGGTGCCACAAGAGAAATCTTC			
RM6818	SSR	F	CGGCGAAGACTTGGAACCT	16582450	16582596	McCouch et al. [20]
		R	CCGTCACAAGGCTCGTCC			redesigned in this study
RM193	SSR	F	CAATCAACCAAACCGCGCTC	18086456	18086578	Temnykh et al. [19]
		R	CGCGGGCTTCTTCTCCTTC			redesigned in this study
KGC6_19.48	Indel	F	GAAGATAGTTAAGGGGTGTAGTGTGA	19483821	19484065	This study
		R	GACCAAAAGTTAAACAACATATTC TTCTAACCTAG			
KGC6_22.19	Indel	F	ACAAAATATGCTTTCTTCGTGCGTA	22191370	22191498	This study
		R	GCACTCAACTGTATCGTCTTTGAAA			

Table 2. Haplotypes around the segregation distortion region on rice chromosome 6 of BC_4F_2 (W1297).

Haplotype	Genotype of DNA Marker [1]								No. of Plants
	KGC6_8.73	KGC6_10.09	KGC6_12.02	KGC6_13.00	RM6818	RM193	KGC6_19.48	KGC6_22.19	
1	H	H	H	H	H	H	H	H	115
2	T	T	T	T	T	T	T	T	59
3	H	H	T	T	T	T	T	T	2
4	T	T	T	T	T	T	T	H	2
5	H	T	T	T	T	T	T	T	1
6	H	H	H	H	H	H	H	H	1
7	W	H	H	H	H	H	H	H	1
8	W	W	H	H	H	H	H	H	1
9	H	H	H	H	H	H	H	W	1
									183

[1] T, H, and W respectively denote homozygotes for T65, heterozygotes, and homozygotes for W1297.

The distorted segregation ratio 1:2:0 can be explained by one pair of recessive lethal genes. If the lethality occurred at the seedling stage, about 25% of seedlings would be expected to die. However, our visual observation did not fit with such a phenomenon. We then speculated that segregation distortion occurred during seed development. If so, seed fertility of the heterozygotes should be lower than that of the T65 homozygotes by about 25%. To test this, we examined the seed fertility of each of the BC_4F_2 plants. In the BC_4F_2 (W1297) population, the heterozygotes for KGC6_12.02 showed lower seed fertility than the homozygotes of the T65 allele (Figure 3). If lower fertility was caused by one recessive gene, many of the sterile seeds were expected to be aborted after fertilization. Therefore, sterile seeds were dehusked to see if sterility occurred before or after fertilization (Figure 4). The proportion of seeds aborted after fertilization for heterozygotes for KGC6_12.02 was higher than that for homozygotes of the T65 allele (Figure 5). These results suggested that homozygotes of the W1297 allele for KGC6_12.02 die at the seed developmental stage.

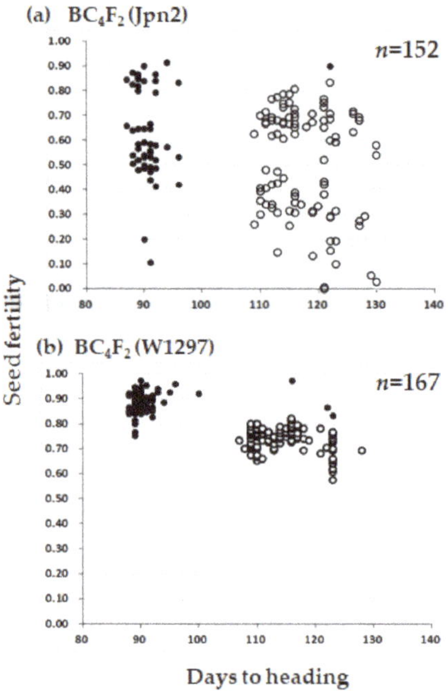

Figure 3. Scatter diagram of days to heading and seed fertility in the two BC_4F_2 populations using T65 as the recurrent parent. Jpn2 and W1297 were respectively used as donor parent in subfigure (**a**) and (**b**). Two classified genotypes were assessed for KGC6_12.02 as indicated: solid circle, homozygous for T65; open circle, heterozygous. In (**b**), plants used for testcross (Table 3) or damaged by birds in Figure 2 were removed in this figure.

Figure 4. Sterile seeds aborted after fertilization (top) and normal fertile seeds (bottom) found in the BC_4F_2 (W1297). One unit of the rightmost scale indicates 1 mm.

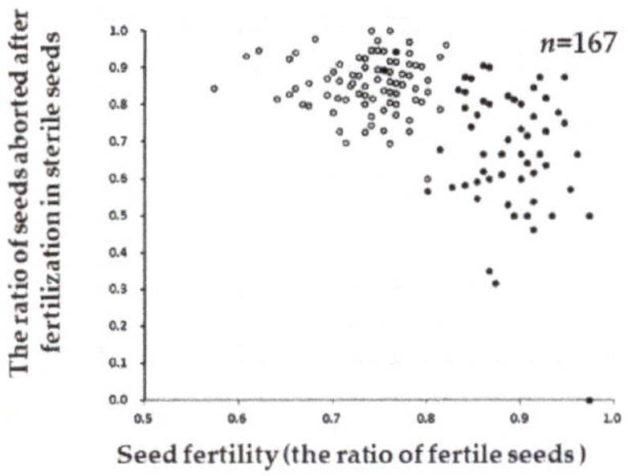

Figure 5. The scatter diagram between seed fertility (the ratio of fertile seeds) (X-axis) and the ratio of seeds aborted after fertilization in sterile seeds (Y-axis) in the BC_4F_2 population (W1297). Two classified genotypes were assessed for KGC6_12.02 as indicated: solid circle, homozygous for T65; open circle, heterozygous. Plants used for testcross (Table 3) or damaged by birds in Figure 2 were removed in this figure.

2.3. Segregation Distortion Caused by Abortion During Seed Development

To confirm this hypothesis, the following experiments were performed. First, pollen fertility was examined for all BC_4F_2 (W1297) plants, with the result that all plants showed more than 90% pollen fertility (data not shown), suggesting that pollen sterility was not the cause of the distorted segregation ratio. Second, reciprocal backcrossing of heterozygotes for KGC6_12.02 to T65 to produce a BC_5F_1 generation was undertaken. The BC_5F_1 from both cross combinations showed the segregation ratio fitted a 1 heterozygote:1 homozygote ratio for the T65 allele (Table 3), indicating that normal gene segregation occurred at both the egg and pollen developmental stage. The BC_4F_2 plants used for backcrossing were also selfed to produce a BC_4F_3 generation. DNA was extracted from the embryo of the fertile seeds. The segregation ratio was largely distorted from 1:2:1 at the KGC6_12.02 locus, and no homozygotes for the W1297 allele for KGC6_12.02 appeared, indicating that segregation distortion occurred during seed development, and was not caused by ungerminated fertile seeds, though the segregation ratio did not fit to a 1:2:0 ratio. (Table 3). The BC_4F_3 plants deriving from selfed seeds of the BC_4F_2 plants heterozygous for the KGC6_12.02 locus also showed distorted segregation, and the ratio fitted a 1: 2: 0 ratio, confirming the other experimental results (Table 3). Taken together, all the experimental results indicated that a recessive lethal gene causing abortion at the seed developmental stage was located between KGC6_10.09 and KGC6_22.19 on chromosome 6 of W1297 (Table 2).

Table 3. Segregation of progeny of BC_4F_2 (W1297) heterozygous for KGC6_12.02 genotype.

BC_4F_2	Genotype of the KGC6_12.02 [1]															
Individual Number	BC_4F_3										BC_5F_1					
	Plant					Embryo of Fertile Seeds					T65 as Pollen Parent			T65 as Egg Donor		
				P					*P*				*P*			*P*
	T	H	W	χ^2 (1:2:1)	χ^2 (1:2:0)	T	H	W	χ^2 (1:2:1)	χ^2 (1:2:0)	T	H	χ^2 (1:1)	T	H	χ^2 (1:1)
1	10	23	0	0.004	0.712	21	26	0	0.000	0.099	10	17	0.178	12	13	0.841
2	12	14	0	0.004	0.166	17	27	0	0.000	0.456	10	17	0.178	8	9	0.808
3	12	20	0	0.004	0.617	16	30	0	0.000	0.835	17	10	0.178	16	7	0.061
4	18	18	0	0.000	0.034	15	33	0	0.000	0.759	12	12	1.000	14	15	0.853
5	10	15	0	0.011	0.480	20	27	0	0.000	0.180	10	14	0.414	15	13	0.705
6	12	18	0	0.005	0.439	21	27	0	0.000	0.126	14	12	0.695	14	9	0.297
7	6	23	0	0.002	0.149	24	24	0	0.000	0.014	8	20	0.023	21	23	0.763
Sum	80	131	0	0.000	0.158	134	194	0	0.000	0.004	81	102	0.121	100	89	0.424

[1] T, H, and W respectively denote homozygotes for T65, heterozygotes, and homozygotes for W1297.

The same segregation distortion was also found in the BC_4F_2 (Jpn2) population (Figure 2). The ratio of 59 homozygotes of T65 allele: 97 heterozygotes at the KGC6_12.02 locus fitted very well to 1:2 (χ^2(1: 2) = 0.556, P = 0.456), and no homozygotes of Jpn2 allele appeared. The seed fertility of heterozygotes of KGC6_12 was lower than that of the homozygote of T65 allele, supporting the view that a recessive lethal gene causing abortion at seed developmental stage was located close to KGC6_12 of Jpn2 (Figure 3). The seed fertility of BC_4F_2 (Jpn2) was highly variable for both homozygotes of the T65 allele and heterozygote at the KGC6_12.02 locus, suggesting that other genetic factor(s) were involved in the large variance of seed fertility. Our preliminary results showed low pollen fertility might be responsible for low seed fertility of some plants (unpublished data). Therefore, the cause of the seed sterility was not investigated further. For Jpn1, both BC_3F_2 (Jpn1) and BC_4F_2(Jpn1) showed that normal gene segregation occurred around the *HD1* locus (Figures 1 and 2).

These results indicated that the two Australian *O. meridionalis* strains, W1297 and Jpn2, carry a recessive lethal gene causing abortion at the seed developmental stage, which was located between the two DNA markers, KGC6_10.09 and KGC6_22.19, spanning 12 Mb on chromosome 6, and that the Australian *O. rufipogon* strain Jpn1 does not carry such a gene.

3. Discussion

There have been many genes conferring hybrid seed sterility, hybrid pollen sterility, and segregation distortion found on *Oryza* chromosome 6 in inter-and intra-specific crosses, most of which *O. sativa* is involved with [21–27]. However, to our knowledge, the segregation distortion caused by seed abortion after fertilization has not been reported in the genus *Oryza*. We name this gene *SEED DEVELOPMENT 1* (gene symbol: *SDV1*), according to the gene nomenclature system for rice [28], because this gene is involved in the early seed developmental stage. In the intraspecific crosses among *O. sativa*, there have been no reports of gene distortion or partial seed sterility phenomena as described above on chromosome 6, though other phenomena have been reported [21–27]. Therefore, all *O. sativa* is thought to share the same functional dominant allele found in T65. This allele was called *Sdv1-s* (s for *sativa*). The homozygotes of the W1297 allele and the Jpn2 allele of this locus do not exist in the T65 genetic background probably because they die at the early seed development stage. W1297 and Jpn2 have originated from different places in Australia: W1297 is from Northern Territory, and Jpn2 is from Queensland. According to Juliano et al. [29], most crosses between Northern Territory and Queensland accessions produced sterile hybrids. Our preliminary results showed the hybrids from the reciprocal crosses between W1297 and Jpn2 were highly sterile (unpublished data). DNA marker-based analyses showed *O. meridionalis* genetic differentiation corresponding to geographic origin [29]. Further, in the CSSL lines of an *O. meridionalis* accession, W1625, chromosomal segments in a T65 genetic background [12], no lines were fixed for the W1625 chromosomal segment on which *SDV1* locus is located (https://shigen.nig.ac.jp/rice/Oryzabase/locale/change?lang=en). The results described above on the whole suggest that all *O. meridionalis* share the recessive abortive allele. This allele was called *sdv1-m* (m for *meridionalis*).

Though carrying the recessive abortive allele in homozygous form at the *SDV1* locus, the *O. meridionalis* accessions can self-fertilize and bear seeds. We speculate that the *SDV1* gene may have been duplicated before the divergence between *O. meridionalis* and the other AA genome *Oryza* species, and that *O. meridionalis* has lost the function of the *SDV1* gene and has kept the function of the other gene while *O. sativa* kept the function of the *SDV1* gene and has lost the function of the other gene (Figure 6). Such duplication and loss of reproductive barrier-related genes has been reported; Yamagata et al. [30] found that the reciprocal loss of duplicated genes encoding mitochondrial ribosomal protein L27, essential for the later stage of pollen development, causes hybrid pollen sterility in F_1 hybrid between *O. sativa* and *O. glumaepatula*. Nguyen et al. [31] reported that the duplication and loss of function of genes encoding RNA polymerase III subunit C4 hybrid causes pollen sterility in F_1 hybrid between *O. sativa* and *O. nivara* (annual form of *O. rufipogon*). Ichitani et al. [32] performed linkage analysis of hybrid chlorosis genes in rice, and found that the causal recessive genes *hca1-1* and *hca2-1* are located on the distal region of the short arm of chromosome 12 and 11, respectively, known to be highly conserved as a duplicated chromosomal segment.

There are other models explaining the hybrid incompatibility (abortion, lethality, or weakness) known as the Bateson–Dobzhansky–Muller (BDM) model (for a review, Bomblies et al. [33]). In the incompatibility caused by the two nonallelic dominant genes, if the one locus is heterozygote or fixed for the incompatibility-causing allele, both heterozygotes and homozygotes of the incompatibility-causing allele of the other locus should show incompatibility. If the one locus is fixed for the normal allele, incompatibility does not occur. In the incompatibility caused by the heterozygote on one locus, only heterozygotes should show incompatibility. Therefore, these models cannot explain the segregation distortion in this study. In the hybrid breakdown model proposed by Oka [34], the combination of the heterozygotes or the homozygotes of recessive alleles at one locus and the homozygotes of recessive alleles at the other locus show incompatibility. This model cannot explain the segregation in this study either. Therefore, the gene duplication model as described above fits the phenomenon in this study best.

As the counterpart of *SDV1*, "the other" putative gene is named *SEED DEVELOPMENT 2* (gene symbol: *SDV2*). *SDV1* and *SDV2* are thought to be derived from duplication. *O. meridionalis* accessions and *O. sativa* accessions should carry the functional allele and the unfunctional allele at the *SDV2* locus, respectively. We name these respective alleles *Sdv2-m* and *sdv2-s*. The presence and chromosomal location of *SDV2* have not been elucidated. We are undertaking the genetic analysis of *SDV2*, tracing back to earlier backcrossing populations.

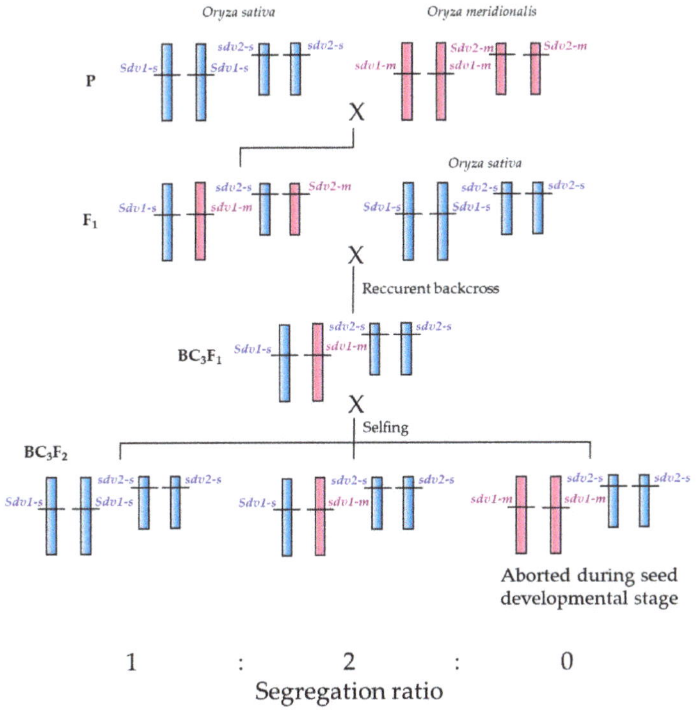

Figure 6. A genetic model that explains segregation distortion and seed sterility assuming gene duplication and loss of gene function for seed development.

If useful genes of *O. meridionalis* for rice genetic improvement are located close to *sdv1-m*, the introgression of these genes into *O. sativa* genetic background should be combined with *Sdv2-m*. Therefore, the chromosomal location of *SDV2* and tightly linked DNA markers to it are urgently needed.

In the frequently cited high-density rice genetic linkage map by Harushima et al. [35], the centromeric region of chromosome six is located between 64.7 cM and 65.7 cM. Some of the DNA marker sequences located on the centromeric region are available in NCBI (https://www.ncbi.nlm.nih.gov). C574 (accession name: D15395) is located at 13,685 kb, and G294 (accession name: D14774) is located at 17,056 kb in Nipponbare genome (Os-Nipponbare-Reference-IRGSP-1.0). Therefore the physical size of the centromeric region is at least 3371 kb. The candidate chromosomal region of *SDV1* encompasses this region (Tables 1 and 2). Recombination events were, in general, highly suppressed around the centromere. Our result is consistent with that. The combination of high resolution linkage analysis with gene expression analysis, gene disruption, and association study will be necessary to identify the *SDV1* gene.

Seed development is dissected into embryogenesis and endosperm development. We are undertaking microscopic observation of seed development of *Sdv1-s sdv1-m* heterozygotes in the T65 background to define the cause of the seed abortion. Several genes required for embryogenesis and endosperm development have been reported [36–39]. Identification of the *SDV1* and *SDV2* genes will contribute to the molecular genetics of seed development.

Direct evidence supporting the gene model was that the DNA from the embryo of aborted seeds deriving from the heterozygote of KGC6_1202 was homozygous for the W1297 allele. In our preliminary experiments, we tried to extract DNA from them, modifying the method below so that the DNA concentration would be higher. A few embryos were homozygous for the W1297 allele. However, PCR failed in most cases. This suggests that embryogenesis stops at an early stage in the homozygotes of *sdv1-m*. One alternative approach might be to extract DNA from developing seeds, not from mature seeds. Combination of microscopic observation of developing embryo and DNA genotyping will contribute to understanding the abortion mechanism caused by the *sdv1-m* gene.

According to the chloroplast genome analyses by Wambugu et al. [40], Yin et al. [41] and Sotowa et al. [42], *O. rufipogon* in Australia carries a chloroplast genome similar to that of *O. meridionalis* rather than that of *O. rufipogon* in Asia and *O. sativa*, probably because of chloroplast capture (introgression). During the process of chloroplast capture, some nuclear genome genes could be shared by *O. rufipogon* in Australia and *O. meridionalis*. However, they might carry distinct alleles on *SDV1* and *SDV2* loci. When the DNA sequences of these alleles of the two loci are uncovered, this information can be applied for the analysis of plants growing in the wild, and possible ongoing hybridization between *O. rufipogon* and *O. meridionalis* can be monitored in the Northern part of Australia, in which the two species are sympatric. Hybrids have been found in the wild and confirmed by molecular analysis [43], but the low frequency of these hybrids and the continued existence of the two distinct AA genome taxa in the northern Australian environment may be explained by these genes that create a reproductive barrier.

4. Materials and Methods

4.1. Plant Material

Three wild rice strains, Jpn1, Jpn2, and W1297, and one cultivated rice cultivar Taichung 65 (T65) were used in this study. W1297 is a strain of *O. meridionalis* collected in Darwin, Northern Territory, Australia, and provided by National Institute of Genetics, Mishima, Japan. Jpn1 and Jpn2 were collected in Australia with the permission from the Queensland government, under the EcoAccess program [42]. Judging from its perennial life history, typical of Australian *O. rufipogon*, and Indel marker genotypes, Jpn1 was classified as *O. rufipogon*. The Australian *O. rufipogon* population at the Jpn1 site has been shown to have a chloroplast similar to that of *O. meridionalis* and a nuclear genome closer to *O. rufipogon* [44] suggesting it may need to be considered as a distinct taxon. Jpn2 was distinct with a short anther, typical of *O. meridionalis*, and perennial life history in its habitat in Queensland, Australia [42]. *O. meridionalis* is now described as including both annual and perennial types [45]. It had Indel marker genotypes that were the same as 18 *O. meridionalis* Core collection accessions. It was treated as a type of *O. meridionalis* based on five Indel DNA markers that reflect varietal differentiation in comparisons, such as Indica–Japonica, temperate Japonica–tropical Japonica with high accuracy [46,47]. Our visual observations indicated that the three wild rice strains each showed a uniform phenotype in the first growing year, suggesting that they had been fixed for at least the loci controlling agronomic traits. Before anthesis, the panicles of the wild rice strains were covered with bags made of glassine paper to force self-fertilization in every generation. The selfed progeny also showed uniform phenotypes. The preliminary analysis of DNA markers covering the whole 12 chromosomes indicated that they were homozygous at all the DNA marker loci. T65 is a Japonica cultivar used frequently in the study of rice genetics, as a recurrent parent of CSSLs, isogenic lines, and the study of induced mutation [12,21,48].

We bred CSSLs in a T65 genetic background incorporating the three Australian wild rice strains, W1297, Jpn1, and Jpn2, chromosomal segments by recurrent backcrossing. First T65 was crossed with W1297, Jpn1, and Jpn2 as pollen parents. One plant per each wild rice strain was used for producing the F_1 generation. Then the F_1 was backcrossed with T65 as a pollen parent in all subsequent backcross generations with some exception described above. A total of 39, 43, and 33 BC_1F_1 plants were obtained using W1297, Jpn1, and Jpn2 as donor parent, respectively. All the BC_1F_1 plants were backcrossed with T65. One BC_2F_1 plant originating from each BC_1F_1 plant was backcrossed with T65 to produce BC_3F_1. One BC_3F_1 plant originating from each BC_2F_1 plant was backcrossed with T65 to produce BC_4F_1. W1297, Jpn1, and Jpn2 have many characters different from T65, such as late heading, red pericarp, long awn, and easy shattering. Some BC_3F_1 plants had such characteristics in T65 genetic background, which was suitable for genetic dissection of these characters. As a model character, late heading was selected. We selected the latest BC_3F_1 plants and collected seeds from these plants to produce a BC_3F_2 generation. As shown above, because the segregation of genes conferring days to heading did not fit the expected Mendelian single gene segregation, we focused on the analysis of the distorted segregation. The BC_4F_2 generations deriving from the late heading BC_4F_1 plants were also examined.

Plant cultivation followed Ichitani et al. [48]. Germinated seeds were sown in nursery beds in a greenhouse. About two weeks after sowing, seedlings were transferred out of the greenhouse. About 30 days after the sowing date, seedlings were planted in a paddy field at the Experimental Farm of Kagoshima University, Kagoshima, Japan. The fertilizers applied were 4, 6, and 5 g/m^2, respectively, for N, K_2O, and P_2O_5. Plant spacing was 15 × 30 cm. Sowing and transplanting were done respectively on May 31 and June 24 in 2015, on May 27and June 28 in 2016, May 25 and July 4 in 2017, respectively. Hybridization was performed as follows: For emasculation, panicles of the egg donor were soaked in hot water at 43 °C for 7 min. For pollination, the upper half of the open spikelets were cut about 30 min after emasculation. All the closed spikelets were cut off. Then pollen of the pollen donor was scattered on them. After pollination, panicles were covered with bags made of glassine paper. At least one panicle was left without pollination to check whether emasculation was complete.

4.2. Trait Evaluation

Heading date was recorded for each plant when the first developing panicle emerged from the leaf sheath of the flag leaf. Heading date was converted into days to heading. Seed fertility was evaluated by collecting 50 seeds from the upper side of each of the three panicles, using a modification of the method of Wan and Ikehashi [49], counting fertile and sterile spikelets on the upper half of 3–4 panicles for each plant. Seeds were scored as fertile or sterile. In the W1297 cross, sterile seeds were dehusked to see if sterility occurred before or after fertilization. The BC_4F_2 plants that produced the BC_5F_1 generation were dug up, and transferred from the paddy field to a glass house a day before pollination. We empirically know that rice plants undergoing such a treatment show lower seed fertility, probably because of root damage. Therefore, we did not evaluate seed fertility of these plants. Panicles of some plants were damaged by birds after heading. This is the reason for the inconsistency in BC_4F_2 plant number among tables and figures.

Pollen fertility of the BC_4F_2 (W1297) population was evaluated using iodine-potassium iodide solution. Panicles were collected about three days after emerging from the leaf sheath of the flag leaf, and dried in paper bags at room temperature. All the anthers in a spikelet collected one day before anthesis were cleaved to gather pollen on a glass slide. Pollen were stained with iodine-potassium iodide solution. More than 200 pollen grains were scored for each individual. Densely stained pollen with a normal size were scored as fertile. The other pollen were scored as sterile.

4.3. DNA Analysis

DNA from leaves and embryo from fertile seeds was extracted according to Ichitani et al. [48] with some modifications: Each leaf tip, 2.5 cm long from a single plant, or embryo from dehusked seeds was put in a well of a 96-deep-well plate. Then 100 μL of extraction buffer (100 mM Tris–HCl (pH 8.0), 1 M KCl, and 10 mM EDTA) was added with a 5-mm-diameter stainless steel ball to the well. After being covered with a hard lid, the plate was shaken hard (ShakeMaster ver. 1.2; BioMedical Science Inc., Tokyo, Japan) for 1 min to grind the leaves or embryos. After centrifuging, the plate was incubated at 70 °C for half an hour, then at room temperature for half an hour. Then 10 μL of the supernatant was recovered and 8 μL of 2-propanol was added. After centrifuging, the supernatant was discarded and the DNA pellet was rinsed with 50 μL of 70% ethanol. The DNA pellet was dried and dissolved in 50 μL of sterilized distilled water. It was very difficult to separate the embryo from the other part of seed completely. However, our preliminary experiment indicated that even if DNA was extracted from the whole dehusked seeds produced by a heterozygote for a DNA marker such as KGC6_12.02 (Table 1), DNA marker segregation was observed, suggesting that DNA from the parts of the dehusked seed other than the embryo was negligible. PCR mixture, cycle, electrophoresis, DNA staining, gel image documentation also followed Ichitani et al. [48].

4.4. DNA Markers

Most published PCR-based DNA markers for *Oryza* are based on an *O. sativa* genome sequence such as Nipponbare (Os-Nipponbare-Reference-IRGSP-1.0, [50]) and 9311 (GCA_0000046551, [51]). However, a preliminary survey comparing the genome of Nipponbare (IRGSP 1.0) and that of *O. meridionalis* accession (GCA_000338895.2. [7]) showed that there were many discrepancies between them, leading to expected failure in amplification from the *O. meridionalis* genome when using *O. sativa* genome-based DNA markers. Our strategy of designing co-dominant DNA markers was that insertion/deletion (indel) polymorphisms ranging from 5 to 100 base pairs were searched for between the Nipponbare and the *O. meridionalis* genomes. Then, the indels found only between *O. meridionalis* and Nipponbare, not between *meridionalis* and two Indica cultivars, 93-11 and HR-12 (GCA_000725085), were selected. The event causing such indels were thought to have occurred in Japonica rice after Japonica-Indica differentiation. T65 is a typical Japonica cultivar. Our preliminary survey showed that T65 shared the banding patterns of Nipponbare in most of the DNA markers examined [52]. Therefore, the indels as described above were expected to show polymorphism between T65 and *O. meridionalis*. The selected indels were screened based on sequence similarity surrounding indels between Nipponbare and the *O. meridionalis* genomes. The primer design followed Busung et al. [53].

Author Contributions: Conceptualization, K.I.; methodology, K.I.; validation, D.T. and M.U.; formal analysis, D.T., M.U., and K.I.; investigation, D.T. and M.U.; resources, S.T., T.S., R.H., R.I., K.I.; data curation, D.T., M.U., K.I.; writing—original draft preparation, D.T. and K.I.; writing—review and editing, D.T., M.U., S.T., T.S., R.H., R.I. and K.I.; visualization, D.T., M.U., and K.I.; supervision, K.I.; project administration, R.I.; funding acquisition, R.I.

Funding: This research was funded by JSPS KAKENHI Grant Number JP16H05777 from the Japan Society for the Promotion of Science.

Acknowledgments: We are grateful to the National Institute of Genetics for their kind provision of W1297 seeds. We thank Mr. Masaaki Ikenoue, Mr. Nishiobino Tsubasa, Ms. Yoko Nakashima and Ms. Asako Kobai for their technical assistance.

Conflicts of Interest: The authors declare no conflict of interest.

References

1. Henry, R.J.; Rice, N.; Waters, D.L.E.; Kasem, S.; Ishikawa, R.; Hao, Y.; Dillon, S.; Crayn, D.; Wing, R.; Vaughan, D. Australian *Oryza*: Utility and conservation. *Rice* **2010**, *3*, 235–241. [CrossRef]

2. Huang, X.; Kurata, N.; Wei, X.; Wang, Z.-X.; Wang, A.; Zhao, Q.; Zhao, Y.; Liu, K.; Lu, H.; Li, W.; et al. A map of rice genome variation reveals the origin of cultivated rice. *Nature* **2012**, *490*, 497–501. [CrossRef] [PubMed]
3. Ohtsubo, H.; Cheng, C.; Ohsawa, I.; Tsuchimoto, S.; Ohtsubo, E. Rice retroposon *p-SINE1* and origin of cultivated rice. *Breed. Sci.* **2004**, *54*, 1–11. [CrossRef]
4. Cheng, C.; Tsuchimoto, S.; Ohtsubo, H.; Ohtsubo, E. Evolutionary relationships among rice species with AA genome based on SINE insertion analysis. *Genes Genet. Syst.* **2002**, *77*, 323–334. [CrossRef] [PubMed]
5. Xu, J.-H.; Kurata, N.; Akimoto, M.; Ohtsubo, H.; Ohtsubo, E. Identification and characterization of Australian wild rice strains of *Oryza meridionalis* and *Oryza rufipogon* by SINE insertion polymorphism. *Genes Genet. Syst.* **2005**, *80*, 129–134. [CrossRef]
6. Zhang, Q.-J.; Zhu, T.; Xia, E.-H.; Shi, C.; Liu, Y.-L.; Zhang, Y.; Liu, Y.; Jiang, W.-K.; Zhao, Y.-J.; Mao, S.-Y.; et al. Rapid diversification of five *Oryza* AA genomes associated with rice adaptation. *Proc. Natl. Acad. Sci. USA* **2014**, *111*, E4954–E4962. [CrossRef] [PubMed]
7. Stein, J.C.; Yu, Y.; Copetti, D.; Zwickl, D.J.; Zhang, L.; Zhang, C.; Chougule, K.; Gao, D.; Iwata, A.; Goicoechea, J.L.; et al. Genomes of 13 domesticated and wild rice relatives highlight genetic conservation, turnover and innovation across the genus *Oryza*. *Nat. Genet.* **2018**, *50*, 285–296. [CrossRef] [PubMed]
8. Naredo, M.E.B.; Juliano, A.B.; Lu, B.-R.; Jackson, M.T. Hybridization of AA genome rice species from Asia and Australia I. Crosses and development of hybrids. *Genet. Resour. Crop Evol.* **1997**, *44*, 17–23. [CrossRef]
9. Naredo, M.E.B.; Juliano, A.B.; Lu, B.-R.; Jackson, M.T. Taxonomic status of *Oryza glumaepatula* Steud. II. Hybridization between New World diploids and AA genome species from Asia and Australia. *Genet. Resour. Crop Evol.* **1998**, *45*, 205–214. [CrossRef]
10. Shim, R.A.; Angeles, E.R.; Ashikari, M.; Takashi, T. Development and evaluation of *Oryza glaberrima* Steud. chromosome segment substitution lines (CSSLs) in the background of *O. sativa* L. cv. Koshihikari. *Breed. Sci.* **2010**, *60*, 613–619. [CrossRef]
11. Hirabayashi, H.; Sato, H.; Nonoue, Y.; Kuno-Takemoto, Y.; Takeuchi, Y.; Kato, H.; Nemoto, H.; Ogawa, T.; Yano, M.; Imbe, T.; et al. Development of introgression lines derived from *Oryza rufipogon* and *O. glumaepatula* in the genetic background of *japonica* cultivated rice (*O. sativa* L.) and evaluation of resistance to rice blast. *Breed. Sci.* **2010**, *60*, 604–612. [CrossRef]
12. Yoshimura, A.; Nagayama, H.; Sobrizal; Kurakazu, T.; Sanchez, P.L.; Doi, K.; Yamagata, Y.; Yasui, H. Introgression lines of rice (*Oryza sativa* L.) carrying a donor genome from the wild species, *O. glumaepatula* Steud. and *O. meridionalis* Ng. *Breed. Sci.* **2010**, *60*, 597–603. [CrossRef]
13. Arbelaez, J.D.; Moreno, L.T.; Singh, N.; Tung, C.-W.; Maron, L.G.; Ospina, Y.; Martinez, C.P.; Grenier, C.; Lorieux, M.; McCouch, S. Development and GBS-genotyping of introgression lines (ILs) using two wild species of rice, *O. meridionalis* and *O. rufipogon*, in a common recurrent parent, *O. sativa* cv. Curinga. *Mol. Breed.* **2015**, *35*, 81. [CrossRef] [PubMed]
14. Yano, M.; Katayose, Y.; Ashikari, M.; Yamanouchi, U.; Monna, L.; Fuse, T.; Baba, T.; Yamamoto, K.; Umehara, Y.; Nagamura, Y.; et al. *Hd1*, a major photoperiod sensitivity quantitative trait locus in rice, is closely related to the arabidopsis flowering time gene *CONSTANS*. *Plant Cell* **2000**, *12*, 2473–2483. [CrossRef]
15. Doi, K.; Izawa, T.; Fuse, T.; Yamanouchi, U.; Kubo, T.; Shimatani, Z.; Yano, M.; Yoshimura, A. *Ehd1*, a B-type response regulator in rice, confers short-day promotion of flowering and controls *FT-like* gene expression independently of *Hd1*. *Genes Dev.* **2004**, *18*, 926–936. [CrossRef] [PubMed]
16. Ichitani, K.; Okumoto, Y.; Tanisaka, T. photoperiod sensitivity gene of *Se-1* locus found in photoperiod insensitive rice cultivars of the northern limit region of rice cultivation. *Jpn. J. Breed.* **1997**, *47*, 145–152. [CrossRef]
17. Ichitani, K.; Okumoto, Y.; Tanisaka, T. Genetic analysis of the rice cultivar Kasalath with special reference to two photoperiod sensitivity loci, E_1 and *Se-1*. *Jpn. J. Breed.* **1998**, *48*, 51–57. [CrossRef]
18. Inoue, H.; Nishida, H.; Okumoto, Y.; Tanisaka, T. Identification of an early heading time gene found in the taiwanese rice cultivar Taichung 65. *Jpn. J. Breed.* **1998**, *48*, 103–108. [CrossRef]
19. Temnykh, S.; Park, W.D.; Ayres, N.; Cartinhour, S.; Hauck, N.; Lipovich, L.; Cho, Y.G.; Ishii, T.; McCouch, S.R. Mapping and genome organization of microsatellite sequences in rice (*Oryza sativa* L.). *Theor. Appl. Genet.* **2000**, *100*, 697–712. [CrossRef]
20. McCouch, S.R.; Teytelman, L.; Xu, Y.; Lobos, K.B.; Clare, K.; Walton, M.; Fu, B.; Maghirang, R.; Li, Z.; Xing, Y.; et al. Development and Mapping of 2240 New SSR Markers for Rice (*Oryza sativa* L.). *DNA Res.* **2002**, *9*, 199–207. [CrossRef]

21. Sano, Y. Genetic comparisons of chromosome 6 between wild and cultivated rice. *Jpn. J. Breed.* **1992**, *42*, 561–572. [CrossRef]
22. Matsubara, K.; Khin-Thidar; Sano, Y. A gene block causing cross-incompatibility hidden in wild and cultivated rice. *Genetics* **2003**, *165*, 343–352. [PubMed]
23. Koide, Y.; Ogino, A.; Yoshikawa, T.; Kitashima, Y.; Saito, N.; Kanaoka, Y.; Onishi, K.; Yoshitake, Y.; Tsukiyama, T.; Saito, H.; et al. Lineage-specific gene acquisition or loss is involved in interspecific hybrid sterility in rice. *Proc. Natl. Acad. Sci. USA* **2018**, *115*, E1955–E1962. [CrossRef] [PubMed]
24. Yanagihara, S.; McCouch, S.R.; Ishikawa, K.; Ogi, Y.; Maruyama, K.; Ikehashi, H. Molecular analysis of the inheritance of the *S-5* locus, conferring wide compatibility in Indica/Japonica hybrids of rice (*O. sativa* L.). *Theoret. Appl. Genet.* **1995**, *90*, 182–188. [CrossRef]
25. Liu, Y.S.; Zhu, L.H.; Sun, J.S.; Chen, Y. Mapping QTLs for defective female gametophyte development in an inter-subspecific cross in *Oryza Sativa* L. *Theor. Appl. Genet.* **2001**, *102*, 1243–1251. [CrossRef]
26. Yang, J.; Zhao, X.; Cheng, K.; Du, H.; Ouyang, Y.; Chen, J.; Qiu, S.; Huang, J.; Jiang, Y.; Jiang, L.; et al. A killer-protector system regulates both hybrid sterility and segregation distortion in rice. *Science* **2012**, *337*, 1336–1340. [CrossRef]
27. Long, Y.; Zhao, L.; Niu, B.; Su, J.; Wu, H.; Chen, Y.; Zhang, Q.; Guo, J.; Zhuang, C.; Mei, M.; et al. Hybrid male sterility in rice controlled by interaction between divergent alleles of two adjacent genes. *Proc. Natl. Acad. Sci. USA* **2008**, *105*, 18871–18876. [CrossRef]
28. McCouch, S.R.; CGSNL (Committee on Gene Symbolization, N. and L., Rice Genetics Cooperative). Gene Nomenclature System for Rice. *Rice* **2008**, *1*, 72–84. [CrossRef]
29. Juliano, A.B.; Naredo, M.E.B.; Lu, B.-R.; Jackson, M.T. Genetic differentiation in *Oryza meridionalis* Ng based on molecular and crossability analyses. *Genet. Resour. Crop Evol.* **2005**, *52*, 435–445. [CrossRef]
30. Yamagata, Y.; Yamamoto, E.; Aya, K.; Win, K.T.; Doi, K.; Sobrizal; Ito, T.; Kanamori, H.; Wu, J.; Matsumoto, T.; et al. Mitochondrial gene in the nuclear genome induces reproductive barrier in rice. *Proc. Natl. Acad. Sci. USA* **2010**, *107*, 1494–1499. [CrossRef]
31. Nguyen, G.N.; Yamagata, Y.; Shigematsu, Y.; Watanabe, M.; Miyazaki, Y.; Doi, K.; Tashiro, K.; Kuhara, S.; Kanamori, H.; Wu, J.; et al. Duplication and loss of function of genes encoding RNA polymerase III subunit C4 causes hybrid incompatibility in rice. *G3 Genes Genom. Genet.* **2017**, *7*, 2565–2575. [CrossRef] [PubMed]
32. Ichitani, K.; Takemoto, Y.; Iiyama, K.; Taura, S.; Sato, M. Chromosomal location of *HCA1* and *HCA2*, hybrid chlorosis genes in rice. *Int. J. Plant Genom.* **2012**, *2012*. [CrossRef] [PubMed]
33. Bomblies, K.; Weigel, D. Hybrid necrosis: Autoimmunity as a potential gene-flow barrier in plant species. *Nat. Rev. Genet.* **2007**, *8*, 382–393. [CrossRef] [PubMed]
34. Oka, H.I. Phylogenetic differentiation of cultivated rice. XV. Complementary lethal genes in rice. *Jpn. J. Genet.* **1957**, *32*, 83–87. [CrossRef]
35. Harushima, Y.; Yano, M.; Shomura, A.; Sato, M.; Shimano, T.; Kuboki, Y.; Yamamoto, T.; Lin, S.Y.; Antonio, B.A.; Parco, A.; et al. A high-density rice genetic linkage map with 2275 markers using a single F_2 population. *Genetics* **1998**, *148*, 479–494. [PubMed]
36. Huang, X.; Peng, X.; Sun, M.-X. *OsGCD1* is essential for rice fertility and required for embryo dorsal-ventral pattern formation and endosperm development. *New Phytol.* **2017**, *215*, 1039–1058. [CrossRef] [PubMed]
37. Huang, X.; Lu, Z.; Wang, X.; Ouyang, Y.; Chen, W.; Xie, K.; Wang, D.; Luo, M.; Luo, J.; Yao, J. Imprinted gene *OsFIE1* modulates rice seed development by influencing nutrient metabolism and modifying genome H3K27me3. *Plant J.* **2016**, *87*, 305–317. [CrossRef] [PubMed]
38. Hara, T.; Katoh, H.; Ogawa, D.; Kagaya, Y.; Sato, Y.; Kitano, H.; Nagato, Y.; Ishikawa, R.; Ono, A.; Kinoshita, T.; et al. Rice SNF2 family helicase ENL1 is essential for syncytial endosperm development. *Plant J.* **2015**, *81*, 1–12. [CrossRef] [PubMed]
39. Hong, S.K.; Kitano, H.; Satoh, H.; Nagato, Y. How is embryo size genetically regulated in rice? *Development* **1996**, *122*, 2051–2058.
40. Wambugu, P.W.; Brozynska, M.; Furtado, A.; Waters, D.L.; Henry, R.J. Relationships of wild and domesticated rices (*Oryza* AA genome species) based upon whole chloroplast genome sequences. *Sci. Rep.* **2015**, *5*, 13957. [CrossRef]
41. Yin, H.; Akimoto, M.; Kaewcheenchai, R.; Sotowa, M.; Ishii, T.; Ishikawa, R. Inconsistent diversities between nuclear and plastid genomes of aa genome species in the genus *Oryza*. *Genes Genet. Syst.* **2015**, *90*, 269–281. [CrossRef] [PubMed]

42. Sotowa, M.; Ootsuka, K.; Kobayashi, Y.; Hao, Y.; Tanaka, K.; Ichitani, K.; Flowers, J.M.; Purugganan, M.D.; Nakamura, I.; Sato, Y.-I.; et al. Molecular relationships between Australian annual wild rice, *Oryza meridionalis*, and two related perennial forms. *Rice* **2013**, *6*, 26. [CrossRef] [PubMed]
43. Moner, A.M.; Henry, R.J. *Oryza meridionalis* Ng. In *The Wild Oryza Genomes*; Mondal, T.K., Henry, R., Eds.; Springer-Verlag: Heidelberg, Germany, 2018; pp. 177–182.
44. Brozynska, M.; Copetti, D.; Furtado, A.; Wing, R.A.; Crayn, D.; Fox, G.; Ishikawa, R.; Henry, R.J. Sequencing of australian wild rice genomes reveals ancestral relationships with domesticated rice. *Plant Biotechnol. J.* **2017**, *15*, 765–774. [CrossRef] [PubMed]
45. Moner, A.M.; Furtado, A.; Chivers, I.; Fox, G.; Crayn, D.; Henry, R.J. Diversity and evolution of rice progenitors in australia. *Ecol. Evol.* **2018**, *8*, 4360–4366. [CrossRef] [PubMed]
46. Ichitani, K.; Taura, S.; Sato, M.; Kuboyama, T. Distribution of *Hwc2-1*, a causal gene of a hybrid weakness, in the World Rice Core collection and the Japanese Rice mini Core collection: Its implications for varietal differentiation and artificial selection. *Breed. Sci.* **2016**, *66*, 776–789. [CrossRef] [PubMed]
47. Muto, C.; Ishikawa, R.; Olsen, K.M.; Kawano, K.; Bounphanousay, C.; Matoh, T.; Sato, Y.-I. Genetic diversity of the *wx* flanking region in rice landraces in northern Laos. *Breed. Sci.* **2016**, *66*, 580–590. [CrossRef]
48. Ichitani, K.; Yamaguchi, D.; Taura, S.; Fukutoku, Y.; Onoue, M.; Shimizu, K.; Hashimoto, F.; Sakata, Y.; Sato, M. Genetic analysis of ion-beam induced extremely late heading mutants in rice. *Breed. Sci.* **2014**, *64*, 222–230. [CrossRef]
49. Wan, J.; Ikehashi, H. Identification of two types of differentiation in cultivated rice (*Oryza sativa* L.) detected by polymorphism of isozymes and hybrid sterility. *Euphytica* **1997**, *94*, 151–161. [CrossRef]
50. Kawahara, Y.; de la Bastide, M.; Hamilton, J.P.; Kanamori, H.; McCombie, W.R.; Ouyang, S.; Schwartz, D.C.; Tanaka, T.; Wu, J.; Zhou, S.; et al. Improvement of the *Oryza sativa* Nipponbare reference genome using next generation sequence and optical map data. *Rice* **2013**, *6*, 4. [CrossRef]
51. Yu, J.; Hu, S.; Wang, J.; Wong, G.K.-S.; Li, S.; Liu, B.; Deng, Y.; Dai, L.; Zhou, Y.; Zhang, X.; et al. A draft sequence of the rice genome (*Oryza sativa* L. ssp. indica). *Science* **2002**, *296*, 79–92. [CrossRef]
52. Ichitani, K.; Taura, S.; Tezuka, T.; Okiyama, Y.; Kuboyama, T. Chromosomal location of *HWA1*and *HWA2*, complementary hybrid weakness genes in rice. *Rice* **2011**, *4*, 29–38. [CrossRef]
53. Busungu, C.; Taura, S.; Sakagami, J.-I.; Ichitani, K. Identification and linkage analysis of a new rice bacterial blight resistance gene from XM14, a mutant line from IR24. *Breed. Sci.* **2016**, *66*, 636–645. [CrossRef] [PubMed]

plants

MDPI

Article

Relationships between Iraqi Rice Varieties at the Nuclear and Plastid Genome Levels

Hayba Badro, Agnelo Furtado and Robert Henry *

Queensland Alliance for Agriculture and Food Innovation, University of Queensland, Brisbane, QLD 4072, Australia; haybaq@yahoo.com (H.B.); a.furtado@uq.edu.au (A.F.)

* Correspondence: robert.henry@uq.edu.au

Received: 16 September 2019; Accepted: 5 November 2019; Published: 7 November 2019

Abstract: Due to the importance of the rice crop in Iraq, this study was conducted to determine the origin of the major varieties and understand the evolutionary relationships between Iraqi rice varieties and other Asian rice accessions that could be significant in the improvement of this crop. Five varieties of *Oryza sativa* were obtained from Baghdad/Iraq, and the whole genomic DNA was sequenced, among these varieties, Amber33, Furat, Yasmin, Buhooth1 and Amber al-Baraka. Raw sequence reads of 33 domesticated Asian rice accessions were obtained from the Sequence Read Archive (SRA-NCBI). The sequence of the whole chloroplast-genome was assembled while only the sequence of 916 concatenated nuclear-genes was assembled. The phylogenetic analysis of both chloroplast and nuclear genomes showed that two main clusters, Indica and Japonica, and further five sub-clusters based upon their ecotype, *indica*, *aus*, *tropical-japonica*, *temperate-japonica* and *basmati* were created; moreover, Amber33, Furat, Yasmin and Buhooth1 belonged to the *basmati*, *indica* and *japonica* ecotypes, respectively, where Amber33 was placed in the *basmati* group as a sister of cultivars from Pakistan and India. This confirms the traditional story that Amber was transferred by a group of people who had migrated from India and settled in southern Iraq a long time ago.

Keywords: rice (*Oryza sativa*); evolutionary relationships; chloroplast genome; nuclear genome; phylogeny

1. Introduction

Rice is grown in a wide range of environments worldwide, however, most of the world's rice is cultivated and consumed in Asia [1–3]. Iraq has favorable agricultural conditions for rice cultivation, where rice is a staple food for the majority of the Iraqi people [4]. In Iraq, rice grows as a summer crop, and there are a number of traditional, introduced and improved rice varieties that are cultivated in the central and southern region, as well as in the valleys of northern Iraq [1].

The variety Amber is the most important local Iraqi rice variety and is characterised by high quality in terms of taste (aromatic character) [1]. It has been cultivated in central and southern Iraq, especially in the marshes, for a long time. Anecdotal evidence suggests that Amber was introduced to the marshlands of southern Iraq when water buffalo breeding was introduced to the region by a foreign group from the south Asia, probably from the Indian subcontinent. This popular view was reported in a study by Al-Zahery et al. [5], that highlighted the paternal and maternal origin of the human population in the marsh areas, and observed marginal influences of Indian origin on the gene pool of an autochthonous population of the region. A number of rice varieties have also been introduced to Iraq since the middle of the last century to improve rice productivity [6]. IR8 was the first variety introduced in 1968 by the International Rice Research Institute (IRRI) (Philippines), it has high yield potential but the grain quality has not been high compared to Amber. Since aroma is one of the key traits in determining grain quality in rice [7], Amber became a control variety in the central and south regions of Iraq to assess the grain quality of introduced varieties [1]. Accordingly, Furat and Yasmin were also introduced from Vietnam to Iraq in the late 20th century because they are aromatic, tolerant to limited water, highly productive, and have high grain quality [4]. An understanding of the origin of local Iraqi rice and the genetic relationships between Iraqi rice and Asian domesticated rice will effectively guide Iraqi rice breeding (the aim of the current study). However, few studies have investigated Iraqi rice in general, and the origin and the evolution of Iraqi varieties, especially Amber, in particular [4,8,9].

Each living organism is the consequence of an evolutionary process [10]; therefore, it is imperative to enrich our perception of the evolutionary history of organisms and the relationships among them to guide their genetic improvement. Methods of determining evolutionary history (Phylogeny) have undergone many stages of development. Morphological markers maybe influenced by environmental factors and growth practices. More recent methods have used molecular markers which are independent of environmental factors [11], including techniques such as RFLP, AFLP, RAPD, SSR and ISSR along with morphological markers, to study phylogenetic relationships [12]. The development of high-throughput sequencing technology has revolutionised the study of genetics and evolutionary relationships. Most recently, through next-generation sequencing (NGS), whole-genome sequencing and re-sequencing have become available, so the investigation of the entire genome, rather than targeting precise regions, is now a real opportunity [13–15].

Every plant cell has three genomes—nuclear, chloroplast, and mitochondrial—that may differ in evolutionary history. The chloroplast genome is a maternal genome which is highly-conserved and not involved in recombination, therefore, it is the most commonly used tool to determine the origin and the evolutionary relationships among plant species [16–19]. However, sometimes, evolutionary analysis based on the chloroplast genome must be supported by nuclear genome-based analysis to achieve the most reliable results because the chloroplast genome can only represent the maternal evolutionary history with a slow evolutionary rate [20,21]. Phylogenetic analysis using the nuclear genome can deliver inconsistent trees due to recombination that may confuse phylogenetic resolution. However, this analysis provides greater insights into evolutionary relationships. Several studies have strongly suggested applying this analysis along with chloroplast phylogenetic analysis [18,19,22]. Many studies have applied phylogenetic analysis at both genome levels [23–25], and the results of most of these studies showed that the nuclear genome followed a different evolutionary history pattern to that of the chloroplast genome.

We reported the whole chloroplast genome sequences for Iraqi rice and compared them with the whole chloroplast sequences of other domesticated Asian rice varieties. This provided an important tool for estimating genetic distance and determining evolutionary relationships between rice accessions; the nuclear genomes also provided further information on the relationships between the varieties studied. The study aimed to determine the origin and evolution of Iraqi rice, especially Amber33.

2. Results

2.1. DNA Sequencing and Data Processing

The sequencing process of the five Iraqi varieties (Table 1) generated about 51 Gb of data containing 337 million of 151-bp paired-end reads. The minimum and the maximum number of reads were about 58 and 93 million reads with sequence depth ranging between 23× and 38× for Buhooth1 and Furat, in turn. When raw data was trimmed at the quality limit of 0.01, an average of 15% of the reads' length and 9% of the number of reads were removed, thus the number of reads and data coverage reduced to the range between 53 and 86 million, and 18× and 30×, respectively (Table S1). In terms of downloaded data (Table 2), the average length of raw reads was 83-bp, and the minimum and the maximum number of reads ranged between 43 and 117 million reads while the sequence coverage fluctuated between 10× and 26×. Finally, the number of reads and the data coverage of each of the data sets were assessed after trimming the raw reads at the quality limit of 0.01 (Table S1).

Table 1. The Iraqi plant materials used in this study.

Varieties	History	Varietal Group	BioProject ID	BioSample Accessions
Amber33	Local (Iraq)	Aromatic, medium grain type	PRJNA576935	SAMN13014963
Furat	Introduced from (Vietnam) in 1996	Aromatic, medium grain type	PRJNA576935	SAMN13014964
Yasmin	Introduced from (Vietnam) in 1998	Aromatic, medium grain type	PRJNA576935	SAMN13014965
Buhooth1	Improved	Non-Aromatic, long grain type	PRJNA576935	SAMN13014966
Amber al-Baraka	Improved	Aromatic, long grain type	PRJNA576935	SAMN13014967

Table 2. Summary of data downloaded for sequence comparisons.

No	Sample Unique ID	Project Accession	Species	Country of Origin	Ecotype *	Alignment Name (in Figures 1 and 2) *
1	B243	ERP005654	*O. sativa*	China	Aus	Ch(Aus)B243
2	CX165	ERP005654	*O. sativa*	China	TmpJ	Ch(TmpJ)CX165
3	CX352	ERP005654	*O. sativa*	China	TrpJ	Ch(TrpJ)CX352
4	CX10	ERP005654	*O. sativa*	China	In	Ch(In)CX10
5	CX368	ERP005654	*O. sativa*	India	Aus	India(Aus)CX368
6	IRIS_313-10670	ERP005654	*O. sativa*	India	Bas	India(Bas)IRIS_313-10670
7	IRIS_313-11153	ERP005654	*O. sativa*	India	TmpJ	India(TmpJ)IRIS_313-11153
8	IRIS_313-11479	ERP005654	*O. sativa*	India	TrpJ	India(TrpJ)IRIS_313-11479
9	IRIS_313-11152	ERP005654	*O. sativa*	India	In	India(In)IRIS_313-11152
10	CX129	ERP005654	*O. sativa*	Indonesia	TrpJ	Indo(TrpJ)CX129
11	CX25	ERP005654	*O. sativa*	Indonesia	In	Indo(In)CX25
12	CX104	ERP005654	*O. sativa*	Iran	Bas	Iran(Bas)CX104
13	CX227	ERP005654	*O. sativa*	Japan	Aus	Jap(Aus)CX227
14	CX140	ERP005654	*O. sativa*	Japan	TmpJ	Jap(TmpJ)CX140
15	IRIS_313-10073	ERP005654	*O. sativa*	Japan	TrpJ	Jap(TrpJ)IRIS_313-10073
16	IRIS_313-10549	ERP005654	*O. sativa*	Pakistan	Aus	Pak(Aus)IRIS_313-10549
17	IRIS_313-11021	ERP005654	*O. sativa*	Pakistan	Bas	Pak(Bas)IRIS_313-11021
18	IRIS_313-11026	ERP005654	*O. sativa*	Pakistan	Bas	Pak(Bas)IRIS_313-11026
19	IRIS_313-8656	ERP005654	*O. sativa*	Pakistan	Bas	Pak(Bas)IRIS_313-8656
20	IRIS_313-11829	ERP005654	*O. sativa*	Pakistan	TmpJ	Pak(TmpJ)IRIS_313-11829
21	IRIS_313-10380	ERP005654	*O. sativa*	Philippines	Aus	Phil(Aus)IRIS_313-10380
22	CX59	ERP005654	*O. sativa*	Philippines	Bas	Phil(Bas)CX59
23	IRIS_313-10373	ERP005654	*O. sativa*	Philippines	TmpJ	Phil(TmpJ)IRIS_313-10373
24	CX243	ERP005654	*O. sativa*	Philippines	TrpJ	Phil(TrpJ)CX243
25	IRIS_313-9505	ERP005654	*O. sativa*	Philippines	In	Phil(In)IRIS_313-9505
26	CX126	ERP005654	*O. sativa*	Philippines	In	Phil(In)CX126
27	IRIS_313-10718	ERP005654	*O. sativa*	Sri Lanka	Aus	SriL(Aus)IRIS_313-10718
28	IRIS_313-9949	ERP005654	*O. sativa*	Sri Lanka	TrpJ	Sril(TrpJ)IRIS_313-9949
29	IRIS_313-11248	ERP005654	*O. sativa*	Thailand	TrpJ	Thai(TrpJ)IRIS_313-11248
30	CX106	ERP005654	*O. sativa*	Vietnam	TrpJ	Viet(TrpJ)CX106
31	B009	ERP005654	*O. sativa*	Vietnam	In	Viet(In)B009
32	CX37	ERP005654	*O. sativa*	Vietnam	In	Viet(In)CX37
33	O.glaberrima-PRJNA13765	SRP038750	*O. glaberrima*	-	-	*O. glaberrima*

32 domesticated Asian rice accessions and one domesticated African rice as an out-group downloaded from SAR-NCBI: their unique ID, species, country of origin, and ecotype was from the study of [26]; the alignment names were generated in this study. * In: *indica* subpopulation, TrpJ: *tropical japonica* subpopulation, TmpJ: *temperate japonica* subpopulation, Aus: *aus* population, Bas: *basmati* population, Ch: China, Indo: Indonesia, Jap: Japan, Pak: Pakistan, Phil: Philippines, Sril: SriLanka, Viet: Vietnam.

2.2. Chloroplast Genome Assembly

Mapping all varieties against the reference, *O. sativa* sub sp. japonica *Nipponbare* "GenBank: GU592207.1", under three various fraction settings clarified the most accurate and reliable mapping setting. The number of mismatches and gaps of each variety was virtually stable in all different settings (Table S2). Indeed, this stability confirms that most of these variations were produced from actual differences between the sequences of samples and reference, not due to using different settings; based on that, setting number two (length fraction (LF) of 1 and similarity fraction (SF) of 0.8) was applied to the other steps of assembly, Improvement process (Imp). Moreover, three different settings of Word "W" and Bubble "B" size in *de novo* assembly generated a satisfactory number of contigs that cover the whole chloroplast genome area, around five large chloroplast-contigs with a length of more than 12 kb produced from each setting. Subsequently, four main regions of the chloroplasts, large single copy

(LSC), inverted repeat A (IR A), small single copy (SSC) and inverted repeat B (IR B), were assembled successfully for all 38 varieties through a *de novo* assembly pipeline. The lengths of these regions were about 80 kb for LSC, 12 kb for SSC and 20 kb for IR A and IR B (Figure S1 shows only Iraqi rice varieties). In manual-curation, the comparison between both sub-approaches of the chloroplast genome assembly pipeline showed no significant differences in terms of the number of variations; however, any minor conflicts were resolved by reference to the reads (Table S3 shows only Iraqi rice varieties). The minimum and maximum lengths of the whole chloroplast for all Iraqi varieties and downloaded accessions were 134,259 and 134,556 bp, respectively, while the coverages ranged from 839× to up to 11,466×, and the average coverage was 3818× (Table 3).

Table 3. The results of the chloroplast and nuclear genome assembly.

Varieties	Chloroplast Genome		Length of Nuclear Genome (bp)
	Length of Genome (bp)	Coverage (×)	
Amber33	134,536	2909	616,371
Furat	134,500	7819	616,190
Yasmin	134,502	5495	616,301
Buhooth1	134,550	4759	616,393
Amber al-Baraka	134,493	3651	616,310
B243	134,497	2203	616,278
CX165	134,542	8818	616,377
CX352	134,553	5305	616,324
CX10	134,503	11,466	616,274
CX368	134,504	3674	616,236
IRIS_313–10670	134,535	1669	616,369
IRIS_313–11153	134,551	1639	616,360
IRIS_313–11479	134,259	2726	616,367
IRIS_313–11152	134,503	1413	616,271
CX129	134,535	6076	616 337
CX25	134,503	5830	616,210
CX104	134,532	6784	616,348
CX227	134,504	4267	616,314
CX140	134,547	5185	616,393
IRIS_313–10073	134,556	2036	616,355
IRIS_313–10549	134,495	1636	616 324
IRIS_313–11021	134,531	1978	616,383
IRIS_313–11026	134,532	1723	616,358
IRIS_313–8656	134,532	2334	616 380
IRIS_313–11829	134,539	4164	616,389
IRIS_313–10380	134,496	1857	616,331
CX59	134,536	5913	616,370
IRIS_313–10373	134,551	1464	616,363
CX243	134,556	4375	616,362
IRIS_313–9505	134,503	968	616,283
CX126	134,503	3510	616,220
IRIS_313–10718	134,531	2332	616,324
IRIS_313–9949	134,532	3041	616 385
IRIS_313–11248	134,413	974	616,336
CX106	134,529	6145	616,339
B009	134,528	839	616,231
CX37	134,503	4836	616,258
O.glaberrima-PRJNA13765	134,542	2567	616,099

The table includes the length of the chloroplast genome, the number of bases of mapped reads, and the coverage of assembled chloroplast genome for five Iraqi varieties and 32 domesticated Asian accessions and one domesticated African rice as an out-group downloaded from SAR-NCBI. This table also shows the length of the nuclear genome.

2.3. Phylogenetic Analysis of the Chloroplast Genome

Two phylogenetic approaches were used to analyse the multiple alignments of thirty-nine chloroplast genomes which had a total length of 134,535 bp. Although the result of both phylogenetic methods showed some minor alterations at the end of some subclades, the content of the main clades and subclades, which followed their ecotype classifications, were identical (Figure 1). Phylogenetic analysis of the chloroplast genome divided the thirty-nine rice accessions into two main clades, an Indica clade and a Japonica clade. The Indica clade (In) included most individuals under *indica* (6 accessions) and *aus* (5 accessions) ecotypes except two individuals, B009 and IRIS_313-10718. The Japonica clade contained two subclades, a main Japonica clade and a Basmati clade; the first subclade which was the main Japonica clade (Jap) included all *japonica* individuals (13 accessions) from the two subpopulations of *japonica* ecotype, *tropical* and *temperate*, while the second subclade, the Basmati clade (Bas), involved all individuals of *basmati* ecotype (6 accessions) and the excluded individuals from the first clade (Indica). Additionally, the Iraqi varieties were distributed as following: Furat, Yasmin and Amber al-Baraka into the Indica clade whereas Amber33 and Buhooth1 into the Japonica clade. Buhooth1 was close to accessions from *tropical japonica* ecotype more than accessions under *temperate japonica* ecotype, and interestingly, Amber33 was located within the Basmati subclade.

The multiple alignments of chloroplast genomes comprised 134,535 bp, the number of identical sites was 134,270 characters (99.8%) while the number of variable bases among all the accessions totaled 265 (0.2%). These 265 variable bases were sorted into 85 variation positions which were in turn grouped into four types of polymorphisms including single nucleotide polymorphism (SNP), multi nucleotide polymorphism (MNP), insertions (Ins) and deletions (Del) (Table 4). The most abundant polymorphism types among all accessions were SNPs. Out of 85 polymorphisms, 83%, 12% and 5% were located in the four main regions of the chloroplast genome, LSC, SSC and IR A and B, respectively (Table S4).

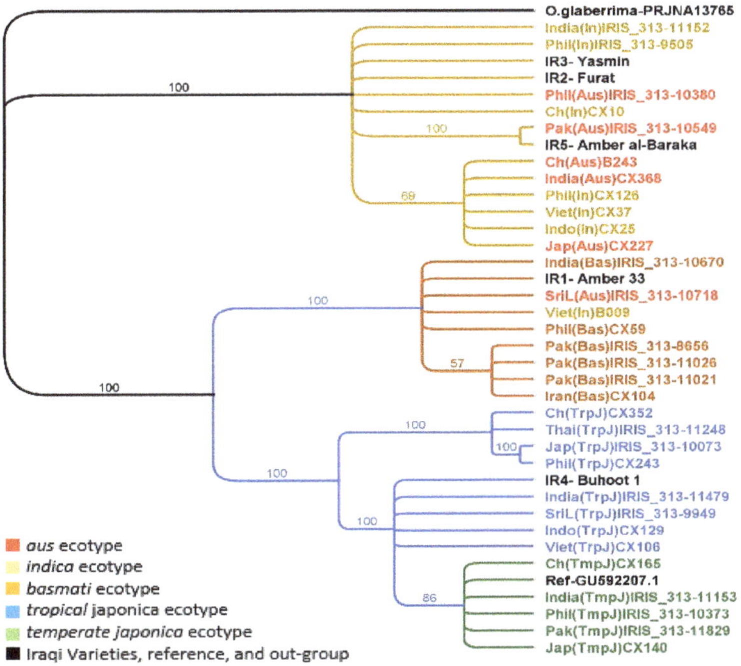

Figure 1. Phylogenetic relationships among chloroplast genomes of thirty-nine rice accessions. Tree topology based on MrBayes software (branch labels represent probability percentage).

Considering the variations identified, all thirty-nine rice accessions were sorted into three main groups: (1) Indica, (2) Japonica and (3) Basmati. As expected, the highest number of polymorphisms among the species studied (255 bases in 76 variant positions) was found in the Indica group, 11 accessions and 3 Iraqi varieties; within 76 variants, there was only one variation (1-bp deletion at position of 75990 bp) between *indica* and *aus* accessions. While the second largest number of variations (55 bases within 21 variant positions) was within the Basmati group, 8 accessions and one Iraqi variety. Part of the Basmati group, 4 accessions, showed unique polymorphisms (2 variable bases (SNPs) within 2 variant positions), three accessions were from Pakistan IRIS_313–8656, IRIS_313–11026, and IRIS_313–11021) and one from Iran (CX104). As expected, the Japonica group, 13 accessions along with the reference (*O. sativa* sub sp. japonica Nipponbare "GenBank: GU592207.1") and one Iraqi variety, possessed the lowest number of polymorphisms (13 bases within 10 variant positions) (Table S4). Most of the polymorphisms in the Japonica group belonged to only four accessions from *tropical japonica* (TrpJ) subpopulation, including CX352, IRIS_313–10073, CX243, and IRIS_313–11248.

Furthermore, a heat-map was drawn according to the number of variable bases (Table S5); in this map, the two main clusters, Indica and Japonica, were clearly distinguished, whereas the Basmati group was comprised within the Japonica group. Within the Japonica group two individuals, 24:IRIS_313–11479 and 27:IRIS_313–11248, clearly showed the greatest distances among the rice accessions. This cluster surprisingly also included two individuals, 33:IRIS_313–10718 and 34:B009, from the *aus* and *indica* ecotype, respectively. There were no variable bases between a number of pairs (dark red in Table S5) such as (3:IRIS_313–11152 and 4:IRIS_313–9505), (9:CX126 and 11:CX37), (9:CX126 and 13:CX25), (17:IRIS_313–10073, and 18:CX243), (19:Ref-GU592207.1 and 20:IRIS_313–11153), (20:IRIS_313–11153 and 21:IRIS_313–10373); and (30:IRIS_313–8656, 31:IRIS_313–11026 and 37:CX104); whereas the highest number of variable bases, 260 bases, was found between (14:CX227 and 24:IRIS_313–11479) (dark green in Table S5). The smallest number of variable bases between Iraqi varieties and other domesticated rice accessions were 1, 3, 1, 6 and 4 bases, those bases were between Iraqi varieties: Amber33, Furat, Yasmin, Buhooth1, and Amber al-Baraka, and the following accessions: 28:IRIS_313-10670, 12:CX10, 3:IRIS_313–11152, 20:IRIS_313–11153 and 5:IRIS_313–10549, respectively (Table S5).

2.4. *Phylogenetic Analysis of the Nuclear Genome*

Within a group of thirty-nine rice accessions, the multiple alignment of 916 concatenated nuclear genes was 621,012 bp in length; the minimum and maximum lengths were 616,099 and 616,393 bp, respectively (Table 3). The nuclear phylogenies using two different methods showed that the two main clusters, Indica and Japonica, and further five sub-clusters were based upon their ecotype, *indica*, *aus*, *tropical japonica*, *temperate japonica* and *basmati* (Figure 2). Unlike the results of the chloroplast phylogeny, the accessions of *indica*, and *aus* ecotypes were represented by two well-resolved subclades within the Indica clade. The Iraqi varieties, Furat and Yasmin, were found in the *indica* subclade while the rest of the Iraqi collection was grouped in the Japonica clade, where Amber33 acted as a sister to all the *basmati* varieties within the *basmati* subcluster, which included all accessions with the *basmati* ecotype. Buhooth1 was part of the *temperate japonica* subcluster that comprised accessions from the *temperate japonica* ecotype and the reference, *O.s japonica* cv. Nipponbare. Amber al-Baraka was a sister to both the Indica and Japonica clades; however, Geneious Tree Builder showed that it was close to the Indica clade, while MrBayes suggested that Amber al-Baraka was closer to the Japonica clade.

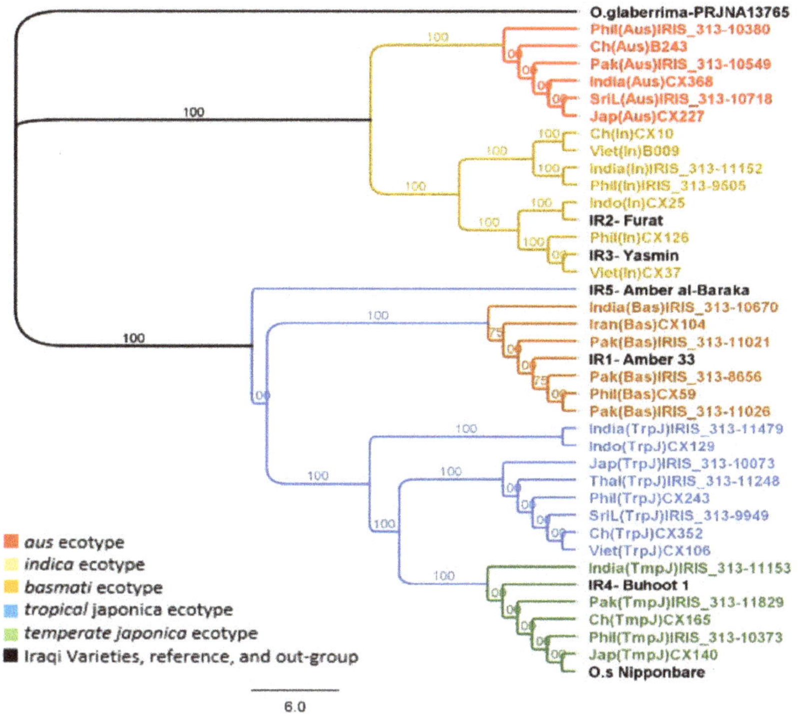

Figure 2. Evolutionary relationships among the multiple alignment of 916 concatenated nuclear genes of domesticated rice. Tree topology based on MrBayes software (branch labels represent probability percentage).

3. Discussion

Rice phylogeny has been extensively studied as a better understanding of the evolutionary relationships among rice species is critical for rice breeding programmes as well as comparative genomics studies. Recent advances in next-generation DNA sequencing (NGS) have improved the phylogenetic reconstruction of any plant species including *Oryza*. In this study, both plastid and nuclear genomes were assembled using NGS reads (whole genome DNA sequencing) to identify the phylogenetic relationships among Iraqi rice varieties and other accessions. According to Sims et al. [27], the accuracy of a genome assembly using NGS reads depends on many factors including sequencing depth (coverage) and the accuracy of the assembly pipeline. Therefore, even after trimming, the sequence coverage of the sequenced and downloaded accessions (Table S1) was enough to ensure coverage of all the chloroplast and most of the nuclear genome, thereby guaranteeing a high-quality assembly.

A dual pipeline was applied to the assembly of the chloroplast genome in this study; this pipeline consisted of two procedures, mapping assembly (MA), and *de novo* assembly (*d*A). The comparison between the sequence of *de novo* and mapping consensus showed no significant differences in terms of the number of variations. Interestingly, the variety Yasmin showed no difference in both approaches with regard to a number of variations (Table S3), but the length of the consensuses was different; this observation indicates that even when the number of copies of an insertion or deletion was similar, the number of bases that were inserted or deleted was diverse. Therefore, in agreement with an earlier study [28], a manual-curation step was critical in resolving any conflicts by reference to the reads. A pipeline of nuclear genes assembly was also developed in this study. This pipeline involved multiple tools on the CLC Genomics Workbench, unlike a previous study [25] that used different software packages to assemble the nuclear genes for phylogenetic analysis at the nuclear genome level. The number of genes selected to represent the nuclear genome in the phylogenetic analysis was only 916 genes with a length of 621,012 bp, considerably lower than that reported previously [25].

Phylogenetic analysis of the chloroplast genome sorted the thirty-nine rice accessions into two main clades, an Indica clade and a Japonica clade (Figure 1). The Indica clade (In) included most individuals under the *indica* and the *aus* ecotypes except for two accessions. Accessions from *indica* and *aus* ecotypes were not clearly distinct but were placed together in one clade; this was confirmed by the results of genetic polymorphism analysis that showed only one variation (1-bp deletion at the position of 75,990 bp; Table S4) between the *indica* and *aus* accessions. The Japonica clade (Jap) contained two subclades, the main Japonica clade (Jap) which included all individuals from the two *japonica* subpopulations, and the Basmati clade (Bas) that included all basmati accessions as well as the individuals excluded from the first clade (Indica). Moreover, the presence of accessions from *aus* ecotype in the Indica clade as well as in the Basmati subclade within Japonica clade agrees with earlier outcomes [18] which indicated that the two different ecotypes, *indica* and *japonica*, might be involved in the origins of the maternal genome in two Korean *aus* landrace rices. This also agrees with the conclusion made by Civáň et al. [29]. which suggested that *aromatic* rice resulted from a hybridization between *japonica* and *aus*. Analysis of genetic polymorphisms at the chloroplast genome level revealed that the most abundant variation types were SNPs, 57% of 85 variants (Table 4). This analysis also showed 255 nucleotide differences within 76 variant positions between the *O.sativa* spp. *indica* and the *japonica* reference (GU592207) in agreement with the previous studies of Brozynska et al. [22] and Wambugu et al. [28].

Table 4. Summary of the number and types of variants in the chloroplast-genomes of thirty-nine domesticated rice-accessions.

Group \ Variation Type	SNPs	MNPs	Del	Ins	Total
Indica	37	5	10	9	61
Japonica	1	0	1	1	3
Basmati	4	1	0	1	6
Indica & Basmati	3	2	3	0	8
Indica and Basmati and Japonica	3	0	1	3	7
Total	48	8	15	14	85

At the nuclear genome level, the phylogenetic analysis using two different approaches sorted accessions from *indica* and *aus* ecotypes into two completely independent subclades within the Indica clade, unlike the result of the chloroplast phylogeny, whilst the second clade was a Japonica clade which included three sub-clusters *tropical japonica*, *temperate japonica*, and *basmati* (Figure 2). Accordingly, the findings of the evolutionary relationship based on nuclear and chloroplast data in the current study aligned with an earlier study by Garris et al. [30] which reported that the closest evolutionary relationships were between *indica* and *aus* groups, and among the *tropical japonica*, *temperate japonica*, and *aromatic* groups. In general, in the present study, the phylogenetic analysis at both genome levels, chloroplast and nuclear, showed relatively comparable evolutionary history patterns with insignificant differences at the end of clades, unlike other studies that recorded significant differences in evolutionary history pattern using both chloroplast and nuclear genomes (regardless of plant materials) [23–25,29]. Furthermore, the phylogenetic trees of both genomes, chloroplast and nuclear, constructed using different methodologies, were highly compatible. However, Amber al-Baraka showed slightly different relationships at the level of the nuclear genome according to the method used; where Geneious Tree Builder software placed Amber al-Baraka close to the Indica clade whereas Amber al-Baraka was closer to the Japonica clade and distant from the Indica clade by MrBayes software. This was unexpected and requires further investigation.

The phylogenetic analysis of both the chloroplast and nuclear genomes indicated that Amber33, Furat and Yasmin, and Buhooth1 belonged to *basmati, indica* and *japonica* ecotypes, respectively. Our results supported that Buhooth1 is an improved cultivar, where the nuclear phylogenies showed a divergent relationship to those deduced from the chloroplast genomes, analogous to *temperate japonica* subpopulation and *tropical japonica* subpopulation, respectively. Furat and Yasmin were introduced to Iraq from Vietnam [4], this was obvious by the results of phylogenetic analysis of the nuclear genome, but their chloroplast genome was closely related to accessions from China, India and Philippines. This may be explained by the breeding history of the genotype.

In this study, Amber33, which is local Iraqi variety, was placed in the *basmati* ecotype group as a sister of cultivars from Pakistan and India by analysing the evolutionary relationship at both levels of the genome. Based on distance analysis, the number of differences in the chloroplast genome between Amber33 and all accessions within the Basmati subclade was in the following order: 28:IRIS_313–10670 O (1 bp), 35:CX59 (1 bp), 30:IRIS_313–8656 (3 bp), 31:IRIS_313–11026 (3 bp), 37:CX104 (3 bp), 33:IRIS_313–10718 (3 bp), 32:IRIS_313–11021 (5 bp), 34:B009 (6 bp) (Table S5); it can be accordingly concluded that Amber33 is closely related to accession from India which is visibly reflected in the observed phylogenetic tree (Figure 1). This confirms the popular tradition that says that the Amber variety was transferred by a group of people who had migrated from India (the Southeast) and settled in southern Iraq a long time ago.

Recently, the term Basmati has been used to indicate a long-grain and high-quality rice, but this name originally refers to *aromatic* rice because it was derived from the Sanskrit words "Vas" and "Matup" which stand for "aroma" and "ingrained from the beginning", respectively, and then both words were combined making 'Vasmati' which changed to become 'Basmati' later on [31,32]. Therefore, the presence of Amber33 within the Basmati subcluster does not necessarily mean that it is a long grain cultivar; indeed, it is an aromatic medium-grain cultivar. Furthermore, Basmati is a group that can be described basically as the fifth isozyme group identified by Glaszmann [33], and it is closer to the *japonica* group than the *indica* [7,30,34]; this group is also phenotypically diverse as it includes both long or medium grain, and aromatic or nonaromatic varieties [7]. In many studies, this group is also known as the "*aromatic*" subpopulation [7,33], but most of the time it is known as "Group V" to avoid confusion. In this study, we refer to this group as "Basmati" according to Wang et al.'s study [26] which is the information resource of the downloaded accessions.

4. Conclusions

In the present study, we have assembled the whole chloroplast genome and the nuclear genome of the five Iraqi rice varieties, together with thirty-three domesticated Asian rice, to find the origin of Iraqi varieties, especially Amber33, and to gain insight into the evolutionary relations between Iraqi and domesticated Asian rice varieties. Our results suggest that the possibility of an Indian and/or Pakistani origin for Amber33; to evaluate this hypothesis, further historical biogeographical analyses are required. Moreover, further study on the chloroplast and nuclear genome in Iraqi rice varieties are required to determine the functional genome annotations that might be useful for future rice breeding programmes in Iraq.

5. Materials and Methods

5.1. Plant Materials

A total of five varieties of *Oryza sativa* were provided and tested by the Office of Agricultural Research, and Directorate of Seed Testing and Certification, Ministry of Agriculture, Baghdad, IRAQ, respectively. Among these varieties, one variety, Amber33, is local and one of the most highly valued varieties in Iraq because of its fragrance, and two varieties, Furat and Yasmin, were introduced from Vietnam; however, they are successfully cultivated in the central and southern regions of Iraq; while the other two, Buhooth1 and Amber al-Baraka, are improved varieties [4]. The plant materials used in this chapter are described in detail in Table 1.

5.2. *Seed Germination and Growth*

About 15 seeds of each individual, a total of 75 seeds, were first dehusked, and then placed in a container with plenty of liquid fertilizer, Flowfeed EX7, that was diluted to half concentration (full concentration is 0.5 g/1L) to break the dormancy phase; this method was the non-heat treatment method. Once the radicle emerged, the germinated seeds were transferred to a petri dish covered with a layer of tissue that was saturated with liquid fertilizer, and planted within three days. All the germination and planting processes were carried out under extremely restricted quarantine conditions in quarantine facilities.

5.3. *DNA Extraction and Sequencing*

After harvesting leaves tissues, total genomic DNA was extracted individually using the modified CTAB protocol described by Furtado [35] with slight modifications. The modifications that were made can be summarised as the following: the mixture of ground plant tissue and nuclear extraction buffer was incubated at 65 °C for 60 min with periodic mixing by inverting the tubes every 5 min; as well as the speed and time of centrifuge were increased to 4000× *g* and 7 min, respectively, after the steps of protein denaturation and DNA precipitation. However, the most vital modification in the DNA extraction procedure was the exclusion of the mixture of phenol:chloroform:isoamyl alcohol (25:24:1). The quality of DNA was assessed by NanoDrop™ 8000 Spectrophotometers (Thermo Scientific, http://www.nanodrop.com) while the DNA quantity was estimated by agarose gel electrophoresis (1%, 120 V for 1 h) based on Furtado's study [35].

The whole genomic DNA of Iraqi rice varieties was sequenced by preparing and indexing five PCR-free libraries separately (one library for each variety), then pooling them together and sequencing over a half lane of an Illumina HiSeq 4000 flow-cell at MACROGEN (Seoul, Korea; http://dna.macrogen.com).

5.4. *Data Downloaded for Sequence Comparisons*

Raw sequence reads of 33 domesticated rice accessions were sourced from the Sequence Read Archive (SRA)-NCBI website (https://www.ncbi.nlm.nih.gov/sra) using "Download/Search for Reads in SRA" tool on CLC Genomics Workbench version 11.0.1 (CLC Bio, a QIAGEN Company, Aarhus, Denmark; www.clcbio.com). All of the species, except one, were Asian rice (*O. sativa*) relatives. *O. glaberrima*, an African rice, was included as an out-group. All related information such as the sample unique ID, project accession, species, country of origin, and ecotype was obtained from an earlier study [26], as shown in Table 2.

5.5. *Data Processing*

The raw reads of both sequenced and downloaded data were subjected to quality control (QC) analysis using the "Create Sequencing QC Report" tool in the CLC Genomics Workbench, which was used to verify the integrity of the data and determine the appropriate trimming score. The low-quality reads were trimmed at a quality limit of 0.01 and a minimum PHRED score of 25 "Trim Sequences" tool on the CLC.

5.6. *Chloroplast Genome Assembly*

A chloroplast genome of the domesticated rice was assembled and validated using a dual pipeline approach: (1) mapping assembly (MA), and (2) *de novo* assembly (*d*A) [36]. In the mapping assembly (MA) pipeline, the trimmed reads were mapped against the reference, which is *O. sativa* sub sp. japonica Nipponbare "GenBank: GU592207.1", using "Map reads to reference" tool at three various fraction settings of length-fraction and similarity-fraction (1) 0.8 and 0.8, (2) 1 and 0.8, and (3) 1 and 0.9, this step was known as "R". Additionally, in an attempt to mend the Cp map, two tools, "InDels and Structural Variants" and "Local Realignment", were applied. This step was named "S". All the analyses of mapping assembly were performed on the CLC Genomics Workbench 11.0.1.

In the *de novo* assembly pipeline, the Fast "F" model was used with combinations of Word "W" and Bubble "B" settings. Contigs generated by *de novo* were blasted against the Cp reference *O. sativa* sub sp. japonica Nipponbare "GenBank: GU592207.1" to select the Cp-exclusive contigs, and they were then updated using the "Map Reads to Contigs" tool on the CLC Genomics Workbench 11.0.1. Lastly, the updated contigs were aligned to a reference sequence to recognise overlaps and gaps using Clone Manager Professional 9.0 (www.scied.com). When non-overlapping contigs were produced, supplemental *de novo* assembly was conducted at various W-and B-settings to plug all gaps by creating additional contigs, and then all the overlapping contigs were subjected to the further analysis.

An additional improvement process was performed on both the mapping and *de novo* assembly pipelines. The improvement (Imp) process was similar to the mapping assembly (MA) pipeline, repeated twice, Imp-1 and Imp-2, with one difference, the consensus generated from each process would be a reference for the following process. The sequences of both improved Cp consensus generated by the mapping and *de novo* improvement processes were compared to identify all mismatches and then were manually corrected by reference to the reads; this step was named "manual-curation" (Figure S2). Eventually, the Cp sequence of each variety was ready for the phylogenetic analysis.

5.7. Phylogenetic Analysis

The consensus chloroplast sequences of the Iraqi rice and the other domesticated rice accessions were used to perform a phylogenetic analysis using Geneious software version 9.1.8 (https://www.geneious.com). The multiple alignment was conducted using the plugin MAFFT Alignment [37] with default parameters; subsequently, to analyse evolutionary relationships; the phylogenetic tree was constructed through software that roots the constructed tree based on the outgroup method: MrBayes [38], and Geneious Tree Builder. The distance between the chloroplast genomes of Iraqi and comparative rice was determined by detecting all the variants using the "variant/SNP detection" tool on Geneious software and then counting the differences (number of bases which are not identical), one of the outputs of the phylogenetic tree construction process.

5.8. Phylogenetic Analysis of the Nuclear Genome

An evolutionary relationship analysis at the level of the nuclear genome was undertaken using the CLC Genomics Workbench 11.0.1 and Geneious software version 9.1.8; this analysis started with the nuclear genome assembly (NGA) pipeline (Figure S3). In NGA pipeline, the "Map Reads to Reference" tool was used to map the trimmed reads of the Iraqi rice (Table 1), and the domesticated rice accessions from Asia and Africa (Table 2) against the reference, which is *O. sativa* sub-spp. Japonica cv Nipponbare "GenBank: IRGSP1.0", applying the following setting: length-fraction of 1 and similarity-fraction of 0.8. After mapping, the consensus sequence of a whole genome for each variety was extracted using the "Extract Consensus Sequence" tool, and from that, the genome and coding sequence (CDS) tracks were generated by the "Convert to Tracks" tool. By investigating the CDS tracks for all varieties, a subset of 916 genes was identified in all varieties, and then the nucleotide sequences of 916 CDS were separately extracted from the genomes using the "Extract Annotations" tool. At the final stage of the nuclear genome assembly (NGA), all the nucleotide sequences of the 916 CDS selected from each genome were concatenated into a super-matrix of 621,012 bp by the "Join Sequences" tool. The super-matrices of all varieties were then aligned using multiple alignments MAFFT [37] on Geneious at default parameters; the alignment output was used in the following phylogenetic inference. A phylogenetic tree was constructed and rooted using the outgroup methods which are MrBayes [38], and Geneious Tree Builder (https://www.geneious.com); the default tree search settings were applied for both methods.

Supplementary Materials: The following are available online at http://www.mdpi.com/2223-7747/8/11/481/s1, Figure S1: The results of *de novo* Assembly on Clone Manager Professional 9.0 software; Figure S2: Illustration of the Chloroplast Genome Assembly Pipeline; Figure S3: Illustration of the Nuclear Genome Assembly Pipeline. Table S1: Summary of the output of sequencing and downloading (Raw Data) and trimming processes; Table S2: Summary of Mapping Assembly process using three different setting of Length fraction and Similarity Fraction; Table S3: Comparison between Mapping and *de novo* assembly in the number of variations in the chloroplast-genome; Table S4: Details of the polymorphisms identified in aligned chloroplast-genomes using the "variant/SNP detection" tool; Table S5: Distance matrix corresponding to the number of non-identical bases in the sequences of domesticated-rice chloroplast-genomes.

Author Contributions: Conceptualization, R.H. and H.B.; Methodology, H.B and A.F.; Software, A.F. and H.B.; Validation, H.B.; Formal Analysis, H.B.; Investigation, H.B.; Resources, H.B.; Data Curation, H.B.; Writing-Original Draft Preparation, H.B.; Writing-Review & Editing, R.H., A.F. and H.B.; Visualization, H.B.; Supervision, R.H. and A.F.; Project Administration, A.F.; Funding Acquisition, H.B.

Funding: A PhD scholarship was provided by HCED Iraq program.

Acknowledgments: We would like to acknowledge the HCED Iraq program for providing PhD scholarship and sincerely thank the Office of Agricultural Research and Directorate of Seed Testing and Certification (Ministry of Agriculture, Baghdad, IRAQ) for providing us with the seeds of five Iraqi varieties for phylogenetic analysis. We acknowledge the University of Queensland Research Computing Centre (UQ-RCC) for providing all the computing resources.

Conflicts of Interest: The authors declare no conflict of interest.

Data Availability Statement: All NGS sequence data as raw data was submitted to NCBI at the Sequence Read Archive (SRA) and is available as SRA Submission# SUB6410326 (under BioProject# PRJNA576935 and BioSample# SAMN13014963, SAMN13014964, SAMN13014965, SAMN13014966, and SAMN13014967 represent Amber33, Furat, Yasmin, Buhooth1, and Amber al-Baraka, respectively).

References

1. Chakravarty, H. *Plant Wealth of Iraq: A Dictionary of Economic Plants*; Botany Directorate, Ministry of Agriculture & Agrarian Reform: Baghdad, Iraq, 1976.
2. Chakravarthi, B.K.; Naravaneni, R. SSR marker based DNA fingerprinting and diversity study in rice (*Oryza sativa* L.). *Afr. J. Biotechnol.* **2006**, *5*, 684–688.
3. Rabbani, M.A.; Pervaiz, Z.H.; Masood, M.S. Genetic diversity analysis of traditional and improved cultivars of Pakistani rice (*Oryza sativa* L.) using RAPD markers. *Electron. J. Biotechnol.* **2008**, *11*, 52–61. [CrossRef]
4. Younan, H.Q.; Al-Kazaz, A.A.; Sulaiman, B.K. Investigation of Genetic Diversity and Relationships among a Set of Rice Varieties in Iraq Using Random Amplified Polymorphic DNA (RAPD) Analysis. *Jordan J. Biol. Sci.* **2011**, *4*, 249–256.
5. Al-Zahery, N.; Pala, M.; Battaglia, V.; Grugni, V.; Hamod, M.A.; Kashani, B.H.; Olivieri, A.; Torroni, A.; Santachiara-Benerecetti, A.S.; Semino, O. In search of the genetic footprints of Sumerians: A survey of Y-chromosome and mtDNA variation in the Marsh Arabs of Iraq. *BMC Evol. Biol.* **2011**, *11*, 288. [CrossRef] [PubMed]
6. Chao, L. *Rice Production*; Report No. 1081 to the Government of Iraq; FAO: Rome, Italy, 1959.
7. Kovach, M.J.; Calingacion, M.N.; Fitzgerald, M.A.; Mccouch, S.R. The origin and evolution of fragrance in rice (*Oryza sativa* L.). *Proc. Natl. Acad. Sci. USA* **2009**, *106*, 14444–14449. [CrossRef] [PubMed]
8. Al-Judy, N. Detecting of DNA Fingerprints and Genetic Relationship Analysis in Local and Improved Rice (*Oryza sativa* L.) Varieties in Iraq. Ph.D. Thesis, College of Science-Baghdad University, Baghdad, Iraq, 2004.
9. Al-Kazaz, A.K.A. Sequence Variation and Phylogenetic Relationships Among Ten Iraqi Rice Varieties Using RM171 Marker. *Iraqi J. Sci.* **2014**, *55*, 145–150.
10. Patwardhan, A.; Ray, S.; Roy, A. Molecular Markers in Phylogenetic Studies—A Review. *J. Phylogenet. Evol. Biol.* **2014**, *2*, 131.
11. Liu, X.; Zhang, H. Advance of molecular marker application in the tobacco research. *Afr. J. Biotechnol.* **2008**, *7*, 4827–4831.
12. Sotowa, M.; Ootsuka, K.; Kobayashi, Y.; Hao, Y.; Tanaka, K.; Ichitani, K.; Flowers, J.; Purugganan, M.; Nakamura, I.; Sato, Y.-I.; et al. Molecular relationships between Australian annual wild rice, *Oryza meridionalis* and two related perennial forms. *Rice* **2013**, *6*, 26. [CrossRef]
13. Nock, C.J.; Waters, D.L.E.; Edwards, M.A.; Bowen, S.G.; Rice, N.; Cordeiro, G.M.; Henry, R.J. Chloroplast genome sequences from total DNA for plant identification. *Plant Biotechnol. J.* **2011**, *9*, 328–333. [CrossRef]

14. Straub, S.C.K.; Parks, M.; Weitemier, K.; Fishbein, M.; Cronn, R.C.; Liston, A. Navigating the tip of the genomic iceberg: Next-generation sequencing for plant systematics. *Am. J. Bot.* **2012**, *99*, 349–364. [CrossRef] [PubMed]
15. Mcpherson, H.; Van Der Merwe, M.; Delaney, S.K.; Edwards, M.A.; Henry, R.J.; Mcintosh, E.; Rymer, P.D.; Milner, M.L.; Siow, J.; Rossetto, M. Capturing chloroplast variation for molecular ecology studies: A simple next generation sequencing approach applied to a rainforest tree. *BMC Ecol.* **2013**, *13*, 8. [CrossRef] [PubMed]
16. Waters, D.L.E.; Nock, C.J.; Ishikawa, R.; Rice, N.; Henry, R.J. Chloroplast genome sequence confirms distinctness of Australian and Asian wild rice. *Ecol. Evol.* **2012**, *2*, 211–217. [CrossRef] [PubMed]
17. Yang, J.-B.; Tang, M.; Li, H.-T.; Zhang, Z.-R.; Li, D.-Z. Complete chloroplast genome of the genus Cymbidium: Lights into the species identification, phylogenetic implications and population genetic analyses. *BMC Evol. Biol.* **2013**, *13*, 84. [CrossRef] [PubMed]
18. Tong, W.; He, Q.; Wang, X.Q.; Yoon, M.Y.; Ra, W.H.; Li, F.; Yu, J.; Oo, W.H.; Min, S.K.; Choi, B.W.; et al. A chloroplast variation map generated using whole genome re-sequencing of Korean landrace rice reveals phylogenetic relationships among *Oryza* sativa subspecies. *Biol. J. Linn. Soc.* **2015**, *115*, 940–952. [CrossRef]
19. Tong, W.; Kim, T.-S.; Park, Y.-J. Rice Chloroplast Genome Variation Architecture and Phylogenetic Dissection in Diverse *Oryza* Species Assessed by Whole-Genome Resequencing. *Rice* **2016**, *9*, 57. [CrossRef]
20. Parks, M.; Cronn, R.; Liston, A. Increasing phylogenetic resolution at low taxonomic levels using massively parallel sequencing of chloroplast genomes. *BMC Biol.* **2009**, *7*, 84. [CrossRef]
21. Moore, M.J.; Pamela, S.S.; Charles, D.B.; Burleigh, J.G.; Douglas, E.S. Phylogenetic analysis of 83 plastid genes further resolves the early diversification of eudicots. *Proc. Natl. Acad. Sci. USA* **2010**, *107*, 4623–4628. [CrossRef]
22. Brozynska, M.; Omar, E.S.; Furtado, A.; Crayn, D.; Simon, B.; Ishikawa, R.; Henry, R.J. Chloroplast Genome of Novel Rice Germplasm Identified in Northern Australia. *Trop. Plant Biol.* **2014**, *7*, 111–120. [CrossRef]
23. Huang, X.; Kurata, N.; Wang, Z.X.; Wang, A.; Zhao, Q.; Zhao, Y.; Liu, K.; Lu, H.; Li, W.; Guo, Y.; et al. A map of rice genome variation reveals the origin of cultivated rice. *Nature* **2012**, *490*, 497–501. [CrossRef]
24. Kim, H.; Jeong, E.; Ahn, S.-N.; Doyle, J.; Singh, N.; Greenberg, A.; Won, Y.; Mccouch, S. Nuclear and chloroplast diversity and phenotypic distribution of rice (*Oryza sativa* L.) germplasm from the democratic people's republic of Korea (DPRK North Korea). *Rice* **2014**, *7*, 7. [CrossRef] [PubMed]
25. Brozynska, M.; Copetti, D.; Furtado, A.; Wing, R.A.; Crayn, D.; Fox, G.; Ishikawa, R.; Henry, R.J. Sequencing of Australian wild rice genomes reveals ancestral relationships with domesticated rice. *Plant Biotechnol. J.* **2017**, *15*, 765–774. [CrossRef] [PubMed]
26. Wang, W.; Mauleon, R.; Hu, Z.; Chebotarov, D.; Tai, S.; Wu, Z.; Li, M.; Zheng, T.; Fuentes, R.R.; Zhang, F.; et al. Genomic variation in 3,010 diverse accessions of Asian cultivated rice. *Nature* **2018**, *557*, 43–49. [CrossRef] [PubMed]
27. Sims, D.; Ian, S.; Nicholas, E.I.; Andreas, H.; Chris, P.P. Sequencing depth and coverage: Key considerations in genomic analyses. *Nat. Rev. Genet.* **2014**, *15*, 121. [CrossRef]
28. Wambugu, P.W.; Brozynska, M.; Furtado, A.; Waters, D.L.; Henry, R.J. Relationships of wild and domesticated rices (*Oryza* AA genome species) based upon whole chloroplast genome sequences. *Sci. Rep.* **2015**, *5*, 13957. [CrossRef]
29. Civáň, P.; Craig, H.; Cox, C.J.; Brown, T.A. Three geographically separate domestications of Asian rice. *Nat. Plants* **2015**, *1*, 15164. [CrossRef]
30. Garris, A.J.; Tai, T.H.; Coburn, J.; Kresovich, S.; Mccouch, S. Genetic Structure and Diversity in *Oryza sativa* L. *Genetics* **2005**, *169*, 1631–1638. [CrossRef]
31. Ahuja, S.C.; Panwar, D.V.S.; Uma, A.; Gupta, K.R. *Basmati Rice: The Scented Pearl, Hisar, Directorate of Publications*; CCS Haryana Agricultural University: Hisar, India, 1995.
32. Akram, M. Aromatic rices of Pakistan: A review. *Pak. J. Agric. Res.* **2009**, *22*, 154–160.
33. Glaszmann, J.C. Isozymes and classification of Asian rice varieties. *Tag. Theor. Appl. Genet. Theor. Angew. Genet.* **1987**, *74*, 21–30. [CrossRef]
34. Caicedo, A.L.; Williamson, S.H.; Hernandez, R.D.; Boyko, A.; Fledel-Alon, A.; York, T.L.; Polato, N.R.; Olsen, K.M.; Nielsen, R.; McCouch, S.R.; et al. Genome-wide patterns of nucleotide polymorphism in domesticated rice. *PLoS Genet.* **2007**, *3*, e163. [CrossRef]
35. Furtado, A. DNA extraction from vegetative tissue for next-generation sequencing. *Methods Mol. Biol.* **2014**, *1099*, 1–5. [PubMed]

36. Moner, A.M.; Furtado, A.; Henry, R.J. Chloroplast phylogeography of AA genome rice species. *Mol. Phylogenet. Evol.* **2018**, *127*, 475–487. [CrossRef] [PubMed]
37. Katoh, K.; Misawa, K.; Kuma, K.; Miyata, T. MAFFT: A novel method for rapid multiple sequence alignment based on fast Fourier transform. *Nucleic Acids Res.* **2002**, *30*, 3059–3066. [CrossRef] [PubMed]
38. Huelsenbeck, J.P.; Ronquist, F. MRBAYES: Bayesian inference of phylogenetic trees. *Bioinformatics* **2001**, *17*, 754–755. [CrossRef]

plants

MDPI

Review

Advances in Molecular Genetics and Genomics of African Rice (*Oryza glaberrima* Steud)

Peterson W. Wambugu [1], Marie-Noelle Ndjiondjop [2] and Robert Henry [3,*]

[1] Kenya Agricultural and Livestock Research Organization, Genetic Resources Research Institute, P.O. Box 30148 – 00100, Nairobi, Kenya; werupw@yahoo.com
[2] M'bé Research Station, Africa Rice Center (AfricaRice), 01 B.P. 2551, Bouaké 01, Ivory Coast; m.ndjiondjop@cgiar.org
[3] Queensland Alliance for Agriculture and Food Innovation, University of Queensland, Brisbane, QLD 4072, Australia
* Correspondence: robert.henry@uq.edu.au; +61-7-661733460551

Received: 23 August 2019; Accepted: 25 September 2019; Published: 26 September 2019

Abstract: African rice (*Oryza glaberrima*) has a pool of genes for resistance to diverse biotic and abiotic stresses, making it an important genetic resource for rice improvement. African rice has potential for breeding for climate resilience and adapting rice cultivation to climate change. Over the last decade, there have been tremendous technological and analytical advances in genomics that have dramatically altered the landscape of rice research. Here we review the remarkable advances in knowledge that have been witnessed in the last few years in the area of genetics and genomics of African rice. Advances in cheap DNA sequencing technologies have fuelled development of numerous genomic and transcriptomic resources. Genomics has been pivotal in elucidating the genetic architecture of important traits thereby providing a basis for unlocking important trait variation. Whole genome re-sequencing studies have provided great insights on the domestication process, though key studies continue giving conflicting conclusions and theories. However, the genomic resources of African rice appear to be under-utilized as there seems to be little evidence that these vast resources are being productively exploited for example in practical rice improvement programmes. Challenges in deploying African rice genetic resources in rice improvement and the genomics efforts made in addressing them are highlighted.

Keywords: African rice; climate change; genomic resources; genetic potential; genome sequencing; domestication; transcriptome and chloroplast

1. Background

African rice (*Oryza glaberrima* Steud) is one of the two rice species that have undergone independent domestication, the other one being Asian rice (*Oryza sativa*). African rice was domesticated about 3500 years ago from its putative progenitor, *Oryza barthii*. These two cultivated species play a vital role in enhancing food security in sub-Saharan Africa where the popularity of rice as a staple food is rising rapidly [1]. Despite this growing popularity of rice, the region is yet to attain self-sufficiency in rice production [1]. In order to achieve self-sufficiency, significant yield increases are required in order to ensure almost complete closure of existing gap between current and potential yields [2]. Climate change is however predicted to be a major threat that is likely to hamper the attainment of these enhanced yields in sub-Saharan Africa [3]. Being part of the *Oryza* primary gene pool and with its wide adaptive potential, African rice presents an important genetic resource that can support the breeding of high yielding climate resilient rice genotypes. Though its production is limited to only a few rice-growing agro-ecologies in West Africa, African rice is of global importance as it is a source of readily available genetic diversity for rice improvement.

Genomic science presents novel tools for exploiting the genetic potential of African rice for accelerated rice productivity. Over the last one decade, there has been tremendous technological advances especially in DNA sequencing which have provided various genomic and genetic tools which have been pivotal in dramatically expanding the frontiers of crop research. Some of these advances include changes in sequencing instruments, chemistry, read length, throughput and bioinformatic tools. In African rice, some of the tools and resources that have been provided by these advances include complete genome reference sequences [4], novel mapping populations [5,6], bacterial artificial chromosome libraries [7] and numerous high throughput molecular markers [4,8]. Other advances include various analytical and bioinformatics tools, resources and platforms. This paper reviews some of the remarkable advances in knowledge that have been witnessed in the last few years in the area of genetics and genomics of this African indigenous *Oryza* species. Challenges in exploiting the immense genetic potential of African rice are highlighted.

2. Genetic Potential and Capacity for Climate Change Adaptation

Literature suggests that African rice possesses important traits that impart great adaptability to various biotic and abiotic stresses as well as climate change adaptation. The superior drought and thermal tolerance capacity of African rice has been reported [9]. This African indigenous rice species may have developed these traits as an adaptive mechanism against the harsh sahelo-saharan climate which is largely characterized by arid conditions. This drought tolerance is achieved through a series of morphological, phenological and physiological responses. Bimpong et al. [10] reported that, compared to Asian rice, some accessions of African rice have capacity to retain more transpirable water when faced with drought stress. These authors suggested that these accessions have capacity to close stomata early enough during periods of drought as a biological survival strategy that ensures effective use of available water. Some varieties of African rice are early maturing and are therefore able to escape terminal drought [11]. It has been found to have thin leaves that roll easily during drought thus reducing transpiration and thin roots which have a high soil penetrative capacity thereby helping in extracting water from the soil [12]. Its leaf and root architecture traits play an important role in enhancing drought tolerance. In a study conducted by Bimpong et al. [13], alien introgression lines derived from a cross between *O. glaberrima* and *O. sativa* had higher yields under drought conditions than the *O. sativa* parent. This demonstrates the potential of transferring drought related traits from *O. glaberrima* to *O. sativa*. About half of the beneficial alleles in the novel drought related quantitative trait loci (QTLs) identified in this study were derived from African rice. In a related study, Shaibu et al. [14] evaluated a total of about 2000 accessions of African rice for drought tolerance and found that some accessions had higher yields under drought conditions than the CG14 *O. glaberrima* drought tolerant check. Though the African rice genotypes were not significantly different from those of the *O. sativa* checks, they provide an important genetic resource for widening the gene pool that can be used to breed for drought tolerance.

Table 1. Important traits on resistance/tolerance to biotic and abiotic stresses found in African rice.

Trait	Reference
Weed competitiveness	[15–18]
Drought tolerance	[15,19,20]
Resistance to nematodes	[21,22]
Resistance to iron toxicity	[23,24]
Resistance to African gall midge	[25]
Resistance to Rice Yellow Mortal Virus	[26–29]
Resistance to bacterial leaf blight (BLB)	[30,31]
Tolerance to lodging	[15,32]
Resistance to green rice leafhopper (*Nephotettix cincticeps* Uhler)	[31]
Tolerance to salinity	[8,33,34]
Tolerance to soils acidity	[19]
Tolerance to submergences	[35]

Natural variation that imparts greater thermal tolerance and adaptation to heat stress compared to *O. sativa* has been identified [9]. This adaptation is particularly important as recent modelling studies have reported potential massive rice yield declines in the West Africa's Sahel region due to high temperature-induced reduction in photosynthesis [3]. African rice therefore possesses valuable genetic diversity for breeding for heat stress in the face of the ever-growing problem of climate change and variability. It has been found to be more tolerant to phosphorus deficiency than Asian rice [36]. Climate change is predicted to lead to increased soil salinity especially in low lying coastal areas and it is expected that this will cause a significant decline in rice yields [37]. Farmers in West Africa where salinity is high have reported that the key strategy they use in mitigating against salinity is planting of tolerant African rice varieties [8]. Owing to its high salt tolerance, African rice seems to be an important source of genes for breeding against salinity. Various types of predictions such as climate modelling show that of all regions, sub-Saharan Africa will be worst hit by climate change [38]. Incidentally, this region has low technological, financial and infrastructural climate change adaptive capacity. It is critically important that rice breeders in sub-Saharan Africa lay concrete strategies for exploiting these important African rice traits for climate change adaptation. Effective deployment of these adapted genetic resources will enhance the resilience and sustainability of rice production systems. In order to leverage the power of genomic tools in taking advantage of these traits, there is need to decipher the loci associated with this adaptive potential or phenotype. Table 1 summarizes the genetic potential of African rice in terms of resistance to a wide range of biotic and abiotic stresses among them drought, soil acidity, iron and aluminium toxicity and weed competitiveness [39].

3. Genetic and Molecular Basis of Important Traits

Genomic research holds the key to greater understanding and unlocking of genetic potential of both wild and domesticated species. In order to leverage the potential of African rice in rice improvement programmes there is need for sound understanding on the molecular underpinning of the functionally important variation. The lack of knowledge on the molecular and genetic basis of important traits acts as a major impediment in the deployment of African rice genetic resources in rice improvement. Aided by the increased availability of genomic resources and other remarkable advances in genomics and molecular genetics, the last couple of years have seen concerted efforts in linking genotypes and phenotypes. These have led to discovery of more loci or causal mutations associated with various traits particularly tolerance to various biotic and abiotic stresses. African rice has superior tolerance to a broad array of nutrient deficiencies and toxicities which are prevalent in most soils. In addition to previously detected QTLs for resistance to iron toxicity which seem stable across genetic backgrounds and environments, seven novel ones were identified [40]. Phosphorus deficiency is a major constraint in rice production particularly in sub-Saharan Africa. A novel allele that is associated with enhanced uptake of phosphorus has been identified in the *OsPSTOL1 (P-Starvation tolerance)* gene which is a major gene controlling the uptake of phosphorus. Candidate genomic regions that are associated with high mineral concentrations among them being key micronutrients have been identified [41]. A genome wide association study based on whole genome resequencing identified genomic regions controlling tolerance to salinity and geographic differentiation, with a total of 28 single nucleotide polymorphisms (SNPs) associated with various salt tolerance traits being identified [8]. This genetic resource of SNP markers is vital for plant breeding and adapting African rice to saline conditions.

Transcriptomic and histological analysis of African rice has identified a set of novel candidate genes for resistance to root knot nematode, *Meloidogyne graminicola*, a pest responsible for major yield losses in *O. sativa* [42]. A second major gene, *RYMV 2*, controlling resistance to *Rice Yellow Mottlel Virus (RYMV)* which is one of the most devastating rice infecting viruses in Africa, has been identified in *O. glaberrima* [26]. Efforts to fine map the *RYMV2* gene led to the identification of a putative loss-of-function one base deletion mutation in one of the candidate genes for *RYMV2*. This low frequency mutation was highly associated with *RYMV* resistance and affected a gene homologous to the *CPR5* defense gene in *Arabidopsis thaliana* [43]. Using *O. sativa* and *O. glaberrima* introgression lines, Gutierrez et al. [44] for the first time identified a major factor QTL controlling Rice stripe necrotic virus located on chromosome 11. These authors also identified a host of other QTLs for various traits, signifying the power of chromosome segment substitution lines (CSSL) as a genetic mapping tool. The continued identification of such locus and alleles is important in rice improvement as it assists in marker assisted selection.

The regulatory mechanisms of key domestication traits are increasingly being unravelled using genomics. Analysis of chromosome segment substitution lines with different genetic backgrounds revealed that the awnless phenotype in African rice was due to a novel recessive allele in the *Regulator of Awn Elongation 3 (RAE3)* gene located on chromosome 6 [45]. Other studies have identified genetic architecture of traits that were selected for by farmers during domestication for adapting the crop to their farming systems. A study by Li et al. [9] has uncovered the QTL responsible for thermal tolerance. Further analysis of this QTL identified a candidate gene, *OsPAB1 (Os03g0387100)*, which was differentially expressed under heat stress and may have been selected for by farmers for adapting African rice to high temperatures. An African rice specific functional SNP, *H99*, in this gene was also identified that may allow the marker assisted introgression of thermal tolerance-enhancing alleles from African rice to other varieties. Despite the growing popularity of genome wide association studies, [8] its application seems limited in African rice as most researchers working on African rice seem to still rely on QTL mapping which has less resolution. Similarly, the application of systems genetics approaches to understand complex traits has been minimal or almost non-existent.

4. Genomic and Transcriptomic Resources

4.1. Genomic Sequences

African rice has one of the smallest genomes in the *Oryza* genus and its assembled reference is about 20% smaller than that of its domesticated counterpart (Table 2). Size differences between the various species are due to lineage-specific expansion and contraction of genes and gene families during the evolutionary process [46]. The first draft genome of African rice was presented by Sakai et al. [47]. This draft genome which was produced through whole genome shot gun sequencing had a size of about 206 Mb, which corresponds to about 0.6X coverage of the African rice genome whose size is estimated to be about 357Mb [48]. Though this genome sequence provided some useful insights on genomic evolution of African rice, it had limited utility as a large portion of the genome was missing. A few years later, a much-improved reference sequence in terms of assembly and annotation was released by Wang et al. [4] under the International *Oryza* Mapping and Alignment Project (IOMAP). Based on the estimated size of the *O. glaberrima* genome, this reference seems incomplete. Recent studies have also reported various assembly errors [49–51]. The CG14 reference sequence was assembled against the *O. sativa* Nipponbare reference sequence and may therefore have missed some *O. glaberrima* specific polymorphisms. The *PSTOL1* locus which is the major gene controlling the uptake of phosphorus was for example found to be missing from the assembled CG14 reference. The *PSTOL1* locus is located within Phosphorus uptake 1 (*Pup1*) which is the major QTL for phosphorus uptake. Aligning the *PSTOL1* locus with unplaced scaffolds revealed that it was present in an unanchored scaffold belonging to chromosome 12. Further analysis revealed that this particular loci and the adjacent sequence of a *Pup1* specific INDEL region spanning about 90 kb is absent in the Nipponbare reference, thus explaining the gap in the CG14 reference in this particular genomic region [51]. Though the identified assembly errors and gaps may hinder its effective utility in rice genetics and genomics, this reference sequence is arguably the most valuable genomic resource for African rice and has opened opportunities for detailed studies on this species. The CG14 reference is also relatively poorly annotated [52,53]. Owing to the importance of this species as a source of readily accessible diversity for rice improvement, there is need for concerted global efforts from the rice scientific community to initiate efforts aimed at improving the quality of this reference sequence.

Table 2. Important assembly and annotation features of selected *Oryza* species.

Species	Feature		Reference
	Genome Size	Gene Count	
O. glaberrima	316 Mb	33,164	[4]
O. sativa	370 Mb	37,544	[54]
O. brachyantha	261 Mb	32,038	[55]
O. barthii	308 Mb	34,575	[56]
O. meridionalis	336 Mb	29,308	[56]
O. punctata	394 Mb	31,679	[56]
O. glumaepatula	373 Mb	35,674	[56]

Genomic studies have revealed that one reference sequence is not enough to represent the full genetic variability present in a species [57,58]. It is against this background that additional varieties of African rice were sequenced. Moreover, as stated, recent studies have identified various errors and gaps in the CG14 reference in addition to its relatively poor annotation [52,53]. These challenges have necessitated additional sequencing efforts to address them. In this regard, sequencing, de novo assembly and annotation of two additional genomes was undertaken. Similarly, using the same de novo approaches, the CG14 genome was also reassembled. These sequencing efforts yielded assemblies which, though smaller and more fragmented, produced better resolution in some loci such as *RYMV1* than the original CG14 reference. As shown in Table 3, they also predicted more protein coding genes than the IOMAP generated reference [50]. A similarly higher number of genes were reported by Zhang et al. [46]. A high-quality reference genome is fundamental for various genetic and genomic applications such as functional and comparative genomics. Lack of quality genomic resources has in some cases limited capacity to validate gene function thereby hindering the unlocking of novel trait variation.

Table 3. Description of various African rice genome assembles.

Feature	CG14 (I-OMAP)	CG14	TOG5681	G22
Assembly size	316 Mb	299 Mb	292 Mb	305 Mb
Gene count	33,164	50,000	51,262	49,662
Scaffold N50	217 kb	10 kb	13 kb	14 kb
Sequencing platform	Roche/454 GS-FLX Titanium Sequencing and Sanger	Illumina	Illumina	Illumina
Assembly approach	Reads aligned to *O. sativa* refseq	De novo	De novo	De novo

Source: [4,50].

In addition to the whole genome sequences, advances in genomics have presented opportunities that have fuelled the development of other types of genomic resources, key among them being high throughput genetic markers. The IOMAP led initiative in which the CG14 variety reference genome was sequenced, generated the first large set of genomic data for African rice. Sequencing 20 diverse accessions of *O. glaberrima* identified a total of 4,447,424 SNPs [4]. Recently, Meyer, et al. [8] generated a genome wide SNP map that contained a total of 2.32 million SNPs by resequencing a total of 93 landraces. Molecular characterization of the *O. glaberrima* accessions conserved in AfricaRice genebank using diversity arrays technology (DArTseq) led to the identification of 3834 polymorphic SNPs [59]. Over 1.4 million Simple Sequence Repeats (SSR) have been identified in the African rice genome [46,47] providing a useful set of genetic markers. The lack of a dedicated set of high throughput markers that can study polymorphisms in interspecific crosses between Asian and African rice has been blamed for the limited exploitation of African rice genetic resources in interspecific breeding [60]. In order to address this gap, Pariasca-Tanaka et al. [60] developed a cost-effective high-throughput genotyping panel comprising of 2015 polymerase chain reaction (PCR)-based SNPs out of which 322 were polymorphic between the two species. These genomic resources provide versatile tools for dissecting the genetic basis of agriculturally important traits, for population genomic studies and other modern breeding applications.

4.2. Chloroplast Genome Sequences

The first chloroplast genome sequences were published by Mariac et al. [61]. Additional chloroplast sequences including data for multiple accessions were later reported by [62] (Figure 1). These authors used a combination of both de novo and read mapping approaches in assembling the genomes. To date, a total of six African rice chloroplast genomes have been released, with the sizes ranging from 132,629–134,661 bp. The significant differences in the sizes of the various released genome sequences can be attributed to the protocol used in retrieving the chloroplast sequences and the assembly approaches used. However, the genome assembled by Mariac et al. [61], with a size of 132,629 bp, appears to be unusually small and to the best of our knowledge is the smallest of all the *Oryza* genomes that have so far been assembled. These genome sequences are providing a versatile tool for use in population genetics, phylogenetic and phylogeographic studies. Wambugu et al. [62] used chloroplast sequences to establish the phylogenetic relationships between African rice and other species constituting the *Oryza* primary gene pool.

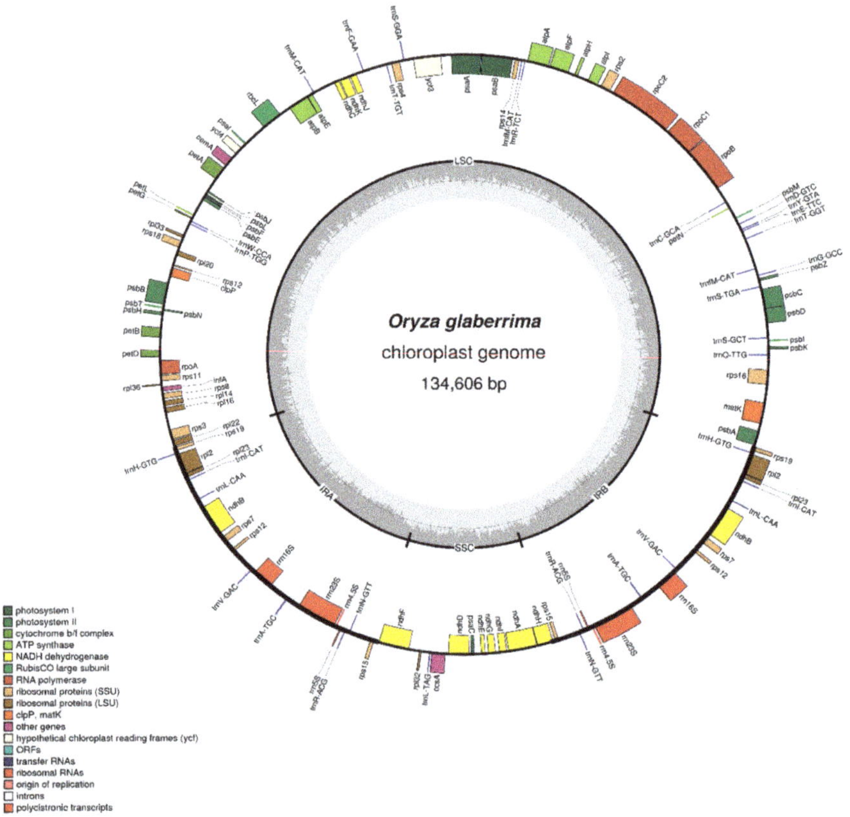

Figure 1. Gene map of *O. glaberrima* chloroplast genome [62].

4.3. Transcriptomic Resources

Transcriptome analysis has played an important role in supporting the assembly, annotation and analysis of the African rice genome and that of other species including *Oryza* wild species.

Wang et al. [4] generated, to our knowledge, the largest multi tissue transcriptomic data for African rice. This RNA sequencing data was used to identify assembly gaps in the CG14 African rice reference. Sequence analysis identified seven genes that were missing in the reference, but RNA data indicated that they were transcribed clearly pointing to genome assembly gaps. In the same study, RNA sequence data was used to conduct comparative analysis of domestication genes in African rice and its progenitor. RNA sequence data was used to confirm the deletion of some genes in African rice among them the ortholog of the *O. sativa* shattering gene (*OsSh1*) which may have been lost during the process of evolution. While transcription for the *O. glaberrima* shattering gene (*OgSh4*) was detected in *O. barthii*, expression level for this gene was found to be limited or absent in African rice. This RNA sequencing analysis together with the analysis of mutation profiles led to the conclusion that African and Asian farmers may have targeted the same traits and genes but sometimes selected different mutations during the domestication process. A similar conclusion was made by Win et al. [63] who used gene expression analysis to unravel the genetic mechanism underlying loss of seed shattering in African rice. Zhang et al. [46] used transcriptome, EST and homology searches to validate predicted gene models during the annotation of de novo assembled *Oryza* genomes. Zhang et al. [46] used transcriptome, EST and homology searches to validate predicted gene models during the annotation of de novo assembled *Oryza* genomes. Further insights into the domestication process were given by Nabholz et al. [64] who used transcriptome sequencing to analyse the genetic diversity of various African rice transcripts. Genetic variation in African rice was reported to be the lowest for all grass species and perhaps for all domesticated crop species [64,65]. As noted by Ndjiondjop et al. [59], it might appear puzzling how a species with such narrow genetic base can possess such unique and exceptional genetic potential in terms of broad resistance to a variety of biotic and abiotic stresses. However, this situation does not seem to be unique as other studies have reported a negative correlation between neutral and functional diversity [66].

Analysis of the transcriptome has been used to decipher the genetic and molecular basis of important morphological, biochemical and physiological traits. African rice and *O. barthii* have uniquely different panicle architectures, but the underlying genetic cause has remained unknown. Comparative RNA analysis of these two African taxa revealed that these differences in panicle morphology are due to expression differences in the miR2118-triggered phased siRNAs [67]. RNA-seq analysis was used to elucidate the cytological and molecular mechanisms of resistance to *M. graminicola* root-knot nematodes, with differentially expressed genes being identified. [42]. Meyer et al. [8] used RNA analysis to identify genes that may be associated with tolerance to salinity based on their gene expression patterns. MicroRNAs are non-coding RNAs that may be involved in regulation of genes involved in response to various biotic and abiotic stresses. In a study analysing miRNAs that are involved in salinity stress response in African rice, Mondal et al. [52] identified a total of 150 conserved and 348 novel miRNAs which may have potential roles in gene expression. A total of 29 known and 32 novel differentially regulated miRNAs were identified suggesting they may have a direct role in response to salinity stress. Additional miRNAs belonging to different gene families have been reported for different *Oryza* species [46,68]. African rice has been found to have less polycistronic miRNA precursors compared to *O. barthii* [69], perhaps as a result of evolutionary and domestication processes. Identification of these important gene expression regulators and their analysis will aid in giving greater insights into their functional and evolutionary roles. This information on the genetic and molecular basis of various traits in African rice is useful to plant breeders as the loci identified can be genetically manipulated in order to impart increased tolerance to biotic and abiotic stresses.

5. Supporting the Conservation and Utilization of African Rice Germplasm Using Genomics

African rice has huge genetic resources which are conserved in various ex situ conservation facilities globally. The largest collection totalling about 3910 accessions is held at AfricaRice Genebank, with the second largest collection of about 2828 accessions being conserved at the International Rice Research Institute (IRRI). As already stated, these collections are a rich reservoir of genes and alleles that is important for rice improvement particularly on tolerance to biotic and abiotic stresses. However, this diversity remains grossly underutilized [23,39]. Over the years, biotechnology-based approaches such as molecular markers have played a key role in genebank management. The current advances in genomics, particularly in DNA sequencing, are offering tools that have capacity for revolutionising the conservation and utilization of plant genetic resources [70,71]. However, compared to other areas of plant science, biodiversity conservation has been slow in embracing these technological advances [72]. Recently, there has been a commendable attempt towards leveraging these genomic-enabled advances in supporting the conservation and utilization of African rice germplasm currently conserved at the AfricaRice genebank. The molecular characterization of this collection using high density molecular markers has recently been reported. A total of 2927 accessions were genotyped with 31,739 DArTseq-based SNP markers. This data has assisted in identification of duplicates, constitution of core and mini-core collections as well as identifying human errors during various genebank operations [59,73]. SNP genotyping is assisting in revealing cases of taxonomic misidentification [74] which is common in genebanks and negatively impacts deployment of genetic resources in plant breeding and other research purposes. This is arguably the largest molecular data collected on this collection and presents a valuable resource for supporting decision making on key conservation aspects. Species SNP diagnostic markers which have capacity for accurately discriminating various *Oryza* species have been developed. Next generation sequencing based approaches have been used to identify duplicates in genebank collections [75] thereby providing a basis for rationalising germplasm collections. The current genomics-enhanced revolution will continue providing novel genomics, analytical and breeding tools that allow more rational and efficient conservation as well as more targeted exploitation of genetic resources.

One of the greatest challenges limiting the use of genetic resources particularly those conserved in genebanks is inadequate understanding of their potential genetic value due to inadequate characterization [72,76]. Genetic diversity especially for genebank samples is usually studied anonymously with little or no efforts to identify the functional diversity [70]. Efforts have been made to analyse the genetic variation of *O. glaberrima* conserved at AfricaRice genebank anonymously using molecular markers [74,77,78]. Genome sequencing has been used to identify functional diversity related to different traits particularly on tolerance to biotic and abiotic stresses [8]. However, in most instances, functional diversity has been studied in only a few genotypes for a particular trait, leaving most of the African rice intraspecific variation largely unknown [23,79]. This limited characterization can largely be attributed to cost related considerations as this remains a major limiting factor. Even with the reduced sequencing and genotyping costs, many labs and researchers particularly in developing countries can still not afford to undertake genomic analysis of large sample sizes. Analysis of bulked samples is becoming a popular approach in genetic mapping and population genetic studies as it allows cost effective analysis of a large number of samples [80,81]. Pool sequencing and whole genome-based bulk segregant analysis are some of the commonly used cutting-edge approaches [82–86]. The lack of quality phenotypic data is increasingly emerging as a major bottleneck in establishing phenotype-to-genotype relationships. The on-going rapid advances in genomics seem to be outpacing capacity to undertake high throughput phenotypic analysis. This calls for an urgent need to invest in human resource capacity and physical infrastructure that will ensure enhanced phenotyping capabilities. Major initiatives aimed at exploiting the vast genetic potential of African rice through intra and inter-specific crossing are currently underway. These initiatives include, Rapid Alleles Mobilization (RAM) and Methodologies and new resources for genotyping and phenotyping (MENERGEP) of African rice species and their pathogens for developing strategic disease resistance breeding programs, both of which are being implemented by AfricaRice and other partners [59].

6. Grain Quality and Its Genetic Control

While priority has been placed on breeding for high yielding crops especially in developing countries, there is also need to ensure that these varieties deliver nutritional security which contributes to human health. Although African rice has potential to contribute genes for improving rice quality [39], this genetic potential has not been deployed in rice improvement and remains poorly studied. However, research interest in the physicochemical and functional properties of starch in African rice is growing [86–93]. Analysis of starch physicochemical properties has revealed that it has unique starch traits [92], a finding that could perhaps explain the renewed interest in starch traits in African rice. Generally, it has higher amylose content (AC) than Asian rice and could be a potential natural source of slowly digestible starch, traits that could confer it potential health benefits [92]. It has been found to have wider diversity of AC than earlier reported [89]. The health benefits of high amylose foods are increasingly being recognised, with such foods being associated with positive gastro-intestinal indices. African rice therefore has potential for use in the development of functional foods [94]. Analysis of *O. glaberrima* introgression lines has revealed that African rice is a novel genetic resource for addressing micro nutrient malnutrition through bio-fortification [41].

Deploying African rice genetic resources in breeding for healthier rice is however constrained by the poor understanding of molecular and genetic mechanisms underlying the unique starch traits, such as AC. Unlike in the case of Asian rice, lack of knowledge on marker-trait associations has hindered the use of marker-assisted selection. Recently, a whole genome based bulk segregant analysis conducted by Wambugu et al. [86] identified genetic markers that are putatively associated with AC. By sequencing bulks of interspecific progenies with low and high AC, this study identified a G/A SNP associated with the *Granule Bound Starch Synthesis (GBSS)* gene located on chromosome 6. Other putative AC associated SNPs were identified in genes encoding the *NAC* and *CCAAT-HAP5* transcription factors located on chromosome 1 and 11 respectively and which have previously been associated with starch biosynthesis. Analysis of natural variation in the *GBSS* locus identified several novel non-synonymous SNPs whose functional importance is still unknown. This study provides useful insights on the genetic control of AC, with the identified candidate genes being novel targets for manipulating AC in African rice.

7. Challenges in Deploying African rice Genetic Diversity in Interspecific Breeding

One of the greatest challenges that have constrained the deployment of African rice diversity in rice breeding is strong and remnant sterility observed in interspecific crosses with Asian rice. This limits rice breeders from taking advantage of heterosis between the two cultivated species. Over the years, there has been intense research efforts on the sterility barriers between the two cultivated species [95–97]. Various approaches have been used in overcoming these barriers, with the first successful cross being achieved about 3 decades ago through the use of another culture and embryo rescue techniques [15]. The use of these conventional biotechnological approaches led to the development of New Rice for Africa (NERICA) varieties, which is arguably the most successful rice improvement program in sub-Saharan Africa. Research has identified a host of loci which are associated with reproductive barriers in cultivated rice [98]. Among these is the S_1 locus, which has a major effect on this interspecific sterility [99]. Despite the huge initial success that was achieved in generating interspecific crosses between *O. sativa* and *O. glaberrima*, the process is still fraught with technical challenges in addition to being tedious and time consuming.

A variety of other methodological approaches have been developed and their effectiveness in addressing these sterility challenges tested. As reported by Lorieux et al. [98], a multi institutional collaborative effort has made efforts to address the challenge of sterility barriers by developing interspecific bridge lines. These are interspecific crosses between *O. sativa* and *O. glaberrima* and are developed through marker assisted selection of progenies that are homozygous for the $S_1{}^s$ allele. Due to the large introgressions of *O. glaberrima* genome in these crosses and by significantly increasing fertility in subsequent crosses with diverse *O. sativa* lines, they ensure effective exploitation of useful *O. glaberrima* genes in conventional breeding programmes. Using mutagenesis, Koide et al. [95] isolated a mutant with an allele in the *S1* locus which is associated with increased fertility. Through this forward genetics approach, these authors were able to create a neutral allele which facilitates crossing these two cultivated species. Another closely related challenge is segregation distortion which has been reported in various genomic regions associated with a sterility locus such as the short arm of chromosome 6 where the *S1* locus is located [44]. Segregation distortion may affect the accuracy of QTL mapping as it may cause the effect of some QTLs to be overestimated. QTLs mapping in regions segregating in non-mendelian fashion should therefore be interpreted with caution.

8. Origin and Domestication of African Rice

Although there has been exceptional interest in studying the domestication and evolutionary history of African rice over the years, this remains unparalleled to that of Asian rice whose domestication is perhaps the most studied of all crop species. Several theories on the origin of African rice have been proposed but the debate rages on. An Asian origin of this species has previously been advanced but has been rejected [65]. Proposals of African rice having been domesticated from Asian rice in West Africa have been put forward but received very little support [100–103]. The dominant theory around which many studies and opinions appear to converge postulates that African rice was domesticated from *O. barthii* in West Africa. This has been supported by studies using gene sequence analysis [65], chloroplast genome based phylogenetic analysis [62] and population genomics [4]. Despite this general consensus, there has been an underlying complexity in understanding the exact location where domestication took place. Whole genome resequencing studies [4,104,105] have provided great insights on the domestication process and especially on the domestication centre but with key studies giving conflicting theories. Using a population genetics approach, Wang et al. [4] were the first authors to map the domestication centre of African rice using whole genome analysis. By resequencing 94 *O. barthii* and 20 *O. glaberrima* accessions as well as comparative genetic analysis of selected domestication genes, these authors mapped the actual domestication centre along the Niger River, consistent with original proposals from Porteres [106] and later supported by Li et al. [65]. Moreover, this resequencing study identified the specific *O. barthii* population from which African rice was domesticated. Recently, analysis of 246 whole genome sequences similarly mapped the Inner Niger Delta as the domestication centre [105]. These findings have however been disputed by Choi et al. [104] who analysed whole genome resequencing data from 286 African rice and *O. barthii* individuals. These authors proposed a non-centric origin of African rice instead of the single origin theory which has been proposed by many previous studies. Moreover, they reported that the progenitor population proposed by Wang et al. [4] lacked genetic differentiation from *O. glaberrima* and had greater resemblance to *O. glaberrima* than *O. barthii*. They therefore concluded that this population may have been misidentified or constitutes a feral weedy population. Rather than settling the debate on the origin of African rice as would have been expected, it appears the era of whole genome data is leading to greater controversies and conflicting theories. The different approaches used in the analysis of whole genome sequences data and the interpretation thereof may be the cause of these contradictory theories and conclusions.

Characterization of domestication genes has enabled deeper understanding of the domestication process of various species. The recently released *O. barthii* reference assembly [56] forms a valuable resource that will allow more insightful analysis of the evolution and domestication of African rice. Genomic analysis is increasing our understanding on the molecular basis of domestication of African rice. While significant progress has been made in the identification and in some cases cloning of domestication genes in Asian rice, relatively little is known about these genes in African rice [45,63,107,108]. The *O. barthii* reference assembly [56] will facilitate identification and analysis of orthologous loci between the domesticate and its progenitor and hence allow in-depth understanding of the target domestication genes. Analysis of selected domestication genes shows that ancient farmers in Africa and Asia targeted the same set of genes during domestication making it an independent and convergent evolution [4,105]. There is increasing evidence indicating that the genetic and molecular basis of the key domestication traits are in some cases conserved between African and Asian rice [4,109] though in other cases the genes and mutation profiles are different [45]. As highlighted earlier, this points to convergent evolution between the two species driven by human selection. The domestication process was associated with major shifts in various morphological traits, among them being grain size where humans showed a strong preference for big seeds. However, in African rice, the selection process appears not to have followed the dominant trend as far as grain size is concerned as the cultivated species typically has smaller seeds than its progenitor. The shift to small seeds has been attributed to a SNP mutation in the *GL4* gene that led to a stop codon. Interestingly, this mutation also led to loss of seed shattering [110]. Analysis of 93 diverse African rice landraces identified *SH3* as an additional gene controlling seed shattering which together with *SH4* led to multiple seed shattering phenotypes [111]. Using association analysis and positional cloning approach, a C/T SNP underlying the loss in seed shattering was identified [63]. The transition from the prostrate growth of *O. barthii* to erect growth in African rice has been attributed to a mutation in the promoter region of the *PROG7* (*PROSTRATE GROWTH 7*) gene which is located on chromosome 7 [107]. A 113kb deletion mutation in the *RICE PLANT ARCHITECTURE DOMESTICATION* (*RPAD*) locus on chromosome 7 has been reported as an additional genetic factor controlling plant architecture in both Asian and African rice [112]. This knowledge is important for plant improvement as important genetic variation can be introduced by targeting these genes and mutations. The domestication process may have been associated with loss of important diversity which may need to be introduced back in a well-targeted manner.

9. Conclusions

In order to meet the food and nutritional requirements of the rapidly growing human population, there is an urgent need to increase per capita rice production. More innovations in rice breeding present an option for achieving the much-needed increases in rice productivity. This can be achieved by the development of super-varieties which have capacity to produce high yields per unit area under low water and nutrient input, in addition to being tolerant to diverse biotic and abiotic stresses. African rice offers a variety of these agronomically important traits. Production of such varieties will require sound knowledge in rice genetics and genomics. There is need to leverage the genomic capabilities that have been presented by cheap genome sequencing technologies to advance their contribution in rice improvement. African resource remains an untapped resource that can play a vital role in the development of novel gene pools. Additional efforts are required in the development of more structural and functional genomic resources. Identification of more functional genetic diversity is also of great value in these efforts. The on-going phenotypic and molecular characterization of African rice genetic resources is also critical in enhancing the utility of these resources in rice improvement.

Author Contributions: P.W.W. wrote the paper; R.H. and M.-N.N. edited an earlier version of the manuscript; all authors read and approved the final manuscript.

Funding: This research received no external funding.

Conflicts of Interest: The authors declare no conflict of interests.

References

1. Seck, P.A.; Tollens, E.; Wopereis, M.C.S.; Diagne, A.; Bamba, I. Rising trends and variability of rice prices: Threats and opportunities for sub-Saharan Africa. *Food Policy* **2010**, *35*, 403–411. [CrossRef]
2. van Ittersum, M.K.; van Bussel, L.G.J.; Wolf, J.; Grassini, P.; van Wart, J.; Guilpart, N.; Claessens, L.; de Groot, H.; Wiebe, K.; Mason-D'Croz, D.; et al. Can sub-Saharan Africa feed itself? *Proc. Natl. Acad. Sci. USA* **2016**, *113*, 14964–14969. [CrossRef] [PubMed]
3. van Oort, P.A.J.; Zwart, S.J. Impacts of climate change on rice production in Africa and causes of simulated yield changes. *Glob. Chang. Biol.* **2018**, *24*, 1029–1045. [CrossRef] [PubMed]
4. Wang, M.H.; Yu, Y.; Haberer, G.; Marri, P.R.; Fan, C.Z.; Goicoechea, J.L.; Zuccolo, A.; Song, X.; Kudrna, D.; Ammiraju, J.S.S.; et al. The genome sequence of African rice (*Oryza glaberrima*) and evidence for independent domestication. *Nat. Genet.* **2014**, *46*, 982–988. [CrossRef] [PubMed]
5. Jacquemin, J.; Bhatia, D.; Singh, K.; Wing, R.A. The International *Oryza* Map Alignment Project: Development of a genus-wide comparative genomics platform to help solve the 9 billion-people question. *Curr. Opin. Plant Biol.* **2013**, *16*, 147–156. [CrossRef] [PubMed]

6. Ali, M.L.; Sanchez, P.L.; Yu, S.-B.; Lorieux, M.; Eizenga, G.C. Chromosome Segment Substitution Lines: A Powerful Tool for the Introgression of Valuable Genes from *Oryza* Wild Species into Cultivated Rice (*O. sativa*). *Rice* **2010**, *3*, 218–234. [CrossRef]
7. Ammiraju, J.S.S.; Luo, M.; Goicoechea, J.L.; Wang, W.; Kudrna, D.; Mueller, C.; Talag, J.; Kim, H.; Sisneros, N.B.; Blackmon, B.; et al. The *Oryza* bacterial artificial chromosome library resource: Construction and analysis of 12 deep-coverage large-insert BAC libraries that represent the 10 genome types of the genus *Oryza*. *Genome Res.* **2006**, *16*, 140–147. [CrossRef] [PubMed]
8. Meyer, R.S.; Choi, J.Y.; Sanches, M.; Plessis, A.; Flowers, J.M.; Amas, J.; Dorph, K.; Barretto, A.; Gross, B.; Fuller, D.Q.; et al. Domestication history and geographical adaptation inferred from a SNP map of African rice. *Nat. Genet.* **2016**, *48*, 1083. [CrossRef] [PubMed]
9. Li, X.-M.; Chao, D.-Y.; Wu, Y.; Huang, X.; Chen, K.; Cui, L.-G.; Su, L.; Ye, W.-W.; Chen, H.; Chen, H.-C.; et al. Natural alleles of a proteasome $\alpha 2$ subunit gene contribute to thermotolerance and adaptation of African rice. *Nat. Genet.* **2015**, *47*, 827. [CrossRef]
10. Bimpong, I.K.; Serraj, R.; Chin, J.H.; Mendoza, E.M.T.; Hernandez, J.E.; Mendioro, M.S. Determination of genetic variability for physiological traits related to drought tolerance in African rice (*Oryza glaberrima*). *J. Plant Breed. Crop Sci.* **2011**, *3*, 60–67.
11. Ndjiondjop, M.N.; Wambugu, P.W.; Rodrigues, S.J.; Karlin, G. The effects of drought on rice cultivation in subSaharan Africa and its mitigation: A review. *Afr. J. Agric. Res.* **2018**, *13*, 1257–1271. [CrossRef]
12. Dingkuhn, M.; Audebert, A.Y.; Jones, M.P.; Etienne, K.; Sow, A. Control of stomatal conductance and leaf rolling in *O. sativa* and *O. glaberrima* upland rice. *Field Crops Res.* **1999**, *61*, 223–236. [CrossRef]
13. Bimpong, I.K.; Serraj, R.; Chin, J.H.; Ramos, J.; Mendoza, E.M.T.; Hernandez, J.E.; Mendioro, M.S.; Brar, D.S. Identification of QTLs for Drought-Related Traits in Alien Introgression Lines Derived from Crosses of Rice (*Oryza sativa* cv. IR64) × *O. glaberrima* under Lowland Moisture Stress. *J. Plant Biol.* **2011**, *54*, 237–250. [CrossRef]
14. Shaibu, A.A.; Uguru, M.I.; Sow, M.; Maji, A.T.; Ndjiondjop, M.N.; Venuprasad, R. Screening African Rice (*Oryza glaberrima*) for Tolerance to Abiotic Stresses: II. Lowland Drought. *Crop Sci.* **2018**, *58*, 133–142. [CrossRef]
15. Jones, M.P.; Dingkuhn, M.; Aluko/snm, G.K.; Semon, M. Interspecific *Oryza Sativa* L. × *O. glaberrima* Steud. progenies in upland rice improvement. *Euphytica* **1997**, *94*, 237–246. [CrossRef]
16. Moukoumbi, Y.D.; Sie, M.; Vodouhe, R.; Bonou, W.; Toulou, B.; Ahanchede, A. Screening of rice varieties for their weed competitiveness. *Afr. J. Agric. Res.* **2011**, *6*, 5446–5456.
17. Fofana, B.; Rauber, R. Weed suppression ability of upland rice under low-input conditions in West Africa. *Weed Res.* **2000**, *40*, 271–280. [CrossRef]
18. Johnson, D.E. The influence of rice plant type on the effect of weed competition on *Oryza sativa* and *Oryza glaberrima*. *Weed Res.* **1998**, *38*, 207–216. [CrossRef]
19. Sano, Y.; Sano, R.; Morishima, H. Neighbour Effects Between two Co-Occurring Rice Species, *Oryza sativa* and *O. glaberrima*. *J. Appl. Ecol.* **1984**, *21*, 245–254. [CrossRef]
20. Ndjiondjop, M.N.; Seck, P.A.; Lorieux, M.; Futakuchi, K.; Yao, K.N.; Djedatin, G.; Sow, M.E.; Bocco, R.; Cisse, F.; Fatondji, B. Effect of Drought on *Oryza glaberrima* Rice Accessions and *Oryza glaberrima* Derived-lines. *Asian J. Agric. Res.* **2012**, *6*, 144–157. [CrossRef]
21. Plowright, R.A.; Coyne, D.L.; Nash, P.; Jones, M.P. Resistance to the rice nematodes *Heterodera sacchari*, *Meloidogyne graminicola* and *M. incognita* in *Oryza glaberrima* and *O. glaberrima* × *O. sativa* interspecific hybrids. *Nematology* **1999**, *1*, 745. [CrossRef]
22. Reversat, G.; Destombes, D. Screening for resistance to Heterodera sacchari in the two cultivated rice species, *Oryza sativa* and *O. glaberrima*. *Fundam. Appl. Nematol.* **1998**, *21*, 307–317.
23. Sikirou, M.; Shittu, A.; Konaté, K.A.; Maji, A.T.; Ngaujah, A.S.; Sanni, K.A.; Ogunbayo, S.A.; Akintayo, I.; Saito, K.; Dramé, K.N.; et al. Screening African rice (*Oryza glaberrima*) for tolerance to abiotic stresses: I. Fe toxicity. *Field Crops Res.* **2018**, *220*, 3–9. [CrossRef] [PubMed]
24. Sahrawat, K.L.; Sika, M. Comparative tolerance of *Oryza sativa* and *Oryza glaberrima* cultivars for iron toxicity in West Africa. *Int. Rice Res. Notes* **2002**, *27*, 30–31.
25. Nwilene, F.E.; Adam, A.; Williams, C.T.; Ukwungwu, M.N.; Dakouo, D.; Nacro, S.; Hamadoun, A.; Kamara, S.I.; Okhidievbie, O.; Abamu, F.J. Reactions of differential rice genotypes to African rice gall midge in West Africa. *Int. J. Pest Manag.* **2002**, *48*, 195–201. [CrossRef]

26. Thiémélé, D.; Boisnard, A.; Ndjiondjop, M.-N.; Chéron, S.; Séré, Y.; Aké, S.; Ghesquière, A.; Albar, L. Identification of a second major resistance gene to Rice yellow mottle virus, *RYMV2*, in the African cultivated rice species, *O. glaberrima*. *Theor. Appl. Genet.* **2010**, *121*, 169–179. [CrossRef] [PubMed]
27. Ndjiondjop, M.N.; Albar, L.; Fargette, D.; Fauquet, C.; Ghesquiere, A. The genetic basis of high resistance to rice yellow mottle virus (*RYMV*) in cultivars of two cultivated rice species. *Plant Dis.* **1999**, *83*, 931–935. [CrossRef] [PubMed]
28. Thottappilly, G.; Rossel, H.W. Evaluation of resistance to rice yellow mottle virus in *Oryza* species. *Indian J. Virol.* **1993**, *9*, 65–73.
29. Albar, L.; Ndjiondjop, M.N.; Esshak, Z.; Berger, A.; Pinel, A.; Jones, M.; Fargette, D.; Ghesquiere, A. Fine genetic mapping of a gene required for Rice yellow mottle virus cell-to-cell movement. *Theor. Appl. Genet.* **2003**, *107*, 371–378. [CrossRef] [PubMed]
30. Djedatin, G.; Ndjiondjop, M.-N.; Mathieu, T.; Cruz, C.M.V.; Sanni, A.; Ghesquière, A.; Verdier, V. Evaluation of African cultivated rice *Oryza glaberrima* for resistance to bacterial blight. *Plant Dis.* **2011**, *95*, 441–447. [CrossRef] [PubMed]
31. Fujita, D.; Doi, K.; Yoshimura, A.; Yasui, H. A major QTL for resistance to green rice leafhopper (*Nephotettix cincticeps* Uhler) derived from African rice (*Oryza glaberrima* Steud.). *Breed. Sci.* **2010**, *60*, 336–341. [CrossRef]
32. Futakuchi, K.; Fofana, M.; Sie, M. Varietal Differences in Lodging Resistance of African Rice (*Oryza glaberrima* Steud.). *Asian J. Plant Sci.* **2008**, *7*, 569–573. [CrossRef]
33. Platten, J.D.; Egdane, J.A.; Ismail, A.M. Salinity tolerance, Na+ exclusion and allele mining of HKT1; 5 in *Oryza sativa* and *O. glaberrima*: Many sources, many genes, one mechanism? *BMC Plant Biol.* **2013**, *13*, 32. [CrossRef] [PubMed]
34. Linares, O.F. African rice (*Oryza glaberrima*): History and future potential. *Proc. Natl. Acad. Sci. USA* **2002**, *99*, 16360–16365. [CrossRef] [PubMed]
35. Watarai, M.; Inouye, J. Internode elongation under different rising water conditions in African floating rice (*Oryza glaberrima* Steud.). *J. Fac. of Agric. Kyushu Univ.* **1998**, *42*, 7.
36. Koide, Y.; Pariasca Tanaka, J.; Rose, T.; Fukuo, A.; Konisho, K.; Yanagihara, S.; Fukuta, Y.; Wissuwa, M.; Yano, M. QTLs for phosphorus deficiency tolerance detected in upland NERICA varieties. *Plant Breed.* **2013**, *132*, 259–265. [CrossRef]
37. Dasgupta, S.; Hossain, M.M.; Huq, M.; Wheeler, D. *Climate Change, Soil Salinity, and the Economics of High-Yield Rice Production in Coastal Bangladesh*; Policy Research Working Paper; No. WPS 7140; World Bank Group: Washington, DC, USA, 2014.
38. Kurukulasuriya, P.; Jain, S.; Mahamadou, A.; Mano, R.; Kabubo-Mariara, J.; El-Marsafawy, S.; Molua, E.; Ouda, S.; Ouedraogo, M.; Sene, I.; et al. Will African Agriculture Survive Climate Change? *World Bank Econ. Rev.* **2006**, *20*, 367–388. [CrossRef]
39. Wambugu, P.; Furtado, A.; Waters, D.; Nyamongo, D.; Henry, R. Conservation and utilization of African *Oryza* genetic resources. *Rice* **2013**, *6*, 29. [CrossRef] [PubMed]
40. Dufey, I.; Draye, X.; Lutts, S.; Lorieux, M.; Martinez, C.; Bertin, P. Novel QTLs in an interspecific backcross *Oryza sativa* × *Oryza glaberrima* for resistance to iron toxicity in rice. *Euphytica* **2015**. [CrossRef]
41. Ohmori, Y.; Sotta, N.; Fujiwara, T. Identification of introgression lines of *Oryza glaberrima* Steud. with high mineral content in grains. *Soil Sci. Plant Nutr.* **2016**, *62*, 456–464. [CrossRef]
42. Petitot, A.-S.; Kyndt, T.; Haidar, R.; Dereeper, A.; Collin, M.; de Almeida Engler, J.; Gheysen, G.; Fernandez, D. Transcriptomic and histological responses of African rice (*Oryza glaberrima*) to Meloidogyne graminicola provide new insights into root-knot nematode resistance in monocots. *Ann. Bot.* **2017**, *119*, 885–899. [CrossRef] [PubMed]
43. Orjuela, J.; Deless, E.F.T.; Kolade, O.; Cheron, S.; Ghesquiere, A.; Albar, L. A Recessive Resistance to Rice yellow mottle virus Is Associated with a Rice Homolog of the *CPR5* Gene, a Regulator of Active Defense Mechanisms. *Mol. Plant-Microbe Interact.* **2013**, *26*, 1455–1463. [CrossRef] [PubMed]
44. Gutierrez, A.G.; Carabali, S.J.; Giraldo, O.X.; Martinez, C.P.; Correa, F.; Prado, G.; Tohme, J.; Lorieux, M. Identification of a Rice stripe necrosis virus resistance locus and yield component QTLs using *Oryza sativa* × *O. glaberrima* introgression lines. *BMC Plant Biol.* **2010**, *10*, 6. [CrossRef] [PubMed]
45. Furuta, T.; Komeda, N.; Asano, K.; Uehara, K.; Gamuyao, R.; Shim-Angeles, R.B.; Nagai, K.; Doi, K.; Wang, D.R.; Yasui, H.; et al. Convergent Loss of Awn in Two Cultivated Rice Species, (*Oryza sativa and Oryza glaberrima*) Is Caused by Mutations in Different Loci. *Genes Genomes Genet.* **2015**. [CrossRef]

46. Zhang, Q.-J.; Zhu, T.; Xia, E.-H.; Shi, C.; Liu, Y.-L.; Zhang, Y.; Liu, Y.; Jiang, W.-K.; Zhao, Y.-J.; Mao, S.-Y.; et al. Rapid diversification of five Oryza AA genomes associated with rice adaptation. *Proc. Natl. Acad. Sci. USA* **2014**, *111*, E4954–E4962. [CrossRef] [PubMed]
47. Sakai, H.; Hamada, M.; Kanamori, H.; Namiki, N.; Wu, J.; Itoh, T.; Matsumoto, T.; Sasaki, T.; Ikawa, H.; Tanaka, T.; et al. Distinct evolutionary patterns of *Oryza glaberrima* deciphered by genome sequencing and comparative analysis. *Plant J.* **2011**, *66*, 796–805. [CrossRef] [PubMed]
48. Kim, H.; Hurwitz, B.; Yu, Y.; Collura, K.; Gill, N.; SanMiguel, P.; Mullikin, J.C.; Maher, C.; Nelson, W.; Wissotski, M.; et al. Construction, alignment and analysis of twelve framework physical maps that represent the ten genome types of the genus *Oryza*. *Genome Biol.* **2008**, *9*, R45. [CrossRef] [PubMed]
49. Wambugu, P.W. Genomic Characterization of African Wild and Cultivated *Oryza* Species. Ph.D. Thesis, University of Queenland, Brisbane, Australia, 2017.
50. Monat, C.; Pera, B.; Ndjiondjop, M.-N.; Sow, M.; Tranchant-Dubreuil, C.; Bastianelli, L.; Ghesquière, A.; Sabot, F. *De Novo* Assemblies of Three *Oryza glaberrima* Accessions Provide First Insights about Pan-Genome of African Rices. *Genome Biol. Evol.* **2017**, *9*, 1–6. [CrossRef] [PubMed]
51. Pariasca-Tanaka, J.; Chin, J.H.; Dramé, K.N.; Dalid, C.; Heuer, S.; Wissuwa, M. A novel allele of the P-starvation tolerance gene *OsPSTOL1* from African rice (*Oryza glaberrima* Steud) and its distribution in the genus *Oryza*. *Theor. Appl. Genet.* **2014**, *127*, 1387–1398. [CrossRef] [PubMed]
52. Mondal, T.K.; Panda, A.K.; Rawal, H.C.; Sharma, T.R. Discovery of microRNA-target modules of African rice (*Oryza glaberrima*) under salinity stress. *Sci. Rep.* **2018**, *8*, 570. [CrossRef] [PubMed]
53. Veltman, M.A.; Flowers, J.M.; van Andel, T.R.; Schranz, M.E. Origins and geographic diversification of African rice (*Oryza glaberrima*). *PLoS ONE* **2019**, *14*, e0203508. [CrossRef] [PubMed]
54. IRGSP. The map-based sequence of the rice genome. *Nature* **2005**, *436*, 793–800. [CrossRef] [PubMed]
55. Chen, J.; Liang, C.; Chen, C.; Zhang, W.; Sun, S.; Liao, Y.; Zhang, X.; Yang, L.; Song, C.; Wang, M.; et al. Whole-genome sequencing of *Oryza brachyantha* reveals mechanisms underlying *Oryza* genome evolution. *Nat. Commun.* **2013**, *4*, 1595. [CrossRef] [PubMed]
56. Stein, J.C.; Yu, Y.; Copetti, D.; Zwickl, D.J.; Zhang, L.; Zhang, C.; Chougule, K.; Gao, D.; Iwata, A.; Goicoechea, J.L.; et al. Genomes of 13 domesticated and wild rice relatives highlight genetic conservation, turnover and innovation across the genus *Oryza*. *Nat. Genet.* **2018**, *50*, 285–296. [CrossRef] [PubMed]
57. Gore, M.A.; Ross-Ibarra, J.; Ware, D.H.; Buckler, E.S.; Chia, J.-M.; Elshire, R.J.; Sun, Q.; Ersoz, E.S.; Hurwitz, B.L.; Peiffer, J.A.; et al. A first-generation haplotype map of maize. *Science* **2009**, *326*, 1115–1117. [CrossRef] [PubMed]
58. Hansey, C.N.; Vaillancourt, B.; Sekhon, R.S.; de Leon, N.; Kaeppler, S.M.; Buell, C.R. Maize (*Zea mays* L.) genome diversity as revealed by RNA-sequencing. *PLoS ONE* **2012**, *7*, e33071. [CrossRef]
59. Ndjiondjop, M.-N.; Wambugu, P.; Sangare, J.R.; Dro, T.; Kpeki, B.; Gnikoua, K. Oryza glaberrima Steud. In *The Wild Oryza Genomes*; Mondal, T.K., Henry, R.J., Eds.; Springer International Publishing: Cham, Switzerland, 2018.
60. Pariasca-Tanaka, J.; Lorieux, M.; He, C.; McCouch, S.; Thomson, M.J.; Wissuwa, M. Development of a SNP genotyping panel for detecting polymorphisms in *Oryza glaberrima/O. sativa* interspecific crosses. *Euphytica* **2014**. [CrossRef]
61. Mariac, C.; Sabot, F.; Santoni, S.; Vigouroux, Y.; Couvreur, T.L.P.; Scarcelli, N.; Pouzadou, J.; Barnaud, A.; Billot, C.; Faye, A.; et al. Cost-effective enrichment hybridization capture of chloroplast genomes at deep multiplexing levels for population genetics and phylogeography studies. *Mol. Ecol. Resour.* **2014**, *14*, 1103–1113. [CrossRef]
62. Wambugu, P.W.; Brozynska, M.; Furtado, A.; Waters, D.L.; Henry, R.J. Relationships of wild and domesticated rices (*Oryza* AA genome species) based upon whole chloroplast genome sequences. *Sci. Rep.* **2015**, *5*, 13957. [CrossRef]
63. Win, K.T.; Yamagata, Y.; Doi, K.; Uyama, K.; Nagai, Y.; Toda, Y.; Kani, T.; Ashikari, M.; Yasui, H.; Yoshimura, A. A single base change explains the independent origin of and selection for the nonshattering gene in African rice domestication. *New Phytol.* **2017**, *213*, 1925–1935. [CrossRef]
64. Nabholz, B.; Sarah, G.; Sabot, F.; Ruiz, M.; Adam, H.; Nidelet, S.; Ghesquière, A.; Santoni, S.; David, J.; Glémin, S. Transcriptome population genomics reveals severe bottleneck and domestication cost in the African rice (*Oryza glaberrima*). *Mol. Ecol.* **2014**, *23*, 2210–2227. [CrossRef] [PubMed]

65. Li, Z.M.; Zheng, X.M.; Ge, S. Genetic diversity and domestication history of African rice (*Oryza glaberrima*) as inferred from multiple gene sequences. *Theor. Appl. Genet.* **2011**, *123*, 21–31. [CrossRef] [PubMed]
66. Yıldırım, Y.; Tinnert, J.; Forsman, A. Contrasting patterns of neutral and functional genetic diversity in stable and disturbed environments. *Ecol. Evol.* **2018**, *8*, 12073–12089. [CrossRef] [PubMed]
67. Ta, K.N.; Sabot, F.; Adam, H.; Vigouroux, Y.; De Mita, S.; Ghesquière, A.; Do, N.V.; Gantet, P.; Jouannic, S. miR2118-triggered phased siRNAs are differentially expressed during the panicle development of wild and domesticated African rice species. *Rice* **2016**, *9*, 10. [CrossRef] [PubMed]
68. Ganie, S.A.; Debnath, A.B.; Gumi, A.M.; Mondal, T.K. Comprehensive survey and evolutionary analysis of genome-wide miRNA genes from ten diploid *Oryza* species. *BMC Genom.* **2017**, *18*, 711. [CrossRef] [PubMed]
69. Baldrich, P.; Hsing, Y.-I.C.; San Segundo, B. Genome-Wide Analysis of Polycistronic MicroRNAs in Cultivated and Wild Rice. *Genome Biol. Evol.* **2016**, *8*, 1104–1114. [CrossRef] [PubMed]
70. Wambugu, P.W.; Ndjiondjop, M.-N.; Henry, R.J. Role of genomics in promoting the utilization of plant genetic resources in genebanks. *Brief. Funct. Genom.* **2018**, *17*, 198–206. [CrossRef] [PubMed]
71. McCouch, S.R.; McNally, K.L.; Wang, W.; Sackville Hamilton, R. Genomics of gene banks: A case study in rice. *Am. J. Bot.* **2012**, *99*, 407–423. [CrossRef]
72. FAO. *Second Report on the World's Plant Genetic Resources for Food and Agriculture*; FAO: Rome, Italy, 2010; p. 299.
73. Ndjiondjop, M.N.; Semagn, K.; Zhang, J.; Gouda, A.C.; Kpeki, S.B.; Goungoulou, A.; Wambugu, P.; Dramé, K.N.; Bimpong, I.K.; Zhao, D. Development of species diagnostic SNP markers for quality control genotyping in four rice (*Oryza* L.) species. *Mol. Breed.* **2018**, *38*, 131. [CrossRef]
74. Orjuela, J.; Sabot, F.; Chéron, S.; Vigouroux, Y.; Adam, H.; Chrestin, H.; Sanni, K.; Lorieux, M.; Ghesquière, A. An extensive analysis of the African rice genetic diversity through a global genotyping. *Theor. Appl. Genet.* **2014**, *127*, 2211–2223. [CrossRef]
75. Singh, N.; Wu, S.; Raupp, W.J.; Sehgal, S.; Arora, S.; Tiwari, V.; Vikram, P.; Singh, S.; Chhuneja, P.; Gill, B.S.; et al. Efficient curation of genebanks using next generation sequencing reveals substantial duplication of germplasm accessions. *Sci. Rep.* **2019**, *650*, 410779. [CrossRef] [PubMed]
76. Wambugu, P.W.; Muthamia, Z.K. *Second Report on the State of Plant Genetic Resources for Food and Agriculture in Kenya*; FAO: Rome, Italy, 2009.
77. Semon, M.; Nielsen, R.; Jones, M.P.; McCouch, S.R. The population structure of African cultivated rice *Oryza glaberrima* (Steud.): Evidence for elevated levels of linkage disequilibrium caused by admixture with *O. sativa* and ecological adaptation. *Genetics* **2005**, *169*, 1639–1647. [CrossRef] [PubMed]
78. Kwon, S.J.; Lee, J.K.; Hong, S.W.; Park, Y.J.; McNally, K.L.; Kim, N.S. Genetic diversity and phylogenetic relationship in AA *Oryza* species as revealed by Rim2/Hipa CACTA transposon display. *Genes Genet. Syst.* **2006**, *81*, 93–101. [CrossRef] [PubMed]
79. Ndjiondjop, M.-N.; Semagn, K.; Gouda, A.C.; Kpeki, S.B.; Dro Tia, D.; Sow, M.; Goungoulou, A.; Sie, M.; Perrier, X.; Ghesquiere, A.; et al. Genetic Variation and Population Structure of *Oryza glaberrima* and Development of a Mini-Core Collection Using DArTseq. *Front. Plant Sci.* **2017**, *8*. [CrossRef] [PubMed]
80. Schlötterer, C.; Tobler, R.; Kofler, R.; Nolte, V. Sequencing pools of individuals-mining genome-wide polymorphism data without big funding. *Nat. Rev. Genet.* **2014**, *15*, 749–763. [CrossRef] [PubMed]
81. Zou, C.; Wang, P.; Xu, Y. Bulked sample analysis in genetics, genomics and crop improvement. *Plant Biotechnol. J.* **2016**, *14*, 1941–1955. [CrossRef] [PubMed]
82. Han, Y.; Lv, P.; Hou, S.; Li, S.; Ji, G.; Ma, X.; Du, R.; Liu, G. Combining Next Generation Sequencing with Bulked Segregant Analysis to Fine Map a Stem Moisture Locus in Sorghum (*Sorghum bicolor* L. Moench): e0127065. *PLoS ONE* **2015**, *10*. [CrossRef]
83. Ramirez-Gonzalez, R.H.; Segovia, V.; Bird, N.; Fenwick, P.; Holdgate, S.; Berry, S.; Jack, P.; Caccamo, M.; Uauy, C. RNA-Seq bulked segregant analysis enables the identification of high-resolution genetic markers for breeding in hexaploid wheat. *Plant Biotechnol. J.* **2015**, *13*, 613–624. [CrossRef] [PubMed]
84. Olasanmi, B.; Akoroda, M.O.; Okogbenin, E.; Egesi, C.; Kahya, S.S.; Onyegbule, O.; Ewa, F.; Guitierrez, J.; Ceballos, H.; Tohme, J.; et al. Bulked segregant analysis identifies molecular markers associated with early bulking in cassava (*Manihot esculenta* Crantz). *Euphytica* **2014**, *195*, 235–244. [CrossRef]
85. Takagi, H.; Abe, A.; Yoshida, K.; Kosugi, S.; Natsume, S.; Mitsuoka, C.; Uemura, A.; Utsushi, H.; Tamiru, M.; Takuno, S.; et al. QTL-seq: Rapid mapping of quantitative trait loci in rice by whole genome resequencing of DNA from two bulked populations. *Plant J.* **2013**, *74*, 174–183. [CrossRef] [PubMed]

86. Wambugu, P.; Ndjiondjop, M.-N.; Furtado, A.; Henry, R. Sequencing of bulks of segregants allows dissection of genetic control of amylose content in rice. *Plant Biotechnol. J.* **2018**, *16*, 100–110. [CrossRef] [PubMed]
87. Falade, K.O.; Semon, M.; Fadairo, O.S.; Oladunjoye, A.O.; Orou, K.K. Functional and physico-chemical properties of flours and starches of African rice cultivars. *Food Hydrocoll.* **2014**, *39*, 41–50. [CrossRef]
88. Gayin, J.; Bertoft, E.; Manful, J.; Yada, R.Y.; Abdel-Aal, E.S.M. Molecular and thermal characterization of starches isolated from African rice (*Oryza glaberrima*). *Starch-Stärke* **2016**, *68*, 9–19. [CrossRef]
89. Gayin, J.; Chandi, G.K.; Manful, J.; Seetharaman, K. Classification of Rice Based on Statistical Analysis of Pasting Properties and Apparent Amylose Content: The Case of *Oryza glaberrima* Accessions from Africa. *Cereal Chem.* **2015**, *92*, 22–28. [CrossRef]
90. Gayin, J.; Abdel-Aal, E.-S.M.; Manful, J.; Bertoft, E. Unit and internal chain profile of African rice (*Oryza glaberrima*) amylopectin. *Carbohydr. Polym.* **2016**, *137*, 466–472. [CrossRef] [PubMed]
91. Gayin, J. Structural and Functional Characteristics of African Rice (*Oryza glaberrima*) Flour and Starch. Ph.D. Thesis, The University of Guelph, Guelph, ON, Canada, 2015.
92. Wang, K.; Wambugu, P.W.; Zhang, B.; Wu, A.C.; Henry, R.J.; Gilbert, R.G. The biosynthesis, structure and gelatinization properties of starches from wild and cultivated African rice species (*Oryza barthii* and *Oryza glaberrima*). *Carbohydr. Polym.* **2015**, *129*, 92–100. [CrossRef] [PubMed]
93. Vargas, C.G.; Costa, T.M.H.; de Rios, A.O.; Flôres, S.H. Comparative study on the properties of films based on red rice (*Oryza glaberrima*) flour and starch. *Food Hydrocoll.* **2017**, *65*, 96–106. [CrossRef]
94. Manful, J.T.; Graham-Acquaah, S. African Rice (*Oryza glaberrima*): A Brief History and Its Growing Importance in Current Rice Breeding Efforts. In *Reference Module in Food Science*; Elsevier: Amsterdam, The Netherlands, 2016. [CrossRef]
95. Koide, Y.; Ogino, A.; Yoshikawa, T.; Kitashima, Y.; Saito, N.; Kanaoka, Y.; Onishi, K.; Yoshitake, Y.; Tsukiyama, T.; Saito, H.; et al. Lineage-specific gene acquisition or loss is involved in interspecific hybrid sterility in rice. *Proc. Natl. Acad. Sci. USA* **2018**, *115*, E1955–E1962. [CrossRef]
96. Koide, Y.; Onishi, K.; Nishimoto, D.; Baruah, A.R.; Kanazawa, A.; Sano, Y. Sex-independent transmission ratio distortion system responsible for reproductive barriers between Asian and African rice species. *New Phytol.* **2008**, *179*, 888–900. [CrossRef]
97. Xie, Y.; Xu, P.; Huang, J.; Ma, S.; Xie, X.; Tao, D.; Chen, L.; Liu, Y.G. Interspecific Hybrid Sterility in Rice Is Mediated by *OgTPR1* at the *S1* Locus Encoding a Peptidase-like Protein. *Mol. Plant* **2017**, *10*, 1137–1140. [CrossRef]
98. Lorieux, M.; Garavito, A.; Bouniol, J.; Gutiérrez, A.; Ndjiondjop, M.N.; Guyot, R.; Pompilio Martinez, C.; Tohme, J.; Ghesquière, A. *Unlocking the Oryza glaberrima Treasure for Rice Breeding in Africa*; CABI: Wallingford, UK, 2013; pp. 130–143.
99. Garavito, A.; Guyot, R.; Lozano, J.; Gavory, F.; Samain, S.; Panaud, O.; Tohme, J.; Ghesquière, A.; Lorieux, M. A genetic model for the female sterility barrier between Asian and African cultivated rice species. *Genetics* **2010**, *185*, 1425–1440. [CrossRef]
100. Nayar, N.M. Origin and Cytogenetics of Rice. In *Advances in Genetics*; Caspari, E.W., Ed.; Academic Press: Amsterdam, Netherlands, 1973; Volume 17, pp. 153–292.
101. Nayar, N.M. Evolution of the African Rice: A Historical and Biological Perspective. *Crop Sci.* **2012**, *52*, 505–516. [CrossRef]
102. Nayar, N.M. The history and genetic transformation of the African rice, *Oryza glaberrima* Steud. (Gramineae). *Curr. Sci.* **2010**, *99*, 1681–1689.
103. Nayar, N.M. Chapter 5—The Origin of African Rice. In *Origin and Phylogeny of Rices*; Nayar, N.M., Ed.; Academic Press: San Diego, CA, USA, 2014; pp. 117–168.
104. Choi, J.Y.; Zaidem, M.; Gutaker, R.; Dorph, K.; Singh, R.K.; Purugganan, M.D. The complex geography of domestication of the African rice *Oryza glaberrima*. *PLoS Genet.* **2019**, *15*, e1007414. [CrossRef] [PubMed]
105. Cubry, P.; Tranchant-Dubreuil, C.; Thuillet, A.C.; Monat, C.; Ndjiondjop, M.N.; Labadie, K.; Cruaud, C.; Engelen, S.; Scarcelli, N.; Rhone, B.; et al. The Rise and Fall of African Rice Cultivation Revealed by Analysis of 246 New Genomes. *Curr. Biol.* **2018**, *28*, 2274–2282. [CrossRef] [PubMed]
106. Porteres, R. Berceaux Agricoles Primaires Sur le Continent Africain. *J. Afr. Hist.* **1962**, *3*, 195–210. [CrossRef]
107. Hu, M.; Lv, S.; Wu, W.; Fu, Y.; Liu, F.; Wang, B.; Li, W.; Gu, P.; Cai, H.; Sun, C.; et al. The domestication of plant architecture in African rice. *Plant J.* **2018**, *94*, 661–669. [CrossRef]

108. Wambugu, P.W.; Nyamongo, D.; Ndjiondjop, M.-N.; Henry, R.J. Evolutionary Relationships Among the *Oryza* Species. In *The Wild Oryza Genomes*; Mondal, T.K., Henry, R.J., Eds.; Springer International Publishing: Cham, Switzerland, 2018; pp. 41–54.
109. Gross, B.L.; Steffen, F.T.; Olsen, K.M. The molecular basis of white pericarps in African domesticated rice: Novel mutations at the *Rc* gene. *J. Evol. Biol.* **2010**, *23*, 2747–2753. [CrossRef]
110. Wu, W.; Liu, X.; Wang, M.; Meyer, R.S.; Luo, X.; Ndjiondjop, M.-N.; Tan, L.; Zhang, J.; Wu, J.; Cai, H.; et al. A single-nucleotide polymorphism causes smaller grain size and loss of seed shattering during African rice domestication. *Nat. Plants* **2017**, *3*, 17064. [CrossRef]
111. Lv, S.; Wu, W.; Wang, M.; Meyer, R.S.; Ndjiondjop, M.N.; Tan, L.; Zhou, H.; Zhang, J.; Fu, Y.; Cai, H.; et al. Genetic control of seed shattering during African rice domestication. *Nat. Plants* **2018**, *4*, 331–337. [CrossRef]
112. Wu, Y.; Zhao, S.; Li, X.; Zhang, B.; Jiang, L.; Tang, Y.; Zhao, J.; Ma, X.; Cai, H.; Sun, C.; et al. Deletions linked to *PROG1* gene participate in plant architecture domestication in Asian and African rice. *Nat. Commun.* **2018**, *9*, 4157. [CrossRef] [PubMed]

MDPI
St. Alban-Anlage 66
4052 Basel
Switzerland
Tel. +41 61 683 77 34
Fax +41 61 302 89 18
www.mdpi.com

Plants Editorial Office
E-mail: plants@mdpi.com
www.mdpi.com/journal/plants

www.ingramcontent.com/pod-product-compliance
Lightning Source LLC
LaVergne TN
LVHW070152120526
838202LV00013BA/1026
* 9 7 8 3 0 3 9 3 6 8 2 6 6 *